Baking Business Sustainability Through Life Cycle Management

João Miguel Ferreira da Rocha
Aleksandra Figurek
Anatoliy G. Goncharuk • Alexandrina Sirbu
Editors

Baking Business Sustainability Through Life Cycle Management

Springer

Editors
João Miguel Ferreira da Rocha
Universidade Católica Portuguesa, CBQF - Centro de Biotecnologia e Química Fina – Laboratório Associado, Escola Superior de Biotecnologia
Porto, Portugal

Anatoliy G. Goncharuk
Hauge School of Management
NLA University College
Kristiansand, Norwa

Aleksandra Figurek
GNOSIS Mediterranean Institute for Management Science, School of Business
University of Nicosia
Nicosia, Cyprus

Alexandrina Sirbu
Constantin Brancoveanu
University of Pitesti
FMMAE Ramnicu Valcea, Romania

ISBN 978-3-031-25026-2 ISBN 978-3-031-25027-9 (eBook)
https://doi.org/10.1007/978-3-031-25027-9

This Springer imprint is published by the registered company Springer Nature Switzerland AG
The registered company address is: Gewerbestrasse 11, 6330 Cham, Switzerland

Preface

It was a great pleasure to hand over the book *Baking Business Sustainability Through Life Cycle Management* to the reader's world. This book has been written to meet the requirements of the scientific community, teachers, students, as well as producers and managers in the baking industry. It also covers the audience's needs for compiling the latest developments in sustainability science applied to this sector. The practical value of this book is very high, especially for the industry, which is interested in how to make their business models more sustainable.

The book differs from the others available in the market, because, to our knowledge, there is not a compilation of technical information about sustainability in the market for the bakery sector. Care has been taken to include new concepts and achievements related to the food industry directly connected with sustainable business models, circular economy, and sustainable uses of agricultural resources.

We reviewed a large number of references during the book's preparation to seek clear concepts, using familiar examples to illustrate various theories and methods, which may enable various types of readers to have a clear understanding and knowledge of different aspects of sustainability in the baking industry. The book provides various analyses connected with the crucial topic today – cereal production, as the main input for the bread industry, as well as the impact of climate change on possible food shortages. It was very important to consider the experiences of different countries in the world which are the biggest cereal producers.

Baking Business Sustainability Through Life Cycle Management combines theories and applications closely and is also convenient for readers' self-study. The book also targets other agriculture/agri-food sectors, environmental specialists, and regulators on recent advances in sustainability applied to the baking industry. The objective was to connect the different sustainability disciplines in a single book to offer the latest developments at theoretical and practical levels about information and communication requirements, reporting standards, regulation, sustainable business models, supply chain management, footprints, and circular economy, among other topics which are presented in the book.

We sincerely express our gratitude to the COST Association and to the team of the Cost action SOURDOMICS – *Sourdough biotechnology network towards novel, healthier and sustainable food and bioprocesses* (CA18101), for the valuable advice, as well as the Springer team for support and assistance for the completion of the book.

Porto, Portugal João Miguel Ferreira da Rocha
Nicosia, Cyprus Aleksandra Figurek
Kristiansand, Norway Anatoliy G. Goncharuk
Ramnicu Valcea, Romania Alexandrina Sirbu

Contents

Part I
Life Cycle Assessment and Product Environmental Footprint in Bread Industry

Chapter 1
Bread Industry Sustainability Life Cycle Assessment (A Proposal of Analysis of Sustainability Assessment Using Environmental and Social Footprints)

Elena I. Semenova and Aleksandr V. Semenov

1.1 Trends in the Bread Market

Currently, the capacity of the global bread market is about 152.5 million tons. However, these data are approximate because, in some territories, the situation cannot be accounted for, and the concept of bread differs from country to country.

The annual growth of global consumption is estimated at 1.3%–1.5%, and its rate is projected to increase. The greatest potential for consumption growth is seen in Africa and Asia (especially in China), while consumption levels stagnate or even decline in the EU and North America, Russia, and many others (Table 1.1).

Let us also note specific trends for individual countries:

- In the USA, there are two main trends in bread consumption – the consumption of cheap products (e.g., Subway sandwiches for $5 or burgers for $1), or the pursuit of a healthy diet, attention to food composition, a preference for low-calorie products that do not contain trans-isomer fatty acids or GMOs, low salt content, and gluten-free products, regardless of price;
- Increasing availability of bread to the population of African countries, a narrow range of products, the lack of special requirements for the product;
- "Europeanization" of bread in the Middle East and Asia, accompanied by increased investment in the industry, the formation of fashion on the consumption of baguettes, croissants, pastries, etc. for the younger generation.

The most significant trends in the global market are as follows:

1. Growth of the share of the large industrial sector (e.g., Bimbo and Yamasaki);
2. Redistribution of production in the direction of network retail;

E. I. Semenova (✉) · A. V. Semenov
Federal Research Center of Agrarian Economy and Social Development of Rural Areas – All Russian Research Institute of Agricultural Economics, Moscow, Russia

J. M. Ferreira da Rocha et al. (eds.), *Baking Business Sustainability Through Life Cycle Management*, https://doi.org/10.1007/978-3-031-25027-9_1

Table 1.1 The main factors common to many countries affecting the consumption of bread

Increasing demand	Constraining demand
a satisfaction from bread – its aroma and flavor; traditional and regional bakery products of daily demand and individual preferences; snack products for meals outside the home or "on the go," for which bread products are the basis or a mandatory accompanying component; Healthy foods, such as whole-grain and multigrain bread, with less salt or increased magnesium and potassium content; Brand-free products – clean label or E-free groups; Globalization of product range.	The existing rather high level of consumption of bread; Economic crises; Overall population decline for North America and the EU; Changes in the preferences of the population – the transition to other foods, including pasta and cereals, products containing animal proteins; The negative image of some products (e.g., pastries and confectionery are considered unhealthy).

Source: Compiled by the authors

3. Development of the range of marginal bread varieties.

Russia belongs to the countries with a medium level of bread consumption. The baking industry in Russia is one of the leading food industries in the agro-industrial complex. Large bread production plants provide mass varieties of bread to most of the population, leading in the production of rye and rye-wheat bread. The share of mass (traditional) varieties of bread is 80%.

The main trends in the global bakery market are also typical for Russia:

- Stabilization of the volume of the bread market, a decrease in the production volume of small businesses with a significant increase in network bakeries;
- Little growth of industrial enterprises with the steady development of foreign players (e.g., Fazer, East Balt, and Lantmannen UniBake);
- Growth in the number of foreign and Russian companies supplying a variety of ingredients for bakery products – "Puratos," "Ireks," CSM, "Backaldrin," "Leipurin Tukku," "Dalnyaya melnitsa," etc.; Increased consumption of baking mixes and product improvers;
- Changes in the traditional structure of the range of products and its globalization, reducing the share of bread from a mixture of rye and wheat flour and increasing demand for varieties and types of bread popular on the European market – multigrain mixes, toasted bread, national types of bread (Wholewheat, ciabatta, baguette, etc.), croissants, etc.
- Reduction in the production of short-term bakery products, development of the market of frozen semi-finished products, sandwiches, and products of long-term storage;
- Growth in volumes of packaged and sliced products, marginal varieties of bread, the bread for preventive, therapeutic, and functional purposes, ethnic types of bread (e.g., Baltic and Belarusian varieties), and portioned bread.

- The tendency to "return to tradition": traditional production technologies, the use of traditional raw materials, the production of products of time-tested taste and quality. The trend is the production of artisan bread – a natural, healthy product prepared according to traditional technology with a long maturation process and baking.

To improve the quality of bakery products and expand the range of products for therapeutic, preventive, dietary, and special purposes, branch institutes of the Russian Academy of Sciences have developed new recipes and innovative technologies:

- With different biological leaven;
- With enrichment with vitamins, organic acids, and other BAA, ensuring microbiological safety during storage;
- With prolonged storage time for different categories and groups of people living in remote areas;
- Bread based on frozen semi-finished products, etc.

1.2 Characteristics of Organizations Engaged in Baking

A number of publications devoted to the problems of small businesses consider the choice of type of activity and their industry affiliation (Almunia & Lopez-Rodriguez, 2018; Benzarti & Carloni, 2019), criteria for the size of enterprises (Skvortsova et al., 2019), and the development of technology, entrepreneurship, and innovation in small business (Linton & Solomon, 2017).

In addressing the issues of efficiency and sustainability of production development in the food industry (Tireuov et al., 2018), the emphasis is placed on the problems of food safety and the peculiarities of quality management (Akhmetova et al., 2017; Poltarykhin et al., 2018).

The paper aims to consider the sustainability of organizations by type of activity 10.71.1 "Production of bread and bakery products of short-term storage."

Initial data were obtained from the TestFirm database (TestFirm, 2021), compiled from the accounting and statistical reports of organizations. During this research, the authors evaluated 256 organizations related to the kind of activity 10.71.1 "Production of bread and bakery products of short-term storage" according to the All-Russian classifier of types of economic activities (OKVED) (Table 1.2).

By organizational and legal form, there prevail joint-stock companies (92.97%); there are 5.08% of consumer societies and 1.95% of cooperatives. The optimal form of business organization in the industry is a limited liability company, of which there are 214 units or 83.59% of the total number of joint-stock companies.

According to the period of activity on the market, among 256 organizations, the largest share of organizations is from 7 to 9 years (39.45%), 10 years and more (27.34%) (Fig. 1.1).

Table 1.2 General characteristics of organizations by type of activity 10.71.1 "Production of bread and bakery products of short-term storage"

Period of operation, years	Number of organizations, units	Share of organizations, %	Organizational and legal form			Revenue per organization in 2020, million rubles	Assets per organization in 2020, million rubles
			Joint Stock Company	Cooperative	Consumer society		
10 years and more	70	27.34	64	1	5	316.35	229.60
From 7 to 9 years	101	39.45	93	1	7	30.98	21.36
From 4 to 6 years	41	16.02	39	2	0	55.48	39.61
From 1 to 3 years	44	17.19	42	1	1	37.30	25.33

Source: Calculated by the authors based on (TestFirm, 2021)

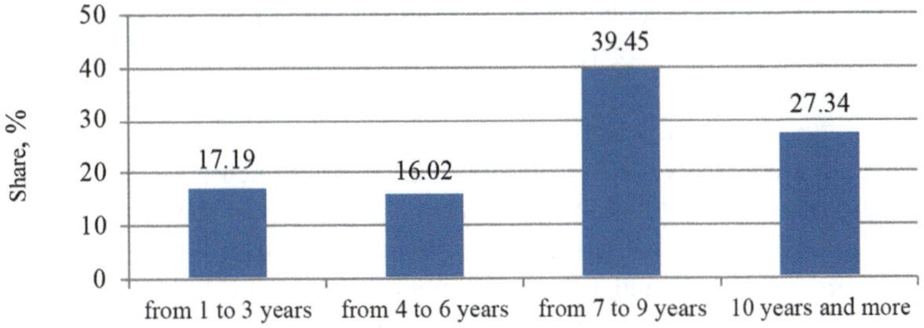

Fig. 1.1 Distribution of organizations by type of activity 10.71.1 "Production of bread and bakery products of short-term storage" by a period of activity in the market. *Source*: Calculated by the authors based on (TestFirm, 2021)

Table 1.3 Characteristics of organizations by type of activity 10.71.1 "Production of bread and bakery products of short-term storage" on the scale of activity

Categories of organizations	Number in the group, units	Share of organizations, %	Period of operation on the market, years	Revenue per organization in 2020, million rubles	Assets per organization in 2020, million rubles	Revenue to assets ratio
Large	2	0.78	10	5215.00	4773.50	1.09
Medium	2	0.78	10	1147.50	1133.50	1.01
Small	24	9.38	8.5	422.75	228.62	1.85
Microenterprises	142	55.47	7.5	42.21	23.46	1.80
Small-scaled	86	33.59	6.1	3.79	3.91	0.97

Source: Calculated by the authors based on (TestFirm, 2021)

According to the Federal law "On the development of small and medium-sized businesses in the Russian Federation" (July 24, 2007 No. 209-FZ) (Russian Federation, 2007), according to the scale of activity – the size of revenues, organizations are divided into medium-sized enterprises (revenues from 800 to 2000 million rubles), small enterprises (revenues from 120 to 800 million rubles), and microenterprises (revenues from 10 to 120 million rubles). Accordingly, enterprises with revenues over 2000 million rubles are considered large, and those with revenues of less than ten million rubles are considered small-scale.

Small- businesses can keep simplified accounting, prepare simplified accounting statements, and quit approving the cash balance limit. The main advantages of small bakeries are the flexibility of production and a large assortment. Small bakeries operate in small towns and rural settlements, thus increasing the level of employment of the rural population, as well as the physical and price affordability of bread.

According to the scale of activity (Table 1.3), micro-enterprises (55.47%) and small enterprises prevail in the industry (33.59%). Small enterprises occupy their segment due to the production of a wider range of products with a larger price range and due to their location within walking distance from the customer. Nevertheless, small enterprises are inferior to large bakeries in terms of technological indicators and product quality because bakeries are equipped with technological laboratories and ensure quality control of raw materials and finished products throughout the technological process.

The volume of bakery products in large and medium-sized enterprises is about 73%, in small enterprises – 16%, and individual entrepreneurs – 11%.

The ratio of revenues to assets is one of the indicators of the efficiency of organizations. Small and microenterprises are the most effective, with the ratio of revenues to assets being 1.85 and 1.80, respectively. The least efficient are small-scale enterprises with revenue to assets ratio of 97 kopecks per ruble.

1.3 Life Cycle Curve and Sustainability of Baking Businesses

According to the lifecycle concept, an industry goes through four stages in its development: market entry, growth, maturity, and decline (Porter, 2005). These stages are determined by changes in the growth rate of industry sales. Therefore, the life cycle curves for activity 10.71.1 "Production of bread and bakery products of short-term storage" are built according to the revenue growth rate.

Lifecycle curves for all organizations by rates of revenue and assets growth for 2011–2020 for activity 10.71.1 "Production of bread and bakery products of short-term storage" are presented in Fig. 1.2, and depending on the scale of their activities – in Fig. 1.3.

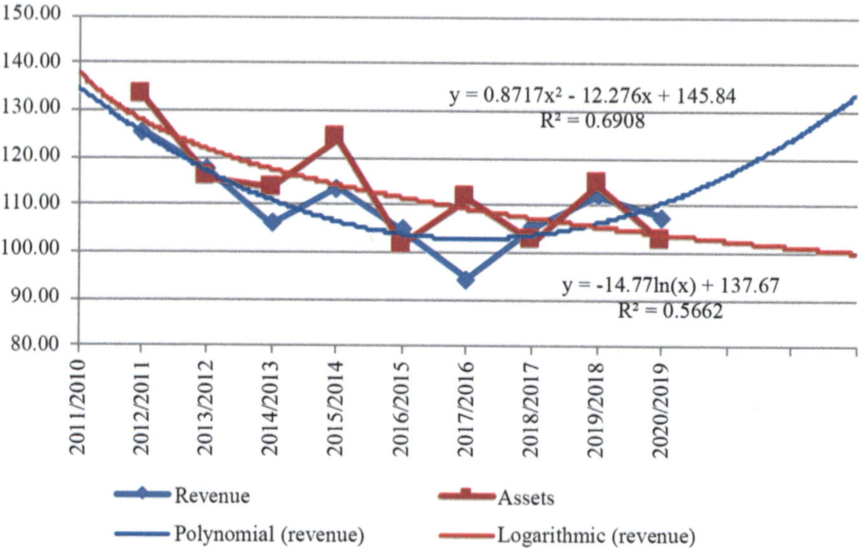

Fig. 1.2 Lifecycle curves for all organizations by rates of revenue and assets growth for 2011–2020 for activity 10.71.1 "Production of bread and bakery products of short-term storage". *Source*: Calculated by the authors based on (TestFirm, 2021)

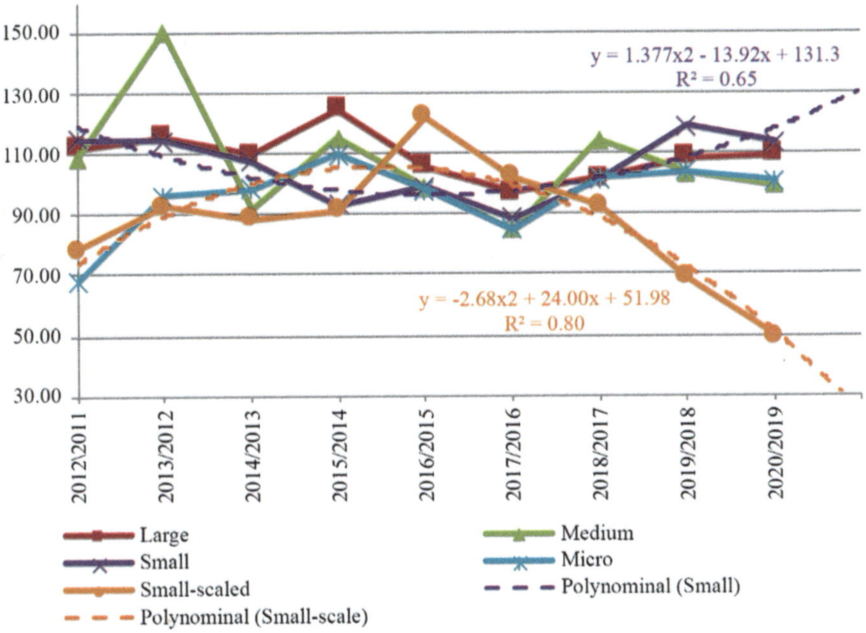

Fig. 1.3 Lifecycle curves of organizations depending on the scale of their activities in terms of revenue growth rate for 2011–2020 by type of activity 10.71.1 "Production of bread and bakery products of short-term storage". (*Source*: Calculated by the authors based on TestFirm, 2021)

The dynamics of the revenue growth rate are expressed as a polynomial curve trend with a confidence level of $R^2 = 0.69$. The dynamics of the growth rate of assets are described by the logarithmic curve with a confidence level of $R^2 = 0.56$. The predicted values on the curve indicate a possible increase in industry revenues and a decrease in the size of industry assets.

Lifecycle curves for organizations by the scale of activity show that organizations are at the maturity stage; growth is observed in small businesses, whose revenue growth rate by 2019 was the highest for the entire period. Small-scale enterprises with revenues of less than 10 million rubles a year are gradually leaving the market; it was shown above that they are less efficient. The change in revenue rates for small-scaled and small businesses are described by polynomial equations with a sufficiently high level of confidence. Forecast values indicate a continued decline for small-scaled businesses and growth for small businesses.

The sustainability of bakery organizations is associated with an increase in production cost due to the impact of prices and reduced efficiency in the use of resources. The growth of production costs is influenced by the growth of energy tariffs, which affects enterprises that use obsolete equipment, furnaces of higher required capacity, underutilized furnaces with a continuous process cycle, as well as organizations that work around the clock. This leads to a loss of energy resources and increases repair and maintenance costs.

Baking production is material-intensive because the share of costs for raw materials ranges from 52% to 60%. Prices for raw materials depend on the type of flour, the delivery period, and the supplier. The stability of production is related to breaches of the technology.

The formation of production cost is influenced by the size and efficiency of the use of resources: fixed assets, current assets, labor resources, and the loading of production facilities. The decline in the profitability of enterprises prevents the attraction of investment for modernization and expanded reproduction, which constrains the innovative development of the industry.

Flour quality is also unstable and has different baking properties. Limited financial resources and rising prices do not allow organizations to buy more expensive flour with higher quality. Small stocks of flour in production (2–3 tons on average) do not allow using flour subsorting at its low quality. The use of low-quality flour of class 4 and 5 in the production of bakery products forces organizations to use improvers, which reduce the fermentation time of dough, improve the appearance of products and their volume, but reduce the taste properties and quality of the crumb. The use of imported expensive additives and scarce dry gluten does not compensate for the negative effects of processing low-gluten and lightweight wheat, leading to a significant increase in the cost of production by an average of 3%–5%.

In Russia, only in two groups of Foreign Economic Activity Commodity Nomenclature (TN VED) in 2019–2021, imports of complex food additives (baking improvers) increased by 10% in volume terms and amounted to 40.7 thousand tons in 2021 (Fig. 1.4). Simultaneously, the cost of improvers increased by 41%, which will increase the cost of flour and bread.

France (29.6%) and Germany (29.1%) are the leading exporters of baking improvers (Fig. 1.5).

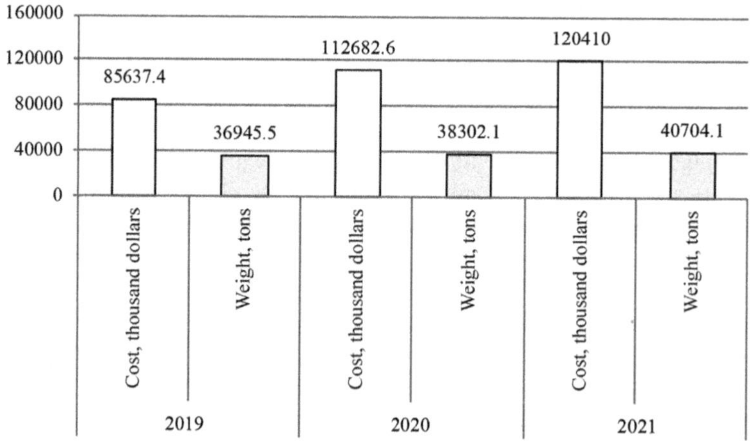

Fig. 1.4 Import of complex food additives – baking improvers into Russia according to TN VED 1901909900 and TN VED 2102300000. (*Source*: Calculated by the authors based on the materials of the Federal Customs Service of Russia (Federal Customs Service of the Russian Federation, n.d.))

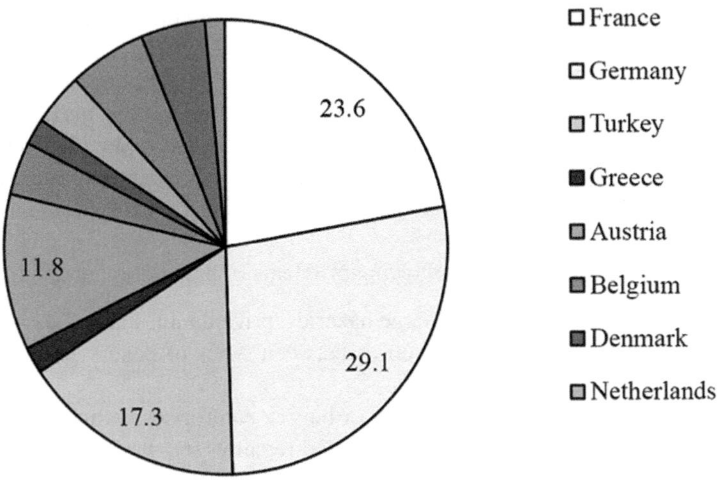

Fig. 1.5 Main countries-exporters of baking improvers to Russia. (*Source*: Code of Foreign Economic Activity Commodity Nomenclature Ltd, n.d.)

Table 1.4 Stability of organizations by type of activity 10.71.1 "Production of bread and bakery products of short-term storage" depending on its scale of activity

Category of organizations	Number in the group, units	Average value	RMS deviation	Fluctuation coefficient	Sustainability coefficient
Large	2	3915.72	935.37	0.24	0.76
Medium	2	1032.49	177.85	0.17	0.83
Small	24	423.33	42.88	0.10	0.90
Microenterprises	142	42.97	8.52	0.20	0.80
Small-scaled	86	11.24	3.38	0.30	0.70

Source: Calculated by the authors based on (TestFirm, 2021)

The uncontrolled use of expensive baking improvers and additives, mostly of imported origin, which is often of dubious quality, poses a threat to the health of bread consumers.

The use of fifth-grade wheat in the milling industry allows for expanding the uncontrolled use of forage wheat for food, which will significantly curb the rise in prices for the bread of dubious quality and for bread in general.

When baking bread, trade organizations often do not expand but rather duplicate the range of bakeries, which does not contribute to the growth of sales.

In statistics, business sustainability is defined as the difference between the unit and the fluctuation (variation) of the dynamics series. Let us calculate the degree of business sustainability by revenue depending on its scale for 2011–2020 (Table 1.4).

The results of the calculations of the stability coefficient show higher stability for small enterprises (0.9) and the lowest for small-scaled enterprises (0.7). The instability of small-scaled organizations in the baking industry, which make up 33.6% of all organizations in the industry, generates not so much financial but rather social risks because the industry has strategic importance for creating conditions for food security.

The instability of small businesses has also been affected by a number of legal regulations. For example, small businesses under the simplified taxation system have an organic number of employees, which leads to underemployment and informal employment. The cessation of the indexation of pensions for working pensioners from 2016 led to their mass layoffs. This exacerbated social problems in the labor market.

Thus, we can distinguish the following problems of bakery organizations:

- Increased competition between large bakeries, private mini-bakeries, production of bread in the retail trade, as well as between types of bakery products (rye-wheat, wheat-rye, wheat);
- Obsolete material and technical base of bakery enterprises, which leads to frequent equipment breakdowns and increased repair costs, the high energy intensity of production, violations of technology, and reduction of product quality;
- Low quality and competitiveness of bakery products produced by small enterprises due to poor quality control of raw materials, production technology, and finished products; unstable and low quality of flour processed by local producers; application of improvers, which leads to higher production costs;
- Differentiation in the cost of bakery products in the range of 9%–14%; imbalances in the formation of capital and resources of organizations;
- Insufficiently high level of professional training, a high proportion of management personnel in enterprises, poor motivation of labor, and low wages of production workers;
- Lack of marketing research of the local market of bakery products, which leads to inconsistency in the formation of a range of retailers; low level of advertising activities, promotion of brands, generation of demand, and sales promotion.

References

Akhmetova, S. O., Fuschi, D. L., & Vasiliunaite, R. (2017). Towards food safety: Quality management peculiarities. *Journal of Security and Sustainability Issues, 6*(3), 513–522. https://doi.org/10.9770/jssi.2017.6.3(15)

Almunia, M., & Lopez-Rodriguez, D. (2018). Under the radar: The effects of monitoring firms on tax compliance. *American Economic Journal: Economic Policy, 10*(1), 1–38. https://doi.org/10.1257/pol.20160229

Benzarti, Y., & Carloni, D. (2019). Who really benefits from consumption tax cuts? Evidence from a large VAT reform in France. *American Economic Journal: Economic Policy, 11*(1), 38–63. https://doi.org/10.1257/pol.20170504

Code of Foreign Economic Activity Commodity Nomenclature Ltd. (n.d.). *Baking improver.* Retrieved from https://kodtnved.ru/podbor/uluchshitel-hlebopekarnyy.html. Accessed 13 May 2022.

Federal Customs Service of the Russian Federation. (n.d.). *Customs statistics.* Retrieved from https://customs.gov.ru/statistic. Accessed 13 May 2022.

Linton, J. D., & Solomon, G. T. (2017). Technology, innovation, entrepreneurship and the small business-technology and innovation in small business. *Journal of Small Business Management, 55*(2), 196–199. https://doi.org/10.1111/jsbm.12311

Poltarykhin, A. L., Suray, N. M., Zemskov, Y. V., Abramov, Y. V., & Glotko, A. V. (2018). Food safety in The Russian Federation, its problems with the solutions. *Academy of Strategic Management Journal, 17*(4), 1–6.

Porter, M. E. (2005). *Competitive strategy: Techniques for analyzing industries and competitors (Transl. From English)*. Alpina Business Books.

Russian Federation. (2007). *Federal law "on the development of small and medium-sized businesses in The Russian Federation"* (July 24, 2007 no. 209-FZ, with amendments and additions). Moscow, Russia.

Skvortsova, T. A., Nikitina, A. A., Pasikova, T. A., & Tagaev, A. V. (2019). Concepts and criteria for the classification of small and medium-sized business in Russia. *International Journal of Economics and Business Administration, 7*(S1), 417–425. Retrieved from https://www.um.edu.mt/library/oar/handle/123456789/46072. Accessed 13 May 2022

TestFirm. (2021). *Rating of organizations by revenue*. Retrieved from https://www.testfirm.ru/rating/01/. Accessed 13 May 2022.

Tireuov, K., Mizanbekova, S., Kalykova, B., & Nurmanbekova, G. (2018). Towards food security and sustainable development through enhancing efficiency of grain industry. *Entrepreneurship and Sustainability Issues, 6*(1), 446–455. https://doi.org/10.9770/jesi.2018.6.1(27)

Chapter 2
Product Environmental Footprint and Bread Industry

Liudmyla Rayichuk, Maryana Draga, and Vira Boroday

2.1 Introduction

The growing consumption of food products with strong environmental impacts associated with economic development, as well as a projected population of 10 billion people (UN DESA, Population Division, 2011), place enormous pressure on the global agri-food system. Coping with the food crisis and tackling sustainable food security requires addressing all key aspects of food production and consumption through a holistic agri-food ecosystem approach, which primarily includes land and resource use, crop production, consumer behaviour and human health (Goucher et al., 2017).

For all of mankind, bread and wheat have been among the most important sources of nutrients for thousands of years in many civilizations. These days, scientific progress in various fields, mechanization and advances in both agriculture and food processing have increased yields and altered the biological and nutritional aspects of many crops and foods (Câmara-Salim et al., 2020).

The World Food Security Committee considers that the main problem today is the lack of growth in wheat yields in many countries around the world, while the world's population is constantly growing, and the issue of food security is becoming more urgent. Therefore, ensuring the production of the required amount of wheat can be mainly due to the increase in crop acreage. Thus, meeting the food needs of a growing population and the associated intensive production technologies as well

L. Rayichuk (✉)
Department of Radioecology and Remote Sensing of Landscapes, Institute of Agroecology and Environmental Management of NAAS, Kyiv, Ukraine

M. Draga · V. Boroday
Department of the Agrobioresources and Environmentally Safe Technologies, Institute of Agroecology and Environmental Management of NAAS, Kyiv, Ukraine

J. M. Ferreira da Rocha et al. (eds.), *Baking Business Sustainability Through Life Cycle Management*, https://doi.org/10.1007/978-3-031-25027-9_2

as an increase in ploughness rate have caused some serious environmental problems that are becoming increasingly global.

According to ISO 14044:2006 Environmental management. Life cycle assessment. Requirements and guidelines (2006), the life cycle is the successive and interconnected stages of the production system, from the acquisition of raw materials or their products from natural resources to the final disposal. Life cycle assessment (LCA) is one of the research methods that contribute to a better understanding of the impacts associated with the products they produce and consume to protect the environment. This helps to identify opportunities to improve the environmental performance of products at different stages of their life cycle. In the baking industry, one of the first stages of the product life cycle is the cultivation of cereals and obtaining high-quality and safe grain.

LCA consists of four stages:

(a) the stage of determining the purpose and scope;
(b) the stage of inventory analysis;
(c) the impact assessment phase;
(d) the stage of interpretation.

In the first and second stages, it is important to set goals and collect data on how to reduce the pesticide load on the cereal cultivation, such as the introduction of pathogens- and pests-resistant varieties, the usage of biological products based on bacteria stimulation that affects plant resistance to environmental factors (both high and low temperatures, lack of moisture, cytotoxic effects of 50 pesticides, damage by pests and diseases), which ultimately contributes to higher yields and improves product quality. To do this, we need to have information on the stability of modern cereal varieties, the biological fertilizers and crop protecting agents market, their application technologies, including their combination with chemical pesticides. Grain is the primary material needed for the production of bakery products (ISO 14044: 2006). One of the mandatory elements of life cycle impact assessment is the selection of impact categories, such as the impact of climate change on the prevalence of pathogens and pests of cereals, leading to overuse of pesticides and, in turn, environmental pollution. The last stage – the interpretation of the life cycle in the study of life cycle assessment covers several elements, namely: identification of significant problems, an assessment that takes into account the verification of completeness, sensitivity and consistency, conclusions, limitations and recommendations.

Each of the stages of the production technological cycle involves a certain environmental load, depletion of natural resources and/or deterioration of their quality. In the vast majority of cases, agricultural production is the first stage in this chain. It usually involves soil degradation, pollution of ecosystems and their components (water, soil, natural vegetation, air) by pesticides, heavy metals, radionuclides, etc., greenhouse gas emissions, reduction of natural biodiversity, ecosystems imbalance, etc.

Multitudinous studies have evaluated the environmental impacts associated with food consumption, expanding the knowledge base on the environmental impact of food (Tukker & Jansen, 2006; Schau & Fet, 2008; Notarnicola et al., 2012). This investigation contributes to the course of research in two ways. First, although bread is among the food products with the least environmental impact, it remains a staple and important food product that is consumed in large quantities and most countries (Braschkat et al., 2003; Roy et al., 2009; Espinoza-Orias et al., 2011; Kulak et al., 2012). For example, in the Nordic countries, the consumption of bread products is linked to food tradition and culture, while in Finland and Denmark there are strong traditions of sourdough rye bread baking (Nordic Ecolabelling, 2013). Secondly, interest in comparing the results of PCF (Product Carbon Footprint) studies is growing in terms of research, but in general is fraught with difficulties (Schau & Fet, 2008; Udo de Haes & Heijungs, 2007; Pulkkinen et al., 2010).

Assessing the environmental footprint in the bread industry, it should first be noted that the scale of the impact of the bakery cycle depends on a number of both global and local factors, such as the impact of climate change, features and scale of production, country and region, type of raw materials and products. It is also worth taking into account the dynamics of regional and global trends in the economic, social and environmental spheres.

2.2 Assessment of the Life Cycle of the Baking Industry. Proposal for the Analysis of Sustainability Assessment Using Environmental and Social Consequences

Plenty of bread products have been studied, including white, rye and wholemeal bread, as well as mixtures of these types. Studies tend to focus on white or mixed bread, but only a few studies specifically evaluating rye bread products (Nielsen et al., 2003; Grönroos et al., 2006; Saarinen, 2012; Jensen & Arlbjørn, 2014).

Similar to other food researches, a distinction is made between traditional and organic production methods, which mainly refers to how the crop was grown and processed (Roy et al., 2009; Hokazono & Hayashi, 2012; Schäfer & Blanke, 2012). Not surprisingly, numerous previous studies have focused on traditional bread, although Braschkat et al. (2003) and Grönroos et al. (2006) also included an assessment of organically produced bread. According to Braschkat et al. (2003), organic bread has a lower carbon footprint (368 g CO_2eq, on average), compared to conventionally produced bread (565 g CO_2eq, on average) and requires less energy use (see also Grönroos et al. (2006)). This distinction is explained by the fact that traditional cereal production also requires the production of mineral fertilizers that are not, or only to a limited degree, used in organic cereal production (Nordic Ecolabelling, 2013). These results indicate that organically produced bread is the best option in terms of greenhouse gas emission, but the production of organic bread does not always lead to a lower environmental impact (DEFRA, 2009; Notarnicola et al., 2012; Salomone & Ioppolo, 2012). The same goes for the comparison of traditional and organic types of grain production: it is difficult to decide which one is "better" due to large differences and uncertainties, particularly when modelling very complex organic production systems (Nordic Ecolabelling, 2013).

Preceding studies have had a shop or local bakery, and homemade cakes. While most researches have focused on industrial bread, Andersson and Ohlsson (1999), Braschkat et al. (2003) focused on comparing bread produced at different scales (Jensen & Arlbjørn, 2014).

The big industrial bakery requires more primary energy and contributes more to acidification, eutrophication and global warming, than the other three systems. The home baking system reflects a relatively high energy requirement; in other circumstances, the differences between the homemade bakery, the local bakery and the small industrial bakery are too small to be important (Andersson & Ohlsson, 1999). At the same time, the value of this distinction largely depends on the country and region and is determined by legislative and economic factors.

A characteristic distinguishing feature of bread production is the geographical region in which production takes place (Iriarte et al., 2010; Ruviaro et al., 2012).

The territory of production with its natural and socio-economic features, traditions, logistics, etc., as well as the peculiarities of the legislation of a certain country or a part of it sometimes affect the magnitude of the ecological footprint impact of products quite significantly. Bread has been studied in various national contexts, but the vast majority of studies have been conducted in Europe. The earliest identified study was from Sweden (Andersson & Ohlsson, 1999), while more recent studies were undertaken in the UK (Espinoza-Orias et al., 2011; Kingsmill, 2012; Sarrouy et al., 2012). The national context influences the PCF. For example, Espinoza-Orias et al. (2011) studied how wheat origins (e.g., in the UK, Canada, France, Germany and USA) affect PCF and determined that purchasing wheat locally or nationally (i.e., in the UK) may be preferable to importing wheat in terms of product quality (Jensen & Arlbjørn, 2014).

The global bread and bakery product market displayed modest, but sustainable growth throughout the period under review, increasing from 122,000 tonnes in 2007 to 129,000 tonnes in 2016.

Based on the results for 2016, the countries with the highest consumption were the U.S. (14.7 million tonnes), China (9.3 million tonnes), Russia (8.7 million tonnes), the UK (6.2 million tonnes), Germany (5.2 million tonnes), Egypt (4.6 million tonnes) and Italy (3.9 million tonnes), together accounting for approximately 41% of global consumption.

The highest annual rates of growth in terms of bread and bakery product consumption from 2007 to 2016 were recorded in China, with a +15.0% growth, the UK and Egypt, with a +10.0% and +10.1% growth, respectively. Consequently, China saw its share of global consumption surged from 2% in 2007 to 7% in 2016.

The market of bread and bakery products in Ukraine is highly competitive and presents many manufacturers. Baking companies offer a wide product range of products with different tastes, nutritional value, weight, and constantly replenish them with new samples. Demand for bakery products is stable, but the industrial output is constantly decreasing. The inconsistency of official statistics with the real capacity of the market complicates the planning processes at baking companies in the medium and long term period and identifying the needs in the flour as the main raw materials for the baking industry at the state level. The problem of providing domestic enterprises with flour of baking relevant properties at a reasonable price is exacerbated with the intensification of the crisis in the country (Kostetska, 2015). In Ukraine in the first half of 2020, the manufacturing of bread and bakery products amounted up to 373 thousand tonnes of bread, which is 14.2% less than in 2019.

There are surprisingly large ranges for the PCF of bread, with results from 256 to 2300 g CO_2eq/kg bread. According to Pulkkinen et al. (2010), this variation is explained by the choice of methodology, climate conditions, and type of energy used. As a rule, the carbon footprint is lower in earlier studies compared to later studies. In particular, studies targeting cradle to retail/ready-to-eat period have shown that production of white bread averages about 675 g CO_2eq/kg of bread, while studies limited from cradle to graves gave results of approximately 1425 g CO_2eq/kg bread on average (see Table 2.1) (Jensen & Arlbjørn, 2014). Wheat is the most widely grown commercial crop worldwide (Mekonnen & Hoekstra, 2010). More than 220 million hectares are under crop annually in many geographic regions of a wide range of climate conditions. Depending on agroclimatic conditions, about 670 million tons are produced annually. Therefore, wheat is one of the most important crops for global food security. Wheat production is roughly equally divided between the developing and the developed world, although production methods may differ (Shiferaw et al., 2013). At the same time, more than 67% of the world's gross grain harvest is provided by ten leading producing countries, including three leaders (China, the USA and India) – almost a half (47.22% of world production in 2019). However, at the same time, the structure of grain harvesting varies depending on the region of the country and the characteristics of the national crop production. Thus, the Asian countries are characterized by the dominance of rice in the structure

Table 2.1 Structure of grain harvest in the top 10 producing countries in 2019 (UNCTAD, 2020)

	Country	Gross grain harvest, million tonnes	The country's share in the world gross grain harvest,%	The structure of the gross harvest of cereals, %					
				wheat	rice	corn	barley	millet, sorghum	others
1	Argentina	70,59	2,38	26,2	1,9	61,6	7,2	2,2	0,9
2	Brazil	103,06	3,48	5,3	11,4	79,8	0,3	2,2	1
3	India	318,32	10,74	31,3	54,2	8,7	0,6	5,2	
4	Indonesia	113,29	3,82		73,3	26,7			
5	Canada	58,1	1,96	54,7		23,9	14,4		7
6	China	610,04	20,69	21,5	34,8	42,2	0,2	0,6	0,7
7	Russian Federation	109,84	3,71	65,7	0,9	10,4	15,5	0,2	7,3
8	United States of America	467,95	15,79	11	2,2	83,9	0,7	2,1	0,2
9	Ukraine	69,11	2,33	35,7	0,1	51,8	10,6	0,4	1,4
10	France	62,74	2,12	57,1	0,1	20,2	17,8	0,6	4,2

of the gross harvest of Mathew Reynolds grains (73.3% in Indonesia and 54.2% in India) (Table 2.1).

The total production of cereal crops in the world for a quarter of a century (1995–2019) has grown by 57%, according to the statistics of the UN FAO, and in the last 10 years, its growth has consistently outstripped the growth in global demand for grain. By 2029, according to the current FAO forecast, world cereal production will increase by about 14% more. More than half of this increase (52%) will be accounted for by corn, 23% – by wheat, 8% – by feed grain.

According to the forecasts, demand for wheat in developing countries is projected to increase and demand for wheat products with differentiated quality will place new demands on market channels; increased economic integration also will change the volume and direction of trade. Wheat demand is increasing in many countries, including the African continent. Demand in developing countries is projected to 60% increase by 2050 (Shiferaw et al., 2013) (Nelson et al., 2010). China, India, Russia and the United States are also world leaders in wheat production (Table 2.2).

Research usually identifies crop cultivation as dominant in PCF and thus marks it as a major hot spot (Table 2.1). Here, emissions of nitrous oxide (N_2O), a potent greenhouse gas, from agricultural land have a significant impact on the climate (Nordic Ecolabelling, 2013). Thus, the creation of sustainable agricultural systems is an important aspect of the development of sustainable food supply chains (Notarnicola et al., 2012; Goucher et al., 2017; Ingrao et al., 2017; Korsaeth et al., 2012).

Table 2.2 Leading wheat production countries[a] (FAOSTAT online database)

	Country	Production, million tonnes	Production per person, kg	Ares, mln. hectares	Yield, t/ha
1	China	131,70	94,48	24,35	5,41
2	India	93,50	69,96	30,23	3,09
3	Russian Federation	73,29	499,02	27,31	2,68
4	United States of America	62,86	191,78	17,76	3,54
5	Canada	30,49	819,24	9,26	3,29
6	France	29,50	438,42	5,56	5,30
7	Ukraine	26,10	617,52	6,21	4,21
8	Pakistan	26,01	128,82	9,14	2,84
9	Germany	24,46	295,67	3,20	7,64
10	Australia	22,27	889,22	11,28	1,97

[a]Aggregate, may include official, semi-official, estimated or calculated data

While there is certainly potential to improve the environmental performance of both pre-farm and off-farm processes, our main focus should be on improving soil and crop management at the farm level.

In addition, the stages of production and consumption of bread are usually identified in studies as the second and third most important stages, except for Andersson and Ohlsson (1999), who identified transport as the second-largest contribution. However, this result is most likely since Andersson and Ohlsson (1999) included consumer transport in retail stores, which are excluded in later PAS-compliant PCF studies (PAS 2050, 2011). Nevertheless, transportation is integral in the life cycle of many products, this is not usually the case in fresh baked goods research (Nordic Ecolabelling, 2013; Jensen & Arlbjørn, 2014).

According to the survey on the determination of carbon footprint in the baking industry in obedience to ISO 14044: 2006 'Environmental management – Life cycle assessment – Requirements and guidelines, one of the main significant problems was the crop growing process (Espinoza-Orias et al., 2011). A similar conclusion was made by Galli et al. (2015), who also called the stage of agricultural cultivation, including the use of nitrogen-containing fertilizers, the primary hotspot in the carbon footprint.

The introduction of the LCA (i.e. Life-Cycle Analysis) model to assess the environmental impact of industrial bread production has shown the importance of starting the chain, namely the supply of raw materials, and identifying possible ways to improve the quality of raw materials. In this context, the authors emphasize that the technological innovation of cultivation systems, including the introduction of GMP standards (i.e. good manufacturing practice), is a particularly important feature that guarantees the eventual result (Galli et al., 2015).

Norwegian scientists have assessed the environmental impact of bread production based on ten impact categories, including the potential for global warming, as well as the impact of growing barley, oats, winter and spring wheat on 93 farms that were representative of Norway's main grain production regions. And it was the latter that had a large share of environmental impacts associated with bread production (Andersson & Ohlsson, 1999; Korsaeth et al., 2012).

The Food and Agriculture Organization of the United Nations (FAO) states that up to 40% of all crops in the world is lost each year from pests and diseases. As a result, the total loss of trade in agricultural products is more than 220 billion US dollars per year (Oerke, 2006). A similar situation is observed in Ukraine, in particular, a significant deterioration of the phytosanitary condition of cereal agrocenoses, which leads to a shortage of about 25% of the crop, and in some years, up to 80% because of epizooty and epiphytosis. According to Mostovyak et al. (2020a), the amount of chemical pesticides is 1.58–1.77 kg as/ha per year, the share of the biological method in the total amount of plant protection products reaches only 5.2%.

In modern technologies of growing cereals in the plant protection system is dominated by the chemical method. However, even with high potential efficiency, such a method is not able to ensure long-term stabilization of the phytosanitary condition of agroecosystems and their environmental safety. At the same time, there is a violation of ecological balance in agrocenoses, their chemical and biological pollution, reduction of product quality, reduction of species biodiversity, the emergence of new species and resistant forms of pathogens, growth of their harmfulness, etc. (Mostovyak et al., 2020b). Seeds of the analyzed varieties of winter wheat, spring barley and oats of intensive and semi-intensive type are contaminated with phytopathogenic fungi of the genera *Alternaria, Fusarium, Nigrospora, Bipolaris, Penicillium, Mucor, Epicoccum, Glicocladium, Drechslera* with a height of eight million biological threat to agrophytocenoses (Mostovyak et al., 2020b). Under favourable conditions of overwintering, warm, moderately humid weather in spring and dry weather in summer, active population and increase in the number of phytophagous insects causes an excess of EPS in crops of cereals by 2.5–5.7 times.

From this perspective, the main factors of destabilization of phytosanitary condition are high ploughness of the Central Forest-Steppe of Ukraine, violation of the scientifically substantiated structure of cultivated areas, intensive and semi-intensive cultivation of grain crops with high ability to stimulate development and accumulation of phytopathogenic microbiome.

2.3 Ecological Trace of the Product and the Bread Industry

An alternative and environmentally friendly method of ensuring high quality and safety standards of agricultural products are to reduce the use of chemical pesticides and the widespread usage of plant protection products of biological origin and resistant crop varieties.

Ukrainian microbiologists have created many microbial specimens based on active strains of nitrogen-fixing, phosphate-mobilizing, growth-stimulating microorganisms. These are Albobacterin, Biogran, Diazobacterin, Microgumin, Polymyxobacterin, Rhizohumin, Hetomik (Institute of Agricultural Microbiology and Agroindustrial Production of National Academy of Agrarian Sciences of Ukraine), Biopolicid, Rhizoactiv, Rhizobophyt (Institute of Agroecology and Environmental Management of National Academy of Agrarian Sciences of Ukraine) Azogran (DK Zabolotny Institute of Microbiology and Virology, National Academy of Sciences of Ukraine) and others. Developed biologies are characterized by high efficiency. Most of them are certified for usage in technologies of organic production of agro-food.

The use of biologicals significantly affects the formation of the root system, its absorptive capacity, the activity of some enzyme systems of the plant, which helps to optimize the absorption of nutrients by the plant. According to experiments with the heavy isotope ^{15}N and isometric studies conducted at the Institute of Agricultural Microbiology and Agricultural Production of the National Academy of Agrarian Sciences of Ukraine, the degree of nitrogen uptake from fertilizers using microbial drugs increases by 20–30%, while reducing the intensity of migration of nutrients to the soil. In general, according to the results of field experiments and industrial tests, the effect of biological on crop productivity is equivalent to the impact of 30–60 kg/ha of mineral nitrogen, 20–40 kg/ha of phosphorus (Volkogon, 2018).

According to Tchaikovsky (2011), under the conditions of the transition of the national economy of Ukraine to a market basis, there was a sharp decrease in agriculture intensification. Due to the dominance of the extensive form of production, soils have lost a significant part of humus and nutrients. The problem of providing plants with phosphorus is especially acute: the results of recent rounds of agrochemical surveys indicate a shortage of mobile phosphorus in the arable soil layer of all soil and climatic zones, including black soils. To preserve and restore soil fertility, it is necessary to introduce some measures that would ensure their rational use and stability of agricultural production (Tchaikovsky, 2011).

Was found that biological products helped to increase the grain yield of spring barley on all backgrounds. However, the most stable effect was observed with the use of Phosphoenterin and Polymyxobacterin. Thus, bacterization of seeds contributed to the growth of grain productivity of barley on unfertilized areas by 7–17%, with use of $N_{30}P_{30}$ – by 11–13%, with estimated dose (N_{53}) – by 7–20% compared to standard treatment. The highest yield growth rates in unfertilized areas were observed using Polymyxobacterin; with use of $N_{30}P_{30}$ the effect of Phosphoenterin and Polymyxobacterin was at the same level, against the calculated dose of fertilizers (N_{53}) the highest yield was obtained using Phosphoenterin. According to the results of the economic evaluation of the use of Phosphoenterine, the additional costs associated with the use of the biological product are repeatedly recouped by the effect of its action. Thus, it was found that when growing spring barley, the economic efficiency of Phosphoenterin in unfertilized areas and with use of $N_{30}P_{30}$ and the estimated dose of fertilizers ($N_{57.0}$) exceeded the control indicators: profit increased by 158, 120 and 188 UAH, profitability – by 9.2, 10.5 and 22.2%; and the cost of grain decreased by 7.2, 7.9 and 15%, respectively. Analyzing the results of three-year studies obtained in the cultivation of winter wheat, it was found that bacterization also has a positive effect on the growth of its grain productivity.

Thus, on all agricultural backgrounds there was an increase in yield growth compared to the control over the use of biological products: on no-fertilized areas – by 8–17%, with using as a background P_{30} – by 17–37%, P_{60} – by 17–32%, P_{90} – at 9–18%. The most stable indicators of the effect of biological products on the yield of winter wheat were obtained with using Polymyxobacterin and Phosphoenterin and P_{30} and P_{60} as a background.

However, a study of the effect of bacterization on grain quality revealed that the highest levels of protein and gluten in the grain were obtained with the use of biological products in comparison with using P_{30} as a background. Thus, with the use of Polymyxobacterin, Albobacterin and Phosphoenterine the protein content in grain respectively was 13.7%, 13.2% and 12.5% (to standard treatment – 9.9%), gluten – 31.7%, 30.0% and 28.0% (to standard treatment – 19,2%) (Tchaikovsky, 2011).

According to Mostovyak et al. (2020a), it is determined that the reduction of herbicide application rates when used in combination with biological drugs and growth regulators, leads to reduction of pesticide load on plants and soil; an increase of photosynthetic productivity of crops and stability of soil microbiota; reducing the cytotoxicity of the soil and increasing its suppression. The high efficiency of pest control for the combination of herbicides, biological preparations and plant growth regulators in tank mixtures has been theoretically substantiated and experimentally confirmed.

The positive synergetic effect of combination herbicides in tank mixtures, biological preparation with fungicidal action and plant growth regulators against pests is proved, which provides reduction of pesticide load on agrocenosis up to 25%, increase of photosynthetic activity of crops by 22% on average. Developed plant protection systems increase the resistance of soil microbiota to pesticides, as evidenced by the growth of the total number of bacteria by 19–54%, a balanced ratio between microorganisms of different ecological and trophic groups, reducing the activity of mineralization processes by 5.1–17.1% and decomposition of organic substances by 2.9–9.0% in the soil. At the same time, the phytotoxicity of the soil decreases by 11.3–43.9% and its suppression increases due to the reduction of the share of phytopathogenic species of micromycetes by 1.2–1.5 times.

The influence of the complex bacterial preparation Azogran on the yield of winter wheat of the Tsarivna variety in the conditions of the Forest-Steppe of Ukraine on grey forest soil was investigated by Korniychuk et al. (2018). The use of the biological preparation with several fertilizer systems contributed to the increase of wheat yield by 0.57–0.62 t/ha for resource-saving plant protection technology and by 0.49–0.55 t/ha for intensive. Bacterization helped to reduce plant damage by root rot, leaf and ear septoria. The content of crude protein and gluten in the obtained grain increased by 0.6–0.9% and 1.2–1.3% for resource-saving and by 0.6–0.9% and 1.1–1.3% for intensive technologies of cultivation accordingly (Korniychuk et al., 2018).

The positive influence of biologization elements on qualitative indicators of crop structure and grain quality has been proved by Burykina et al. (2014). The enhanced elements of the biologized technology of growing winter wheat allowed to obtain high-quality grain with a protein content of 14.5%, gluten – 27.6% and the first-class grain.

The ability of new plant forms of Triticale to maintain high adaptive potential and seed productivity under abiotic stress factors, which is a manifestation of the mechanism of genetic heterogeneity, as evidenced by the heterogeneity of phenetic markers and high ecological plasticity of *Secale cereal* L., *Triticum trispecies* Shulind., some varieties of *Triticum aestivum* L. (Ariivka, Yuvivata 60, etc.). During inoculation of plant seeds of varieties and lines of the Triticale tribe with nitrogen-fixing and phosphate-mobilizing strains of microorganisms, the manifestation of the mechanism of additivity was noted, as evidenced by the increase in seed productivity (0.6–1.3 t/ha).

According to the criteria of adaptability mechanisms, ecologically adaptive varieties and lines of *T. aestivum* L. and *T. trispecies* Shulind. of Forest-Steppe and Polissia ecotypes were created, 9 of which were entered into the Genetic Bank of Plants of Ukraine (Zoryana Nosivska, Chayan, L 41/95, L 3-95, KS 16 -04, KS 7-04, L 22-04, L 4639/96, KS 59-95), Yuvivat 60 – to the State Register of Plants of Ukraine, and for varieties Vivate Nosivske and Nosshpu 100 received certificates of authorship.

For the first time on the example of ecological testing and introduction of highly adaptive and stable plant forms of the *Triticale* tribe into the conditions of Polissia and Forest-Steppe, the way to solving the problems of reducing genetic diversity and pesticidal pressure on the environment manifestations (Moskalets & Moskalets, 2015).

References

Andersson, K., & Ohlsson, T. (1999). Life cycle assessment of bread produced on different scales. *The International Journal of Life Cycle Assessment, 4*, 5–40. https://doi.org/10.1007/BF02979392

Braschkat, J., Patyk, A., Quirin, M., & Reinhardt, G. A. (2003). *Life cycle assessment of bread production – a comparison of eight different scenarios*. In: Proceedings of the 4th international conference on life cycle assess in the agri-food sector, Bygholm, Denmark, 6–8 October 2003

Burykina, S. I., Smetanko, A. V., & Pilipenko, V. N. (2014). Harvest and quality of winter wheat in the steppe zone of Ukraine. *Soil Science and Agrochemistry. Scientific Journal (Minsk), 1*(52), 210–226.

Câmara-Salim, I., Almeida-García, F., González-García, S., et al. (2020). Life cycle assessment of autochthonous varieties of wheat and artisanal bread production in Galicia, Spain. *Science of the Total Environment, 713*(136720), 136720. https://doi.org/10.1016/j.scitotenv.2020.136720

Department for Environment, Food and Rural Affairs [DEFRA]. (2009). *Greenhouse gas impacts of food retailing*. DEFRA. Project code FO0405.

Espinoza-Orias, N., Stichnothe, H., & Azapagic, A. (2011). The carbon footprint of bread. *The International Journal of Life Cycle Assessment, 16*, 351–365.

Food and Agricultural Organization of the United Nations. FAO Statistical Database [FAOSTAT]. http://www.fao.org/faostat/en/#data

Galli, F., Bartolini, F., Brunori, G., et al. (2015). Sustainability assessment of food supply chains: An application to local and global bread in Italy. *Agricultural and Food Economics, 3*(21). https://doi.org/10.1186/s40100-015-0039-0

Goucher, L., Bruce, R., Cameron, D., et al. (2017). The environmental impact of fertilizer embodied in a wheat-to-bread supply chain. *Nature Plants, 3*(17012). https://doi.org/10.1038/nplants.2017.12

Grönroos, J., Seppala, J., Voutilainen, P., et al. (2006). Energy use in conventional and organic milk and rye bread production in Finnland. *Agriculture, Ecosystems & Environment, 117*(2–3), 109–118.

Hokazono, S., & Hayashi, K. (2012). Variability in environmental impacts during conversion from conventional to organic farming: A comparison among three rice production systems in Japan. *Journal of Cleaner Production, 28*(1), 101–112.

Ingrao, C., Licciardello, F., & Pecorino B at al. (2017). Energy and environmental assessment of a traditional durum wheat bread. *Journal of Cleaner Production, 171*, 1494–1509. https://doi.org/10.1016/j.jclepro.2017.09.283

Iriarte, A., Rieradevall, J., & Gabarrell, X. (2010). Life cycle assessment of sunflower and rapeseed as energy crops under Chilean conditions. *J Clean Prod, 18*(4), 336–345.

ISO 14044: 2006. *Environmental management. Life cycle assessment. Requirements and guidelines*. https://www.iso.org/standard/38498.html

Jensen, J. K., & Arlbjørn, J. S. (2014). Product carbon footprint of rye bread. *Journal of Cleaner Production, 82*, 45–57.

Kingsmill. 2009/2012. Kingsmill and the environment. http://www.kingsmillbread.com/fresh-thinking/environment/carbon-footprint/

Korniychuk, O. V., Plitnikov, V. V., Gilchuk, H. H., et al. (2018). Influence of complex bacterial preparation Azogran on winter wheat yield. *Agricultural Microbiology, 27*, 67–73. http://nbuv.gov.ua/UJRN/smik_2018_27_12

Korsaeth, A., Jacobsen, A. Z., & Roer AG at al. (2012). Environmental life cycle assessment of cereal and bread production in Norway. *Acta Agriculturae Scandinavica, Section A – Animal Science, 62*(4), 242–253. https://doi.org/10.1080/09064702.2013.783619

Kostetska, N. I. (2015). The market of bread and bakery products of Ukraine: State and prospects of development. *Galician Economic Bulletin, 48*(1), 26–31.

Kulak, M., Nemecek, T., Trossard, E., et al. (2012). *Ecodesign opportunities for a farmer's bread. Two case studies from north-western France*. In: Proceedings from the 8th international conference on LCA in the agri-food sector, Saint-Malo, 1–4 October 2012

Mekonnen, M. M., & Hoekstra, A. Y. (2010). *A global and high-resolution assessment of the green, blue and grey water footprint of wheat*. Value of water research report series 42, UNESCO-IHE Institute for Water Education: Delft, The Netherlands. http://www.unesco-ihe.org/Value-of-Water-Research-Report-Series/Research-Papers

Moskalets, T. Z., & Moskalets, V. V. (2015). Autecological and democological manifestations of modification ability of genotypes of Triticale tribe of forest-steppe and Polissya ecotypes. *Ecology and noospherology, 26*(1–2), 61–72.

Mostovyak, I. I., Demyanyuk, O. S., & Borodai, V. V. (2020a). Formation of phytopathogenic fond in agrocenoses of cereals of the right-bank Forest-steppe of Ukraine. *Agroecologial Journal, 1*, 28–38.

Mostovyak, I. I., Demyanyuk, O. S., Parfenyuk, A. I., et al. (2020b). Variety as a factor in the formation of stable agrocenoses of grain crops. *Bulletin of the Poltava State Agrarian Academy, 2*(97), 110–118.

Nelson, G. C., Rosegrant, M. W., & Palazzo, A., et al. (2010). Food security, farming, and climate change to 2050 scenarios, results, policy options. IFPRI research monograph. International Food Policy Research Institute (IFPRI). https://doi.org/10.2499/9780896291867

Nielsen, P. H., Nielsen, A. M., & Weidema, B. P., et al (2003). LCA Food Database. www.lcafood.dk

Nordic Ecolabelling Annual Report. (2013). https://www.svanemerket.no/PageFiles/9831/Nordic_Ecolabel_2013_web.pdf

Notarnicola, B., Tassielli, G., & Renzulli, P. (2012). Modeling the agri-food industry with life cycle assessment. In *Life cycle assessment handbook* (pp. 159–183). https://doi.org/10.1002/9781118528372.Ch.7

Oerke, E. C. (2006). Crop losses to pests. *The Journal of Agricultural Science, 144*, 31–43.

Publicly Available Specification [PAS] 2050. (2011). https://shop.bsigroup.com/upload/shop/download/pas/pas2050.pdf

Pulkkinen, H., Katajajuuri, J. M., Nousiainen, J., et al. (2010). *Challenges in the comparability of carbon footprint studies of food products*. In: Proceedings from the 7th international conference on LCA in the agri-food sector, Bari, 22–24 September 2010

Roy, P., Nei, D., Orikasa, T., et al. (2009). LCA on some food products. *Journal of Food Engineering, 90*(1), 1–10.

Ruviaro, C. F., Gianezinia, M., Brandãoa, F. S., et al. (2012). Life cycle assessment in Brazilian agriculture facing worldwide trends. *Journal of Cleaner Production, 28*, 9–24.

Saarinen, M. (2012). *Nutrition in LCA: Are nutrition indexes worth using?* In: Proceedings from the 8th international conference on LCA in the agri-food sector, Saint-Malo, 1–4 October 2012

Salomone, R., & Ioppolo, G. (2012). Environmental impacts of olive oil production: A Life Cycle Assessment case study in the province of Messina (Sicily). *J Clean Prod, 28*, 88–100. https://doi.org/10.1016/j.jclepro.2011.10.004

Sarrouy, C., Davodson, J., & Lillywhite, R. (2012). *Product energy use within the agri-food supply chain*. In: Proceedings from the 8th Int. conference on LCA in the agri-food sector, Saint-Malo, 1–4 October 2012

Schäfer, F., & Blanke, M. (2012). Farming and marketing system affects carbon and water footprint – a case study using Hokaido pumpkin. *Journal of Cleaner Production, 28*, 113–119. https://doi.org/10.1016/J.JCLEPRO.2011.08.019

Schau, E. M., & Fet, A. M. (2008). LCA studies of food products as background for environmental product declarations. *The International Journal of Life Cycle Assessment, 13*(3), 255–264.

Shiferaw, B., Smale, M., Braun, H. J., et al. (2013). Crops that feed the world 10. Past successes and future challenges to the role played by wheat in global food security. *Food Security, 5*, 291–317.

Tchaikovsky, L. O. (2011). The effectiveness of the combined use of biological products based on phosphate-mobilizing bacteria and mineral fertilizers in the cultivation of cereals in the south of Ukraine. *Agricultural Microbiology, 13*, 52–58.

Tukker, A., & Jansen, B. (2006). Environmental impact of products: A detailed review of studies. *J Ind Ecol, 10*(3), 159–182.

Udo de Haes, H. A., & Heijungs, R. (2007). Life-cycle assessment for energy analysis and management. *Applied Energy, 84*(7), 817–827. https://doi.org/10.1016/j.apenergy.2007.01.012

United Nations Conference on Trade and Development [UNCTAD]. (2020). *Report 2020*. https://unctad.org/system/files/official-document/tdr2020_en.pdf

United Nations, Department of Economic and Social Affairs [UN DESA], Population Division. (2011). *World population prospects: The 2010 revision, vol. I: Comprehensive tables*. ST/ESA/SER.A/313

Volkogon, V. V. (2018). Agricultural microbiology in Ukraine: Achievements, problems, prospects. *Bulletin of Agrarian Science, 11*, 20–27.

Chapter 3
Life Cycle Assessment and Product Environmental Footprint: Recommendations for Integral Optimization of Economic and Environmental Performance

Liudmyla Rayichuk, Maryana Draga, and Vira Boroday

3.1 Introduction

Mostovyak et al. (2020) contribute to the solution of an important ecological problem of modern domestic agricultural production – reduction of chemical load and negative impact on the agroecosystem, improvement of an ecological and phytosanitary condition of agrocenoses. To stabilize the phytosanitary condition of crops in the Central Forest-Steppe of Ukraine, they recommended:

- adjustment of the cultivated area structure, namely reduction of the share industrial crops by 50%, incl. sunflower;
- cultivation of varieties that can inhibit the intensity of sporulation of phytopathogenic micromycetes;
- cultivation of cereals after the predecessors of perennial grasses or legumes (peas);
- combination of herbicides, biologicals and growth regulators in the plant protection system.

In the technology of growing cereals it is recommended to use tank mixtures of herbicides, biofungicides and plant growth regulators, which provide: increase the photosynthetic potential of crops, plant resistance to pathogens and phytophagous insects, reducing the chemical load on agroecosystem up to 30%.

Reduction of pesticide load up to 30% on agrophytocenosis of spring cereals (barley, wheat) can be achieved by applying herbicides in tank mixtures at lower

---

L. Rayichuk (✉)
Department of Radioecology and Remote Sensing of Landscapes, Institute of Agroecology and Environmental Management of NAAS, Kyiv, Ukraine

M. Draga · V. Boroday
Department of the Agrobioresources and Environmentally Safe Technologies, Institute of Agroecology and Environmental Management of NAAN, Kyiv, Ukraine

J. M. Ferreira da Rocha et al. (eds.), *Baking Business Sustainability Through Life Cycle Management*, https://doi.org/10.1007/978-3-031-25027-9_3

rates with biological products and plant growth regulators. This combination of agricultural chemicals in the plant protection system creates favourable conditions for plant growth and development, the formation of photosynthetic activity of crops and grain yields, reduces the negative effects of chemical treatments and stress on the plant and soil microbiocenosis caused by pesticides.

The combination of chemical and biological preparations with plant growth regulators in the system of their protection against pests was introduced in 32 farms of Kyiv, Vinnytsia, Kirovohrad and Cherkasy regions on an area of 2412 ha and provided a 22% yield increase in average.

To improve the ecological and phytosanitary condition of agrocenoses of grain crops in the Central Forest-Steppe of Ukraine and to reduce the pesticide load on the agroecosystem, it is necessary to widely implement ecologically safe pest control measures, which include:

- obligatory adjustment to the scientifically substantiated norms of the structure of sown areas and observance of crop rotation in crop rotation;
- cultivation of varieties that during the growing season inhibit the sporulation of phytopathogenic fungi at an ecologically safe level (<1 million spores/ml) (for example, winter wheat – Aurora Myronivska, Podolyanka, spring barley – MIP Sharm, oats – Treasure of Ukraine), or are stable to the settlement of phytophagous and the defeat of viral, fungal and mycoplasmal diseases (for example, winter wheat – Chornyava, Zorepad, Lars, Litanivka, Voloshkova);
- sowing of winter wheat should be carried out in the III decade of October, spring barley – in the 1 decade of April after the predecessors of legumes (peas) or perennial grasses;
- apply chemical plant protection products with less pesticide active substance required for effective control of pests, low potential for pollution of the environment (soil, water sources, etc.) and impact on biodiversity; compatibility with biological drugs and growth regulators; to conduct constant monitoring of pests in agrocenoses and use agrometeorological forecasts to develop an effective plant protection system.

The technology of growing spring barley recommends a cost-effective and environmentally friendly control system of weeds and fungal pathogens, which combines in a tank mix herbicide Caliber 75 v.g. (application rate 40 g/ha) or Lintur 70 WG v.g. (100 g/ha), with a biological product with fungicidal action Agate 25-K, etc. (20 ml/ha) and growth regulator Agrostimulin (10 ml/ha) during the treatment of vegetative plants at the tillering phase, which provides reduced to 30% rates of application of herbicides to preserve grain yield at 19–25%.

The economic analysis proved that the maximum conditional net profit and profitability of 94–97% was provided by the technology with the predecessor of black steam, application of mineral fertilizers by calculation method and pre-sowing seed treatment with biological products Rizoagrin, FMB, Planriz. The use of the Trichodermin helped to reduce the cost with the formation of profitability of 132.7%. Energy intensity increased to 7.06 GJ/t when growing winter wheat after the predecessor peas, applying a full dose of mineral fertilizers under the main tillage and without the use of biological products for seed treatment before sowing. The maximum energy factor at the level of 2.89–3.64 was obtained in the version with the predecessor of black steam, the use of calculated doses of nitrogen, phosphorus and potassium fertilizers and seed treatment with biological products (Burykina et al., 2014).

3.2 The Most Common Groups of Fertilizers and Plant Growth Regulators Used for the Spring Wheat in the Forest-Steppe Zone of Ukraine

Plants need nutrients to grow which they absorb from the soil via the plant's root system. Fertilizers provide the major nutrients (nitrogen, phosphorus, potassium and important secondary elements) that plants need. Unless the nutrients are replenished, the soil's productive capacity declines with every harvest.

Nitrate-based fertilizers are the most used straight fertilizers in Europe. The main products are nitrate-based fertilizers such as ammonium nitrate (AN) and calcium ammonium nitrate (CAN), which are well suited to most European soils and climatic conditions, and urea and urea ammonium nitrate (UAN) aqueous solutions, which are widely used in other parts of the world.

The most common phosphate fertilizers are single superphosphate (SSP), triple superphosphate (TSP), monoammonium phosphate (MAP), diammonium phosphate (DAP) and ammonium polyphosphate liquid.

Potassium is also available in a range of fertilizers, which contain potassium only whether two or more nutrients and include Potassium chloride (KCl), Potassium sulphate (K_2SO_4) or sulphate of potash (SOP), Potassium nitrate (KNO_3), known as KN.

Calcium (Ca), magnesium (Mg) and sulphur (S) are essential secondary plant nutrients. They are not usually applied as straight fertilizers but in combination with the primary nutrients N, P, and K.

Sulphur is often added to straight N fertilizers such as ammonium nitrate or urea. Other sulphur sources are single superphosphate (SSP), potassium sulphate (SOP) and potassium magnesium sulphate (Kainite), the latter also containing magnesium.

Kieserite is a magnesium sulphate mineral that is mined and also used as fertilizer in agriculture, mainly to correct magnesium deficiencies. Calcium is mainly applied as calcium nitrate, gypsum (calcium sulphate) or lime/dolomite (calcium carbonate), of which calcium nitrate is the only readily plant-available source of calcium.

Crop residues, animal manures and slurries are the principal organic fertilizers. Although they have varying nutritional values, they are generally present on the farm and the nutrients and the organic carbon they contain are recycled. Animal manures and slurries cover a wide range of nutrient sources with different physical properties and nutrient contents. Furthermore, their nutrient content varies regionally and depends on the type of livestock and the farm management system.

Cereals are the main product of human food; in addition, they are concentrated feed for animals. Spring wheat is a reserve for high-quality grain, which has high baking and cereal qualities, contains more protein than winter wheat grain (14–16% – soft and 15–18% – hard) and gluten (28–40%).

Spring wheat grain is used for baking quality bread, bakery products, production of the best varieties of pasta, vermicelli, semolina. Spring wheat is also of fodder value; in particular, it can be used for the production of compound feed, bran as a concentrated feed, straw and chaff – as roughage. In addition, it is a valuable insurance crop for reseeding dead winter wheat crops.

One of the most reliable and cost-effective factors for increasing the gross harvest is the use of new high-yielding varieties most adapted to modern cultivation technologies.

To date, 55 varieties of spring wheat have been entered in the State Register of Plant Varieties Suitable for Distribution in Ukraine, of which only in 2020 six new varieties of soft wheat (spring) were registered – Lycamer, Yaryna, MIP Solomiya, PS Petra, SHTRU093736s4, MIP Dana and two varieties of durum wheat – Remarque and MIP Ksenia. Therefore, the choice of varieties is quite wide.

Some of them are offered for sale by seed and nursery entities of the Khmelnytsky region. Thus, in 2020, spring wheat crops of the varieties Elegiya Myronivska, Struna Myronivska, MIP Svitlana, MIP Raiduzhna and Simkoda Myronivska were certified in the region. These varieties belong to the selection of the V.M. Remeslo Myronivka Institute of Wheat of NAAS.

3.3 General Characteristics of Organomineral Fertilizers, Plant Growth Regulators and Microbiologic Preparations, Used in Agrotechnologies for Cereals in Ukraine

In all soil-and-climatic zones of Ukraine, fertilizers, usually, improve the quality of the wheat harvest, increase the content of protein, gluten in grain and improve baking qualities. Doses of fertilizer and terms of their introduction influence primarily. With increasing doses of nitrogen and their application closer to the earing phase, the amount of total and protein nitrogen in the grain increases. The effect of fertilizers also depends largely on soil and climatic conditions. Application of large doses of fertilizers to cereals on soils poor in natural fertility before sowing or in the early stages of growth and development is not always accompanied by increased yields (Vasylenko & Stadnyk, 2019). This is because at low nitrogen content in the soil all the nitrogen additionally introduced into the soil is used by plants to increase the vegetative mass, and to build gluten protein is not enough.

Gumisol is a preparation with a high content of humic substances. The fertilizer has high bactericidal and fungicidal properties, environmentally friendly, contains in dissolved and physiologically active state all the components of biohumus:

humates, fulvic acids, amino acids, vitamins, natural phytohormones, micro- and macronutrients, spores of soil microorganisms. Studies conducted in the Kyiv region, when spraying wheat crops with **Gumisol** in different phases of development, confirmed that the highest yields were obtained when spraying at the end of the tillering – the beginning of the tube phase. Seed spraying with fertilizer before sowing increased yields by 14.4%. Seed treatment with subsequent spraying of crops contributed to the growth of grain yield by 18.1–26.7%.

Pre-sowing seed treatment and the use of Gumisol during the vegetation increases the level of plant life and their resistance to adverse environmental factors. Pre-sowing application of the preparation also increases seed germination and the number of productive stems.

In general, the average use of Gumisol in wheat crops during pre-sowing seed treatment increases grain yield by 18.6% (when spraying – 29.7%), protein content in the grain – by 0.35% (during the growing season – 0.67–0, 75%), protein yield from 12 ha – by 23.2%. The highest increase in protein content is observed when processing seeds before sowing and 2-time spraying of crops.

Vitalyst is a liquid organo-mineral fertilizer. It is intended for the pre-sowing treatment of seeds and planting material as well as foliar fertilization of crops.

Vitalyst contains 3.4% of ammonium nitrogen, 6.5% of phosphorus, 7.9% of potassium, 0.53 of copper, 0.36 of boron, and 0.12% of molybdenum. In addition, it contains humic, fulvic acids, biologically active substances with anti-stress activity. The main advantage of **Vitalyst** is that both macro- and microelements are in a physiologically active organo-mineral form, which allows for low doses to provide a significant increase in the growth and development of crops. The latter is accompanied by increased productivity and crop quality. Unlike traditional mineral fertilizers, much of which is bound in the soil after application, Vitalyst is fully absorbed when applied to seeds and foliar fertilization during the growing season.

Organo-mineral fertilizer **Oasis** is a fertilizer intended for foliar fertilization. The main advantage of **Oasis** is that macro- and microelements are in a physiologically active organo-mineral form, which allows at low doses to provide a significant increase in the growth and development of crops. This is accompanied by increased productivity and crop quality. Unlike traditional mineral fertilizers, a significant part of which binds in the soil after application. Oasis composition: macronutrients: nitrogen – 20.6% (including nitrate – 0.6%, amide – 20.0%), potassium oxide – 4.5%; trace elements (sulfur – not less than 0.1%, boron – 0–0.071%, cobalt – 0.006–0.0084%, copper – 0.014–0.2%, zinc – 0.026–0.001 mg/l, iron – 0–0.08, manganese – 0–0.079, molybdenum – 0–0.018, magnesium – not less than 2.0). Thus, from the dose of **Oasis** 2.5 l/ha, the grain yield of spring wheat increased by 0.35 t/ha. The dose of fertilizer 5.0 l/ha increased grain yield by 0.44 t/ha. Further increase of the dose to 50 l/ha did not increase the yield.

Organo-mineral fertilizer (OMF) **Dobrodiy** is a composite, highly effective, multifunctional fertilizer, which includes humic and fulvic acids, potassium, nitrogen, biogenic microelements in chelated form, as well as a plant growth regulator that has phytohormonal activity.

Dobrodiy is intended for the processing of seed, planting material and foliar feeding of plants. Magnesium (up to 20 g/l) contained in the fertilizer makes it possible to increase not only the productivity of many crops but also quality – increases the accumulation of vitamins, carbohydrates, improves protein and phosphorus metabolism, which accelerates the process of fruit formation and maturation of field crops. According to its composition, **Dobrodiy** promotes:

- increasing germination energy and seed germination;
- growth of total, and especially root biomass of plants;
- frost and drought resistance;
- increase the immune system of plants;
- chemoprotective action in plant damage by pesticides, heavy metals and radionuclides.

Organo-mineral fertilizer **Dobrodiy** was used for vegetation in late May – the early of June (first decade).

Endofit is a water-alcohol solution of the products of vital functions of fungi – endophytes. It contains ethyl alcohol – 95%, water – 4.5% and a complex of physiologically active substances – 0.26–0.52%. The latter include natural plant growth regulators: auxins, gibberellins, cytokinins, and others. Toxins and harmful substances are completely absent.

The biopreparation has a broad spectrum of action. Ristregulating, anti-stress, immunostimulating types of activity prevail. Once in the plant, the preparation ensures its harmonious growth and development, which increases yields. Under the influence of the preparation both respiration and photosynthesis increase, which, in turn, provides the plant with nutrients (regulatory activity). When applying the solution to seeds or vegetative plants, it significantly increases its resistance to disease (immunostimulatory activity). In addition, **Endofit** combines well with pesticides and herbicides.

During pre-sowing treatment of spring wheat grain with **Endofit**, the protein content in the grain increased by 0.56–0.73%.

The content of basic nutrients in the product, among other things, depended on the use of **Endofit**. On wheat crops with the standard treatment, nitrogen was 2.66%, phosphorus – 0.85 and potassium – 0.61%, in straw, respectively, 0.62, 0.18 and 0.71%. Processing of crops and grain slightly increased the content of nutrients as in grain.

Thus, the use of growth stimulant **Endofit** on wheat crops increased grain yield by 22.9%. The optimal dose was 10 ml/ha. Seed treatment before sowing gave a grain yield increase of 28.7%.

Ecostym – a water-alcohol solution of analogues of phytohormones (auxins, cytokinins and gibberellins), amino acids, carbohydrates, vitamins, fatty acids, trace elements and other biologically active substances obtained from the products of metabolism of fungi-endophytes. Toxic and harmful substances are completely absent. The preparation has the following main properties: increases germination, seed germination energy and photosynthesis of plants stimulates root formation, growth and development of plants, increases immunity, changes the content of proteins, sugars and vitamins, stimulates flowering plants.

The solution significantly increases the resistance of spring wheat plants to damage by major diseases, in particular, rust, root rot and others (Vasylenko & Stadnyk, 2019). The highest yield is formed when spraying wheat crops at the end of the tillering – beginning of the tube phase. Application of **Ecostym** 10 l/ha (standard) increased the yield of spring wheat grain by 33.0–45.6%, protein content increased by 0.40–1.40%, gluten content – by 2.0–5.2%.

With an increasing dose of the preparation from 12.5 ml/ha to 50 ml/ha, the yield and protein content increased. Thus, when applying **Ecostym** wheat for vegetation at doses of 12.5, 25.0 and 50 ml/ha, the yield increased by 0.60, 0.83 and 0.82 t/ha respectively in comparison with the standard treatment and on 0.18–0.40 g/ha – compared to the standard; the protein content increased by 0.40%, 0.90 and 1.40% respectively. Protein yield with **Ecostym** increased by 0.74–1.12 c/ha, and gluten yield by 2.34–2.70 c/ha (65.1–75.0%).

Ecostym is compatible with all herbicides, insecticides and fungicides, which makes it possible to apply it together with other drugs without disrupting the technological cycle, and does not require additional costs.

The preparation can be used for processing grain (seeds) and used during the growing season. The norm of the drug during pre-sowing seed treatment is 25 ml.

Foliar feeding is the most effective way to increase yields. The first spraying is best done in the phase of the end of tillering – the beginning of the tube.

Neofit is a biological preparation with phytohormonal action, which is obtained as a result of the isolation of microorganisms from the products of metabolism. The composition of the biopreparation includes amino acids, both macro- and microelements, water-soluble proteins, vitamins, substances with auxin, cytokinin and gibberellin activity. It can be used for seed treatment by spraying crops for the growing season both individually and with pesticides (pesticides, insecticides, herbicides).

The application of **Neofit** during the growing season at a dose of 12.5 ml/ha increased the grain yield by 0.23 t/ha, the protein content – by 1.04% compared to the standard treatment (Vasylenko & Stadnyk, 2019). The optimal dose of **Neofit** when used for vegetation on wheat crops was 50 ml/ha. The increase in yield from using Neophyt was 0.29 t/ha, the protein content increased by 1.64%. The use of the plant growth regulator Neophyt on wheat crops increased the gluten content in the grain by 1.8–3.0%. In general, the use of growth regulator **Neofit** for vegetation on wheat crops increased grain yield by 0.23–0.29 t/ha and protein content in the grain – by 1.04–1.69%.

Vegestym is a phytohormonal activity-balanced composition of growth regulators of microbial and synthetic origin. The preparation has high cytokinin and auxin activity. Also, it promotes germination energy and field germination of seeds, accelerates the processes of photosynthesis, chloroplastogenesis, increases the immune status of plants, resistance to diseases and stressors. The use of the preparation before sowing increased grain yield by 21.3–31.5%, protein content – by 1.23–1.88% (Vasylenko & Stadnyk, 2019).

From a dose of **Vegestym** 250 ml/t when processing grain before sowing, grain yield increased by 28.2% to standard treatment and 16.4% to standard (Emistym). The protein content increased, respectively, by 1.57% to control and 1.43% to standard. At a dose of 300 ml/ha, the increase in grain yield was 31.5% to standard treatment and 19.3% to standard; the protein content increased by 1.88% to control and 1.74%, respectively. Accordingly, the optimal dose is 300 ml/ha.

Noostym is a composition of growth regulators of natural origin and synthetic analogues of phytohormones of auxin and cytokinin nature and organic compounds – polyethylene glycols. Helps increase germination energy and field germination of seeds, creating a strong root system and developed leaf surface, increases the immune status of plants, which reduces the loss of pesticides and fungicides, increases plant resistance to disease and stressors.

Treatment of wheat plants with **Noostym** at a dose of 200 ml/t gave an increase in grain yield by 38.1%, protein content – by 0.7% and gluten – by 2.8% to standard treatment. Increasing the dose to 300 ml/t caused slightly increased yields (Vasylenko & Stadnyk, 2019). The protein content did not change, the gluten content increased slightly.

Thus, the treatment of spring wheat grain with **Noostym** plant growth regulator increased grain yield by 38.1–46.6% to standard treatment, protein content – by 0.7%, gluten – by 2.8–3.4%. The optimal dose for seed treatment before sowing is 300 ml/t, for spraying crops – 300–400 ml/ha.

From the application of **Agrostym** at a dose of 12.5 ml/ha, the increase in grain yield to standard treatment was 7.3%, from a dose of 25 ml/ha – 10.6%, and from a dose of 50 ml/ha – 13.1%. The highest increase in grain yield was obtained from the dose of 75 ml/ha – 14.7%.

The quality of wheat grain also increased from the use of the preparation. The protein content increased by 0.29–1.55% and gluten – by 2.0–3.0% comparably to standard treatment. The highest protein content was at doses of 50–75 ml/ha.

Protein yield in wheat crops increased with increasing dose of the preparation by 10.0–30.4%. The highest yield was from a dose of 75 ml/ha. Gluten yield also increased with increasing dose of **Agrostym** by 8.0–24.5% to control and by 4.5–7.4% to standard. The highest increase in gluten yield was also from the dose of Agrostym 75 ml/ha. The use of **Agrostym** on vegetation at the tillering – the beginning of the tube phase increased the grain yield to 14.7%, protein content – by 1.3–1.55% and gluten content by 2.8–3.0% (Vasylenko & Stadnyk, 2019).

Biological and microbiological preparations and fertilizers play a significant contribution to increasing yields and improving the quality of grain products. Soil fertility, i.e. its ability to release nutrients, to accumulate and retain moisture and air, has been created by soil microorganisms for billions of years. Depending on which species of microorganisms predominate in the soil, it will be more or less fertile. Because of the interaction of soil microflora and plants, nutrients in the soil are scattered and most often adsorbed on solid soil particles.

Azotovit – is an environmentally friendly natural component of the microflora of healthy soil. It obtained based on soil nitrogen-fixing microorganisms. This is a highly effective biological fertilizer to increase crop yields. The drug is especially effective on depleted and contaminated soils. The active substance is a complex of biologically active substances of microbial origin and pyridine derivatives - synthetic plant growth regulators with phytohormonal activity. The drug has a high auxin activity, promotes the rapid formation of symbiotic bonds of plants with the nitrogen-fixing microflora of the rhizosphere. The drug accelerates the processes of photosynthesis, chloroplastogenesis, activates the activity of cellular enzymes (nitrogenase, RNA polymerase) and plant enzyme systems, increases plant resistance to disease.

The use of **Azotovit** has a positive effect on increasing grain yield and some improvement in its quality.

Application of **Azotovit** at a dose of 0.4 l/ha increased grain yield by 26.0% comparably to standard treatment. A dose of **Azotovit** of 0.8 l/ha gave an increase in wheat grain yield of 36.0% to standard treatment. The protein content in the grain increased with increasing dose; it was maximum value at a dose of 0.8 l/ha.

Azotovit increases crop yields, improves product quality, reduces the development of diseases such as scabies, root rot, head rot, septoria, brown rust, rhizomes and others. The action of **Azotovit** is based on the ability to fix atmospheric nitrogen in the soil by microorganisms that are part of it. Bacteria use enzymes to convert atmospheric nitrogen into protein compounds, and when these compounds are decomposed, nitrogen is released mainly in the form of ammonium and assimilated by plants.

These bacteria synthesize growth substances – heteroauxins, vitamins and antibiotics that have fungicidal properties. The latter actively stimulate the growth and development of terrestrial numbers of phytopathogenic fungi and bacteria. They secrete growth-regulating substances, B-group vitamins and synthesize an antibiotic that inhibits the development of harmful microflora of the plant system. In addition, Azotovit restores the natural microflora of the soil, increases crop yields, improves product quality, reduces the development of diseases such as scab, root rot head, septoria, brown rust, root borer, etc.

Embionic – a microbiological fertilizer, which includes lactic acid, photosynthetic, nitrogen-fixing bacteria, yeast – the products of microorganisms.

Application of the biopreparation once at 1 l/ha increased the yield of spring wheat by 0.84 t/ha, protein – by 1.23%, comparably to standard treatment. Application of **Embionic** twice on soil and crops with a dose of preparation 1 l/ha (total 2 l/ha) increased the yield of wheat by 40.7%, protein content by 1.57% in comparing with standard treatment. Increasing the dose of the drug to 4 l/ha increased the yield of wheat by only 0.02 t/ha, the protein content – by 0.31% compared with the dose of 2 l/ha.

Thus, the optimal dose for grey forest soil is 2–4 l/ha. From this dose, wheat yield increases by 40.7–41.7%, protein content – by 1.57–1.88%.

For **Humisol**, the optimal dose was 6 l/ton for seed treatment and double spraying during the vegetation dose was 6 l/ha (the first one – on the phase of the end of tillering – the beginning of the tube and the second – before heading phase. The increase in protein content from the application of Humisol during the vegetation was up to 34% when seeds treatment – up to 17.5% and when seeds treatment with subsequent spraying during vegetation – up to 24.3–33.9%.

The best results on spring wheat were obtained with the use of Ecostym for the treatment of crops during vegetation at a dose of 50 ml/ha: yield increased by 46%

and protein content – by 14% in comparing to standard treatment; with the use of **Embionic** at a dose of 4 l/ha, the yield increased by more than 40%, and the protein content – up to 18% in comparing to standard treatment (Fig. 3.1).

Regarding the content of microelements in spring wheat grains when using organo-mineral fertilizers, growth regulators and biological products (Fig. 3.2), the

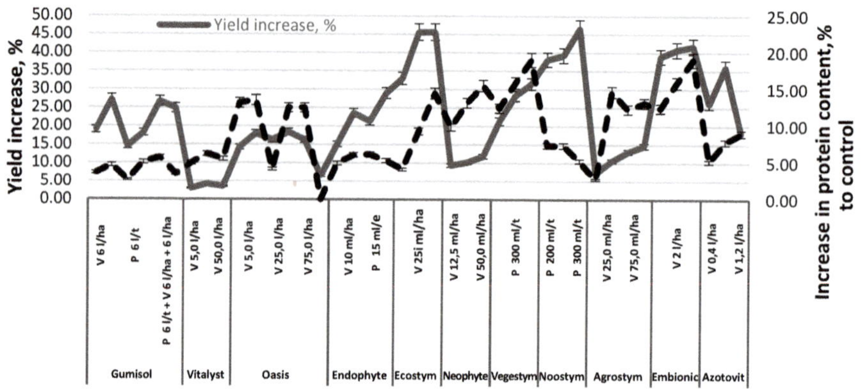

Note: P – pre-sowing seed treatment; V – vegetation treatment

Fig. 3.1 Yield and quality of spring wheat grain depending on the use of organo-mineral fertilizers, plant growth regulators and biological products (Forest-Steppe of Ukraine, grey forest soil)

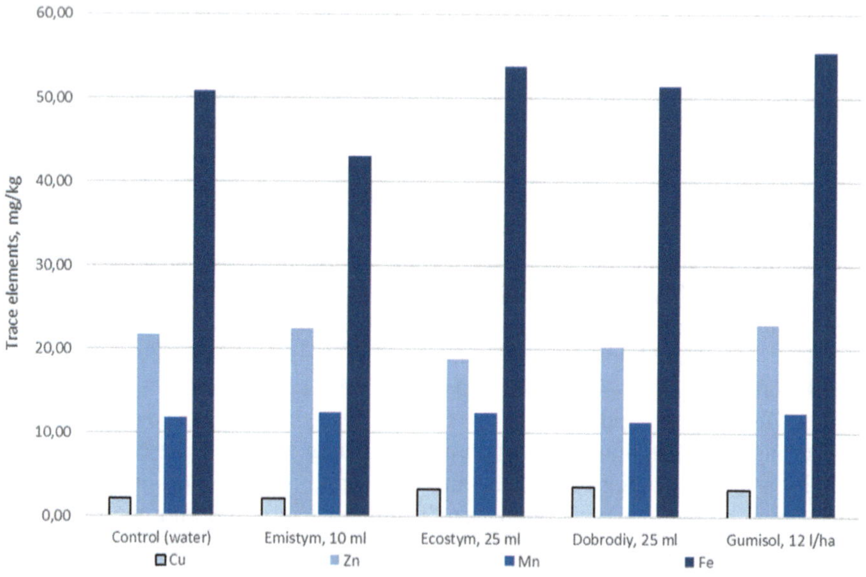

Fig. 3.2 The content of microelements in the grain of spring wheat, depending on the use of organo-mineral fertilizers, growth regulators and biological products (Forest-Steppe of Ukraine, grey forest soil)

tendency to increase iron comparably to standard treatment when using **Dobrodiy** and **Gumisol** was shown, as well as and the tendency to increase copper content when using **Ecostym, Dobrodiy** and **Gumisol**.

3.4 Conclusion

For both Ukraine and Europe in terms of opportunities and threats the analysis of the main tasks and key components of the European Green Deal, which includes, in particular, the new EU's Farm to Fork Strategy, is extremely relevant, especially taking into account economy aspects and food industry. In particular, for Ukraine, which is an agrarian state and has large areas under cereals, it is very important to assess the life cycle of the baking industry. The market of bread and bakery products in Ukraine is characterized by stable demand for products. The enterprises of the bakery industry in the process of functioning have significant ecological impacts on the environment. The problem of the impact on the environment of the life cycle of the domestic baking industry is becoming an increasingly important issue in Ukraine.

Therefore, the use of environmentally friendly technologies for growing cereals and protection systems against pests and diseases for the use of environmentally-friendly preparations in the first link of the life cycle of bakery production in Ukraine is an important task in the Agrarian Sector. The use of organo-mineral fertilizers, growth regulators and biological products helps to increase plant resistance to adverse environmental factors, improve grain quality and reduce the burden on the environment.

References

Burykina, S. I., Smetanko, A. V., & Pilipenko, V. N. (2014). Harvest and quality of winter wheat in the steppe zone of Ukraine. *Soil Science and Agrochemistry. Scientific Journal (Minsk), 1*(52), 210–226.

Mostovyak, I. I., Demyanyuk, O. S., & Borodai, V. V. (2020). Formation of phytopathogenic fond in agrocenoses of cereals of the right-bank Forest-steppe of Ukraine. *Agroecologial Journal, 1*, 28–38.

Vasylenko, M. G., & Stadnyk, A. P. (2019). *Agroecological substantiation of application of new domestic fertilizers and plant growth regulators in agroecosystems*. LLC 'TVORY', Vinnytsia.

Part II
Sustainability of the Baking Industry

Chapter 4
Ensuring Sustainability of Baking Industry in North Macedonia

Mishela Temkov, Elena Velickova, and Elena Tomovska

4.1 Defining Sustainability in the Baking Industry in North Macedonia

The bakery industry usually refers to the grain-based food industry. Grain-based foods may include breads, cakes, pies, pastries, baked pet treats, and similar foods. Along with bakeries, the industry consists of the equipment, supplies, and delivery drivers used to support the industry. According to Euromonitor baked goods are the largest category of staple foods in North Macedonia (Euromonitor, 2021a, b), indicating the importance of the industry in the country. The market is dominated by small artisanal producers, as nearly 2.7% of total retail capacities in the country are specialised bakeries (State Statistical Office of N. Macedonia, 2016). On an industrial scale, the Elbisco corporation is the dominant player on the bakery market. The most common bakery goods in North Macedonia is white bread, but lately the market has accepted a wider variety of bakery products, including sourdough breads. Sourdough bread made with local, organic flour is perhaps the most sustainable bread of all. It entirely avoids the chemical fertilisers used in producing commercial wheat and yeast, as well as the carbon used in transporting this wheat and yeast hundreds of miles from source to factory to retail outlets. On the other hand, prepacked breads and frozen bakery goods are also gaining popularity due to their convenience. Raw materials for the industry are sourced nationally, but also

M. Temkov · E. Velickova (✉)
Department of Food Technology and Biotechnology, Faculty of Technology and Metallurgy, University Ss. Cyril and Methodius, Skopje, RN, Macedonia
e-mail: mishela@tmf.ukim.edu.mk; velickova@tmf.ukim.edu.mk

E. Tomovska
Department of Textile Engineering, Faculty of Technology and Metallurgy, University Ss. Cyril and Methodius, Skopje, RN, Macedonia
e-mail: etomovska@tmf.ukim.edu.mk

J. M. Ferreira da Rocha et al. (eds.), *Baking Business Sustainability Through Life Cycle Management*, https://doi.org/10.1007/978-3-031-25027-9_4

imported. While the country produces around 250,000 t of grain annually, this satisfies only 80% of internal consumption. The gap is bridged by imports from the region, with Serbia as the largest supplier of wheat and grain.

Defining sustainability in any industry poses a question of where to set the boundaries of the system within the supply chain, particularly in regards with the suppliers. While a small artisanal baker can find economic benefits in sustainable practices such as reduced energy consumption and be induced to apply them, they have no control over the grain production. For instance, the carbon impact of wheat production is around 1 kg CO_2 to produce 1 kg of wheat, which is considered as low. When breaking down the total carbon footprint of wheat production, 26% comes from soil carbon, 50% from N_2O emissions and the remaining 24% of carbon impact is direct energy, fertiliser manufacturing, and seed and pesticide manufacturing (Buttenham, 2021). A second major issue to be considered is the number of intermediaries in the supply chain, as longer supply chains imply increased environmental impact of transport. Although fresh bakery goods are in general locally manufactured, reconfiguring the distribution networks to create shorter supply channels will always reduce the overall environmental impact.

While understanding the complexity of upstream supply is important, immediate action within an industry should be taken at the producer level, thus a more focused producer/consumer approach is a good starting point to tackle the issue. Figure 4.1 illustrates the connection between bakery products, human health and wellbeing and ecosystem health. Consumers are currently focused on healthy foods that are safer, convenient and easy to use. These requirements increased the demand of fresh

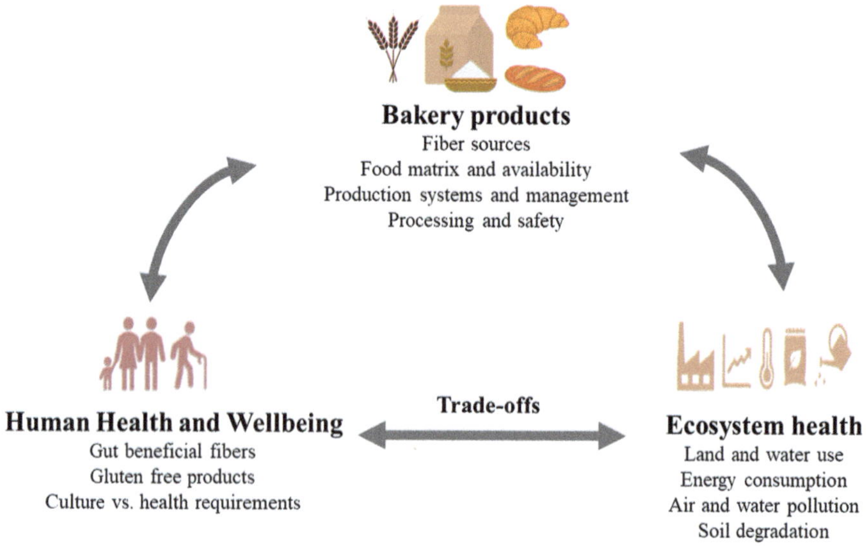

Fig. 4.1 Sustainability of the baking industry as an intersection of human health and wellbeing and ecosystem health

or pre-baked products, devoid of synthetic chemical preservatives, such as sourdough and functional breads, biscuits, crackers, snacks etc. (Velickova et al., 2014). Nowadays, consumers' health concerns have forced the snack foods industry to produce gluten-free, low-carb or low-fat baked goods while keeping their traditional flavour and texture characteristics. For this reason, research aimed at reducing the gluten, carbohydrate or fat content in snacks has gained importance, either by modifying process conditions or by the use of flour and fat substitutes. Food waste such as grape pomace, tomato pulp, olive oil pomace etc. can be revalorised as substitute flours (Temkov et al., 2021). Different oleogels or inulin based emulsions can be used as fat substitutes (Temkov & Mureşan, 2021). Therefore broadening the bakery products food matrix as a source of gut beneficial edible fibres and bioactive compounds is a way to improve nutrition. It is always a challenge to align nutritional requirements and environmental implications towards a safe operating space for food systems (Willett et al., 2019).

There will always be trades off regarding consumption needs and the environmental impact of production on land, water and air ecosystem health. Innovations in production systems and management are required in order to overcome these issues.

In 1987, the United Nations Brundtland Commission defined sustainable as "meeting the needs of the present without compromising the ability of future generations to meet their own needs." Today the 17 points of sustainable development as defined by the UN are widely acted upon. When considering food sustainability there is a strong interconnection between Goal 12 – Responsible consumption and production with the primary drivers of no poverty (Goal 1), zero hunger (Goal 2) and health and wellbeing (Goal 3).

Responsible production and consumption encompasses several factors affecting economic growth and environmental sustainability, energy consumption and waste. In 2016 North Macedonia reported an overall increase of 21.2% in resource productivity and 25% increase in energy productivity compared to 2012 (State statistical office of N. Macedonia, 2019), however energy consumption and waste management are seldom adequately addressed.

The energy consumption of bakery products varies according to product type. For instance, for producing 1 kg of bread the cumulative energy demand ranges from 9 MJ/kg to 32.9 MJ/kg depending on flour variety (Notarnicola et al., 2017). With approximately 250 kg of bread consumed yearly by a Macedonian household, the national annual consumption of energy related to baking products can be approximated to 96 GWh, or 27, 300 tons of CO_2 emissions. Baking ovens are the largest on site energy consumers so introducing better baking practices (new bread formulations, sourdough, etc.) can lead to significant reduction in energy consumption. Recent project by the Carbon Trust's UK found that lowering the core temperature while baking to 90 °C will lead to energy reductions during the baking process.

4.2 Food Waste: A Challenge for Bakery Industry Sustainability

Food waste presents the greatest challenge in the baking industry sustainability in North Macedonia. Food waste refers to the food that is not consumed, but removed from the consumer supply chain together with its inedible parts. Attention should be paid to the nutritive content of waste, as this food may or may not be usable. When considering baking goods, product shelf life drastically varies. Introducing practices to sort and collect bakery waste and reuse it as feed or dried bread crumbs can lead to reduced waste. On the other hand, food loss comes from harvesting, slaughter, and processing activities prior to consumption. These can be generated during production, distribution, retail and in households. Wheat crops grown in moderate climates are particularly prone to food loss. Research and innovation can be utilised to find new products to reutilise food loss. The husks generated during the rafination of cereals can also be further used instead of wasted. Depending on the type of the crops, the husks can represent a rich source of bioactive compounds that can be extracted, concentrated and used for fortification of other food products. Over the last decade extensive scientific research has been dedicated to reutilizing husks either in food or other industries, as can be seen in Grafi and Singiri (2022), Khan et al. (2021), Ibrahim et al. (2019). Food waste is usually mixed with other municipal solid waste, or it might be composted, combusted or incinerated.

Food waste represents a modern problem of rich countries, but appears in all the countries of different economic development (Stangherlin & de Barcellos, 2018). According to the Food Waste Index Report 2021 by United Nation Environment Program one-third (931 million tons) of world's produced food becomes waste out of which 61% (570 million tons) comes from households. On the other hand, food services and retail impart 26% (225 million tons) and 13% (121 million tons), respectively (United Nations Environment, 2021). At the same time, 120 million people suffer from hunger, malnutrition and poverty. Moreover, it causes significant impact to the environment by increasing CO_2 emissions to 3.3 billion tons or increasing greenhouse gases through anaerobic fermentation. The United Nations are acting on the problem by introducing a program called Sustainable Development Goal Target 12.3, which aims to reduce the food loss and waste within the food supply chain by half per capita by 2030. In raw food production, food loss can occur from severe climate or pest infection so called pre-harvest loss. Food can be wasted while being processed due to inadequate temperature or humidity as well as microbiological spoilage. Great amount of food waste is generated in retail due to the strict policies they have towards the food with expired shelf life, which in many cases is still very edible. Some of the big chain markets will not even consider putting bruised or spotted fruit/vegetable on the shelf. Finally, consumers are discarding large quantities of food due to changed texture, appearance, odour or passed expiration date.

In the Republic of North Macedonia, total food waste generated in each step of the food chain is reported as a part of the municipal waste. Its total amount in 2020 was reported as 913,033 tons (452 kg per capita) by the State Statistical Office out of which 83.4% was a mixed waste. A reduction of waste by 0.4% was observed compared to the total quantity collected in 2019 (1% reduction per capita). However, compared to 2008, the total waste increased by an enormous 21.8% (22.7% increase per capita). The greatest waste generators were households producing 522,954 tons (83%), whereas the rest was collected as commercial waste. A big discrepancy between generated (913,033 tons) and collected (630,086 tons) waste is reported. Almost all of the collected quantity is discarded in landfills (Stojanovska, 2021). Since the issue with food waste and food loss is a multi-layered problem, the solution should as well be complex and multidimensional covering many aspects. Some of the aspects that should be considered involve good logistics and planning, improved and innovative packaging, developing a plan and guideline for the supermarkets to donate the food about to be expired to the social categories. Many of the measures are implemented on a voluntary level so that the food supply chain remains sustainable. However, prior to undertaking any measures to increase sustainability precise numbers of generated food waste should be known and classified by types of waste. Thus far there is no public data on the quantity and type of waste generated by the food industry in general including the baking industry.

Nonetheless, adopting good industrial practices already employed by other countries can be beneficial. Many countries realised that food waste is a big economic loss and therefore launched their individual programs while battling food waste with different approaches. The state government of Australia took the initiative to invest in research on food waste reduction and this led to several innovative solutions published in National food waste strategy: Halving Australia's food waste by 2030 (Australian Government, 2017). On the other hand, France was the first country to pass a legislation that forbade retail to dispose of unsold food. As an alternative, supermarkets could donate the food to charities. Another approach is to sell the products soon to expire at a discounted price. France set the goal to reduce food waste by half by 2025 (Mourad, 2015). Recently, the Government of Singapore implemented Zero Waste Masterplan under which the daily waste should be reduced up to 30% in which the private sector also co-joined (Mica, 2022). United Kingdom is fighting food waste using aggressive media campaign to raise public awareness through government-funded "Waste & Resources Action Programme", "Love Food, Hate Waste" campaign and "War on waste" dedicated to take off "best before", create new packaging sizes, recycle "on-the-go" and installing more anaerobic digestion plants (Irving, 2021; WRAP, 2018).

In North Macedonia the initiative to solve the problem with the food waste came from the organisation "Ajde Makedonija – Let's go Macedonia". While 450,000 persons, of which 60,000 are children, are impoverished, the country is in the top-level countries (eighth place) that generate waste per capita (Ajde Makedonija, Lets go Macedonia, 2011). This organisation is influencing the passing of the law for food surplus donation. They recommend establishing a coordinative body that will manage the food surplus from the food chain and will redirect it to donation. As measures to reduce the waste they suggest introducing separation of food waste for different end-users, adjusting the product shelf life and facilitating the use of the products after this period if the food is safe, or donating the surplus of the food. As a benefit from donations, they recommend donors should be exempted from responsibilities; moreover, donated food should be VAT free. They also recommend establishment of a simple and fast administrative procedure for donating food surplus, supporting food donation with errors on labelling, packaging, weight, but is safe to be used and limitation of return of the food with expired shelf life. Food waste reduction would be advantageous for socio-economic improvement, decreasing energy consumption, quenching world hunger, mitigating emissions of greenhouse gases in the atmosphere and improving food security and nutrition.

4.3 Reducing Food Waste as a Way to Sustainability

The complex issue of food waste cannot be tackled using just one simple method, but employing several different symbiotic axes from concerned participants in the food supply chain. If we consider only food waste without food loss, the generated waste can be divided as:

- industrial waste produced as side stream by product,
- waste created in retail due to the expiration of the date labelled as "expiry date"
- waste produced in households as leftovers (edible waste), skins, peels and bones (inedible waste) or expired whole product.

Therefore, the reduction of the waste of each of the different chain links should have a specific approach.

Industrial food waste has high transportation costs and limited landfill disposal sites, two important criteria to drive the incentive towards finding green technologies for its valorisation. Vast amount of research has been published for industrial stream recycling and reuse in some other products for human consumption, source of nutrients, feed, biofuels, compost fertiliser, nutrient source for heterotrophic microalgae cultivation, chemicals or heat and power (Usmani et al., 2021). For this purpose, many bio-based techniques were employed including enzymatic or ultrasonic pretreatment, fermentation, anaerobic digestion, composting etc. (Kopsahelis et al., 2018; Papadaki et al., 2018). The barriers for implementation of many of these techniques are their complexity, capital and operational cost, production of post-processing residue or difficult configuration. High effort and investment is needed in transferring the technology from lab-scale to pilot or plant scale.

On the other hand, the waste generated in retail and households is potentially influenced by consumers' personal and behavioural factors. This side of the problem might be overcome by educating and creating anti-wastage social norms by a systematic approach to food storage (Stangherlin & de Barcellos, 2018). First, there should be very clear and well defined regulation and policies concerning food waste. If institutions in charge of collecting waste create better infrastructure, it might be a driving force for bigger effort from consumers in the actions of recycling as a way for food waste reduction. Household parties should be communicated for waste reduction by awareness campaigns. They can be informed about the differences of the terms "best before" and "expiry date" and educated on sustainable food. Stockpiling of food items and over-purchasing can be limited by planning the meals in advance or reuse leftovers. There is a necessity for consumers to be appropriately informed on how to read labelling or how to store perishable food. Retailers can have a key role in waste prevention with proper sustainable strategies. In some countries, they are already legally obliged to follow the concept of "zero-waste" followed by penalization if they fail to do so. In this line, retailers sell the suboptimal food close to expiration date on reduced prices or create different categories of products. In this particular approach the retail sector has the liberty to create innovative solutions and promote sustainable strategies for marketing the suboptimal product. Any unsold food that is still edible would be donated to charitable organisations or food banks instead of being destroyed. Food that is not edible, but still nutritious and does not impose a risk should be used as animal feed with a proper regulation. Another important parameter that should be observed is food packaging and labelling. The easiest improvement in this domain is to have appropriate portion sized containers to give the choice to the consumer not to overbuy. Another measure is to mend the expiration dates for many of the products such as salt, vinegar, sugar, sweets etc. that don't require "best before" dates on the label, so they would not be unfortunately rejected by the retail chain. The most promising solution for waste reduction is the use of active and intelligent packaging for better storability. This innovative packaging uses different types of sensors that change colour and give real-time information about the food quality or the history of storage/ transportation.

As listed above, there are ways to create a more sustainable food and baking industry. However, its implementation requires a network of various societal stakeholders. Rather than creating individual policies in areas of agriculture, food and environment a harmonised multi sector policy is required to achieve sustainability. Figure 4.2 summarises the mix of policies on which such legislation should be grounded. Currently, in North Macedonia there is a national initiative to prepare the legislation for enabling sustainability in all sectors, including the food industry and shifting towards a circular economy. The procedure is still at a stage of mapping national requirements.

4.4 Conclusion

Despite the importance of food sustainability, when it comes to the baking industry it is easier said than done. As experience teaches us if the environmental impact is overridden by economic restraints, strict legislation is the only way forward to implementing change. Population growth, projected dietary transitions, increased global food waste, and global supply chain turbulence experienced in the 2020s require urgent effort and innovation in order to sustainably increase agricultural production, decrease food waste and loss and improve access to nutritious food.

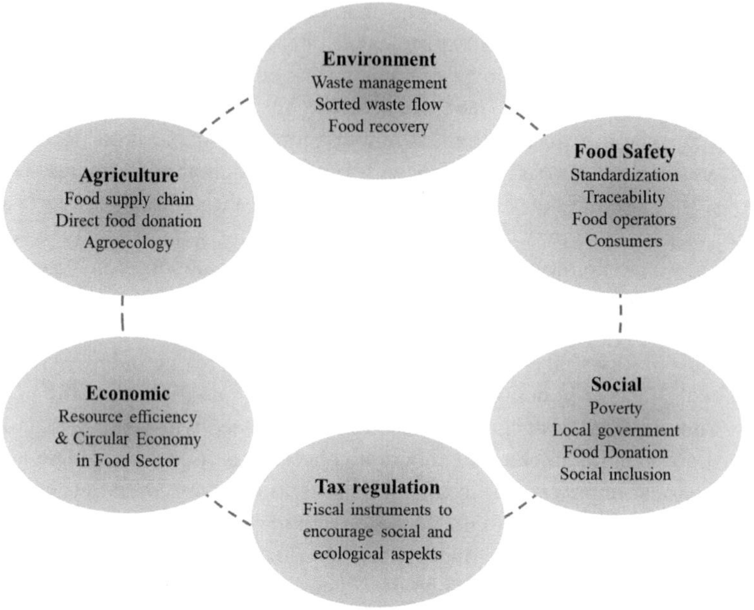

Fig. 4.2 Policy framework for establishing food sustainability

This can only be achieved by a comprehensive, multi level framework involving all relevant stakeholders to ensure environmental, agricultural, food safety, social, economic and financial policy contributing to a sustainable food and bakery sector.

References

Ajde Makedonija, Lets go Macedonia. (2011). *Policy and option impact analysis -draft version.*

Australian Government. (2017). *National food waste strategy: Halving Australia's food waste by 2030.* Retrieved from https://www.awe.gov.au/sites/default/files/documents/national-food-waste-strategy.pdf

Buttenham, M. (2021). *Carbon foot print 101.* Retrived from https://ontariograinfarmer.ca/2021/04/01/carbon-footprints-101/

Euromonitor. (2021a). *Staple foods in North Macedonia, market report.*

Euromonitor. (2021b). *Baked Goods in North Macedonia, market report.*

Grafi, G., & Singiri, J. R. (2022). Cereal husks: Versatile roles in grain quality and seedling performance. *Agronomy, 12*, 172. https://doi.org/10.3390/agronomy12010172

Ibrahim, M. I. J., Sapuan, S. M., Zainudina, E. S., & Zuhria, M. Y. M. (2019). Potential of using multiscale corn husk fiber as reinforcing filler in cornstarch-based biocomposites. *International Journal of Biological Macromolecules, 139*, 596–604. https://doi.org/10.1016/j.ijbiomac.2019.08.015

Irving, H. (2021). *Renewing the war on waste.* History and Polocy. Retrieved from https://www.historyandpolicy.org/policy-papers/papers/renewing-the-war-on-waste

Khan, A. A., Adeel, S., Azeem, M., et al. (2021). Exploring natural colorant behavior of husk of durum (*Triticum durum* Desf.) and bread (*Triticum aestivum* L.) wheat species for sustainable cotton fabric dyeing. *Environmental Science and Pollution Research, 28*, 51632–51641. https://doi.org/10.1007/s11356-021-14241-6

Kopsahelis, N., Dimou, C., Papadaki, A., Xenopoulos, E., Kyraleou, M., Kallithraka, S., Kotseridisa, Y., Papanikolaoua, S., & Koutinas, A. A. (2018). Refining of wine lees and cheese whey for the production of microbial oil, polyphenol-rich extracts and value-added co-products. *Journal of Chemical Technology & Biotechnology, 93*(1), 257–268. https://doi.org/10.1002/jctb.5348

Mica, H. W. L. (2022). *Food wastage statistics in Singapore. bestinsingapore.* Retrieved from https://www.bestinsingapore.co/food-wastage-statistics-singapore/

Mourad, M. (2015). *France moves toward a national policy against food waste*. Retrieved from Paris: https://www.nrdc.org/sites/default/files/france-food-waste-policy-report.pdf

Notarnicola, B., Tassielli, G., Renzulli, P. A., & Monforti, F. (2017). Energy flows and greenhouses gases of EU (European Union) national breads using an LCA (Life Cycle Assessment) approach. *Journal of Cleaner Production, 140*(2), 455–469. https://doi.org/10.1016/j.jclepro.2016.05.150

Papadaki, A., Fernandes, K. V., Chatzifragkou, A., Aguieiras, E. C. G., da Silva, J. A. C., Fernandez-Lafuente, R., Papanikolaoua, S., Koutinas, A., & Freire, D. M. G. (2018). Bioprocess development for biolubricant production using microbial oil derived via fermentation from confectionery industry wastes. *Bioresource Technology, 267*, 311–318. https://doi.org/10.1016/j.biortech.2018.07.016

Stangherlin, I. d. C., & de Barcellos, M. D. (2018). Drivers and barriers to food waste reduction. *British Food Journal, 120*(10), 2364–2387. https://doi.org/10.1108/bfj-12-2017-0726

State Statistical Office of N. Macedonia. (2016). *Census of retail trade sale capacities in the republic of Macedonia*. Retrieved from https://www.stat.gov.mk/Publikacii/8.4.17.04.pdf

State Statistical Office of N. Macedonia. (2019). *Sustainable development goals* Retrieved from https://www.stat.gov.mk/publikacii/2019/Odrzhliv-Zhvillimit.pdf

Stojanovska, S. (2021). *Municipal waste, 2020*. Retrieved from https://www.stat.gov.mk/pdf/2021/9.1.21.01_mk.pdf

Temkov, M., & Mureșan, V. (2021). Tailoring the structure of lipids, oleogels and fat replacers by different approaches for solving the *trans*-fat issue – A review. *Foods, 10*, 1376. https://doi.org/10.3390/foods10061376

Temkov, M., Velickova, E., Stamatovska, V., & Nakov, G. (2021). Consumer perception on food waste management and incorporation of grape pomace powder in cookies. *Journal of Agricultural Economics and Rural Development, 21*(1), 753–762.

United Nations Environment. (2021). *United Nations Environment Programme Food Waste Index Report*. Retrieved from Nairobi: https://www.unep.org/resources/report/unep-food-waste-index-report-2021

Usmani, Z., Sharma, M., Awasthi, A. K., Sharma, G. D., Cysneiros, D., Nayak, S. C., Thakurg, V. K., Pandeyjk, R. N. A., & Gupta, V. K. (2021). Minimizing hazardous impact of food waste in a circular economy – advances in resource recovery through green strategies. *Journal of Hazardous Materials, 416*, 126154. https://doi.org/10.1016/j.jhazmat.2021.126154

Velickova, E., Winkelhausen, E., & Kuzmanova, S. (2014). Physical and sensory properties of ready to eat apple chips produced by osmo-convective drying. *Journal of Food Science and Technology, 51*(12), 3691–3701. https://doi.org/10.1007/s13197-013-0950-x

Willett, W., et al. (2019). Food in the Anthropocene: The EAT-Lancet Commission on healthy diets from sustainable food systems. *Lancet, 393*(10170), 447–492. https://doi.org/10.1016/S0140-6736(18)31788-4

WRAP. (2018). *Your food isn't rubbish*. Retrieved from https://www.lovefoodhatewaste.com/

Chapter 5
Sustainability Approach of the Baking Industry Along the Food Supply Chain

Alexandrina Sîrbu

5.1 Introduction

Humankind history proves that the health and well-being of any society are strongly related to its economic growth; the reason for which increment of the population worldwide and human prosperity demands a stable economy that we wish to become more and more but flourishing. Unfortunately, human activities and the development of the economy came also with a price paid successively when an industrial wave came one after another. Also, as Perrings and Ansuategi (2000) released, environmental consequences become global because an ecological issue solved apparently in a particular country can be displaced to affect other distant-geographically areas or populations, or future generations. So, anthropogenic factors have been influencing our environment; and the obvious conclusion in the second part of the twentieth century referred to finding and following new pathways, technologies, innovations, and business opportunities aimed at lasting development assurance. But sustainability should focus on balancing economic growth, environmental preservation, and social aspects.

Environmental changes, which started to affect health and social welfare more visible for us in the last decades, imposed new approaches to reshape sustainable the global economy since the last part of the twentieth century, even towards a circular one. In this aim, the international and regional policies, strategies, agreements, and laws about environmental issues were put in force and periodically updated to mitigate human actions on climate change as possible. Different tools and programs were applied to plan, perform, check and improve the activities in order to ensure sustainability and achieve the main goals of lasting development at various levels – from local or national to international ones. Additionally, organizations can use

A. Sîrbu (✉)
Constantin Brancoveanu University of Pitesti, FMMAE Ramnicu Valcea,
Ramnicu Valcea, Romania

J. M. Ferreira da Rocha et al. (eds.), *Baking Business Sustainability Through Life Cycle Management*, https://doi.org/10.1007/978-3-031-25027-9_5

standards to manage environmental issues, assess sustainability impacts, report the ecological performance or eco-efficiency, support regulatory compliance, prove and improve their environmental claims.

Indeed, a long distance is between theory and practice. In this respect, the proof is the circular economy, from the conceptual approach to its ecological outputs and eco-efficiency, several differences and paradigms being registered (as Korhonen et al. (2018) observed).

This paper aims to conceptually approach the sustainability model along the agri-food supply chain and describe a synopsis of the sustainability aspects of the baking sector along a food supply chain.

5.2 Sustainability: A Conceptual Approach

Sustainability has been conceptual developed, having its origins in the Brundtland Report of 1987 (Kuhlman & Farrington, 2010). Afterward, sustainable development became part of various international and regional policies in order to reform our society, reshape the markets, focus on information and technology advancements, and improve social systems. Even now, sustainability remains, at the same time, a topic for scientific studies and policy-makers debates.

In the beginning, the interest has oriented to economic development and welfare from the standpoint of resource preservation for future generations, and environmental concerns increased. In this respect, attention has been paid mainly to the economic and environmental dimensions, but later the social component was also integrated into debates on defining sustainability (Eizenberg & Jabareen, 2017). Hopwood et al. (2005) mapped different approaches to sustainable development to understand better the sustainability meanings and its usage in various policies.

The environment should not be considered synonymous with sustainability but is only a part of the overall approach to the sustainability concept. Nowadays, all three dimensions of sustainable development, namely the economic, environmental, and social, are acknowledged. However, increasing socio-economic well-being, equity, and social equality seem to be related to long-term environmental sustainability. As Hopwood et al. (2005) found out, views on sustainable development changed in the last period from environmental concerns focused on technologies and innovations to eco-centered movements with in-depth social insights.

Based on cited literature and professional experience, starting from the conceptual approach to the implementation framework, a sustainability model is proposed further, with the specific case of the agri-food supply chain, as depicted in Fig. 5.1.

As shown in Fig. 5.1, an in-depth assessment should proceed toward sustainability on each of its dimensions, i.e., economic, social, and environmental. Many interactions occur, even apparent overlaps, between these dimensions, the reason for which a sustainable food supply chain supposes a holistic approach.

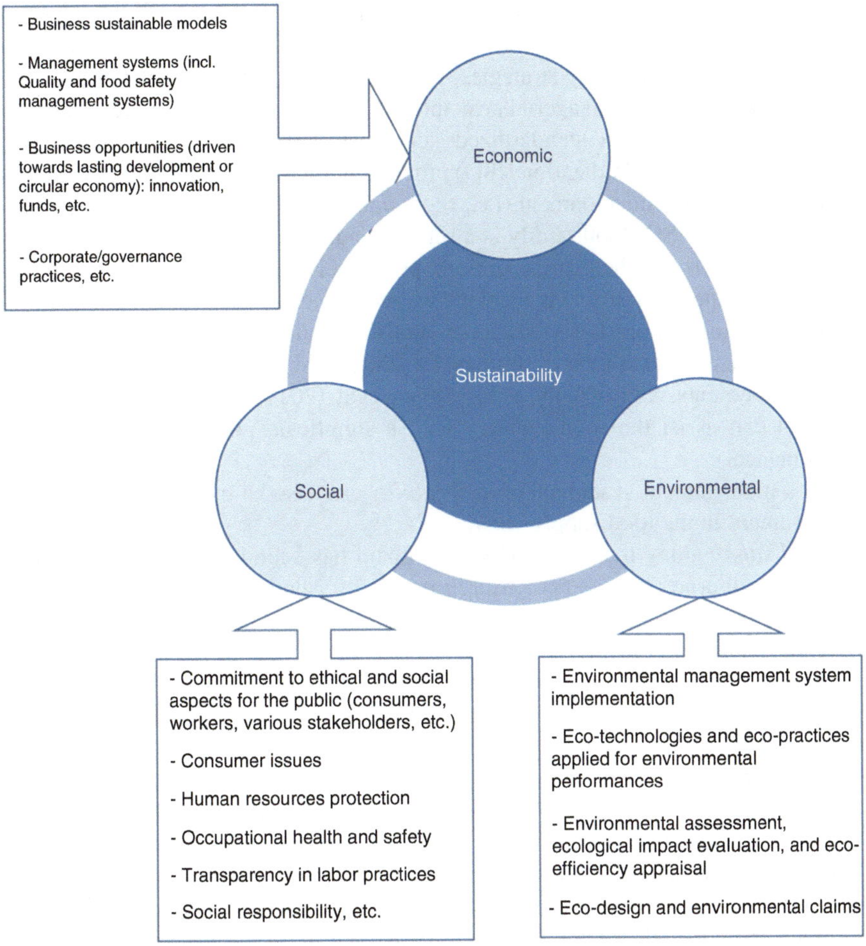

Fig. 5.1 Sustainability model within agri-food supply chain

5.2.1 Economic Dimension of Sustainability Along the Agri-food Supply Chain

Sustainable economic growth or integrative solutions for economic fluctuations, society issues, and environmental concerns have supposed choosing adequate *business models* and developing new ones aimed at contributing companies to sustainable development (Bocken et al., 2013; Schaltegger et al., 2016; Abdelkafi & Tauscher, 2016; Nosratabadi et al., 2019; Goni et al., 2021). However, sustainability should be approached beyond a company level, addressing the whole supply chain. For instance, if we refer to food systems and food supply chain, Nosratabadi et al. (2019) observed that only between 2016 and 2018, at least 100 papers and 1% of subject topics of articles discussed *sustainable business models* within the agri-food supply chain.

As part of their business strategies, many food companies integrated *management systems* based on managerial principles PDCA (Plan-Do-Check-Act), and several systems have been standardized since the 1990s. However, International Organization for Standardization (ISO) provides adequate tools for all three dimensions of sustainable development (i.e., economic, environmental, and social).

For example, ISO 9000 family is a set of standards that deal with *quality management systems* and help organizations proactively manage internal or external quality assurance. Quality management system implementation enhances communication on quality, promotes efficiency and financial performance, and offers an additional trust towards their customers, i.e., the meaning that companies meet specific requirements. Additionally, as Pipatprapa et al. (2017) concluded, quality management can assist the food industry with a significant positive effect on green performance.

As a management systems tool, *traceability* can also be used for sustainability advancement in the food supply chain.

Food safety along the agri-food supply chain has been emphasized globally to prevent foodborne diseases and ensure safe food for human consumption. Besides the social dimension of health and well-being, food safety is approached and should be implemented in a technical and managerial (incl. organizational) way for agri-food sectors. All firms acting within the agri-food supply chain must comply with regulatory requirements on food safety. In this respect, a legal framework for food safety has been reinforced. Any company must identify and keep the food safety hazards under control following the specific steps; different tools for food safety management based on HACCP (Hazard Analysis Critical Control Points) method have been developed worldwide. Any food safety management system based on the HACCP method helps food business operators supply the food market with as possible safe products for consumption. Also, several private certification schemes are foreseen under the Global Food Safety Initiative (GFSI) umbrella to prove that legal provisions and customer specifications regarding food safety have been considered and implemented within certified food companies.

Although many differences between quality and food safety management systems exist, in the literature sometimes appear confusions or overlaps between quality management system (QMS, according to ISO 9001/9004) and food safety management system (FSMS, according to ISO 22001), even with *risk management system* (RMS, according to ISO 31001) applied in the food supply chain. It is important to note that all these systems have a few similar elements concerning communication, resource allocation, some processes, and managerial principles. Still, they are distinct regarding their eligibility, scope, objectives, systemic approach, and even certification.

Generally speaking, management systems are implemented with the main purpose of fostering businesses in the agri-food market. In this regard, firms pay attention without limited to:

- efficiency improvement (based on various economic indicators, such as yield, costs, etc.);
- process control for adequate management of all processes and resources (with higher efficacy, quality control, etc.);
- increase the returns or market share;
- better valorization of resources (inclusive food by-products).

Moreover, quality and food safety management systems are also oriented toward clients and other stakeholders. Indeed, they address to trending demands of the end-consumers with social and environmental consequences. Consumers' needs refer to food safety aspects, sensory attributes, nutrition, organic food, clean label, shelf life, food commodities' diversification, food ethics, sustainable consumption, and the bucket list may continue.

ISO established requirements targeting also other kinds of management systems that can integrate under the same umbrella (Başaran, 2018). Besides the quality or food safety (QMS, FSMS), an integrated management system can be designed to accomplish several organizational objectives within sustainable concepts, such as organizational sustainability (OS), occupational health and safety (OHS), or environmental management (EMS) addressing to all sustainability dimensions. The experience proved that integrated management systems are more efficient than individual systems because it is easier to allocate resources (financial, human, material), considering multifaceted objectives and a holistic business strategy from the beginning. This statement is true if the organization decides or needs to implement more than an independent system to address many requirements and a more extensive action framework. In this respect, an integrated management system is more sustainable than many individual ones, implemented together or in parallel.

Innovation has been considered an essential element of sustainability. Nevertheless, innovation became a *business opportunity* for managers because it drives a competitive advantage in the agri-food market and adds value to the food supply chain. Innovativeness can solve various issues or challenges for food companies with noticeable effects on profitability or sustainability. Pipatprapa et al. (2017) pointed out that innovativeness is critical for green performance and the sustainable food industry. Indeed, various innovations have been implemented along the agri-food supply chain in order to cut costs, introduce new technologies, obtain higher food quality, develop new products, follow the best practices for manufacturing or hygiene, ensure a cleaner production, optimize the processes, implement lean manufacturing, high valorize of food by-products, recycle the food waste, recovery the energy, etc.

Bigliardi and Galanakis (2020) reviewed innovation classification in the food industry in another train of thought. They identified a few innovation models for the food industry, emphasizing sustainability-driven foods innovation cases (e.g., food waste recovery, packaging materials, or modern sanitation). On the other hand, Gimenez-Escalante and Rahimifard (2019) acknowledged the importance of best choosing the technologies following each food company's needs and its micro-economic environment. They proposed Distributed and localized food manufacturing (DLM) as an innovative strategy for this aim.

Besides innovation, other business opportunities are driven towards lasting development or a circular economy (e.g., *incentives, external funds*).

In the last decade, a new trend appeared regarding the *corporate sustainability* concept. Consequently, even food companies aware of the sustainability role should integrate sustainability concepts into corporate governance practices by considering their implications in business management. But, during the 2010s, these steps were still in the beginning. As Krechovská and Procházková (2014) showed, most organizations did not integrate sustainability into their corporate governance. According to their research performed among companies in the Czech Republic in 2012, the enterprises perceived the whole picture of the sustainability concept. Still, those companies did not report their sustainability performance accordingly to combine sustainability within the corporate strategic management or corporate governance.

As mentioned, when approaching sustainability, sometimes it is challenging to trace a line apart between all its conceptual dimensions because a specific topic can address at the same time many goals. For instance, food safety is a technical and organizational issue for the businesses ongoing within the food supply chain with specific objectives but having the social dimension as the primary goal regarding health, well-being, and consumer protection. On the other hand, innovation has an extensive framework, from business – management and marketing to technological advancements and new product development. The consequences of innovation applied within agri-food supply chain are evident not only in financial terms or as marketing aspects, which indeed ensure a competitive advantage, but also in improvement of social dimension (e.g., commitment to social aspects for consumers, workers, and other stakeholders) or developing eco-practices with a positive effect on environmental factors as water, waste, energy, or greenhouse gases.

5.2.2 Social Dimension of Sustainability Along the Agri-food Supply Chain

The social dimension of sustainability has been debated multi- and inter-disciplinary by politicians, experts, policy-makers, researchers, and academicians from a large sphere – geography, urbanism, anthropology, sociology, legal sciences, standardization, etc. Many international and regional organizations, various movements, independent scientists, lawyers, economists, sociologists, ecologists, and other stakeholders contribute to building social sustainability, and different models were proposed. For example, Eizenberg and Jabareen (2017) designed a conceptual framework for social sustainability, in which *equity* is a crucial component. They considered intergenerational and intra-generational equity with three dimensions, i.e., redistributive, recognition, and participation. *Social safety* as a dimension of sustainability is also related to other concepts regarding the security of people in relation to human vulnerability, uncertainties, and risks that should be mitigated for equity (justice). Social agendas also consider developing sustainable urban or rural forms, even eco-prosumption promotion.

Nowadays, an increasing *social responsibility* is also envisaged. The *commitment to ethical issues* is becoming more and more linked to social responsibility for all public (consumers, workers, various stakeholders, etc.). In a first instance, business ethics indicates the behavior of companies as regards how compliant they are with moral and ethical values. Still, ethical issues have been discussed in businesses as a valuable asset for improving the overall performance of the companies and building customer value (McMurrian & Matulich, 2006). Later, Torelli (2021) commented on ethics, companies' behavior, and their links to corporate social responsibility and the concept of sustainability. So, ethics cannot be excluded when discussing environmental responsibility and sustainability. On the other hand, one must appreciate the impact of ethics on society or broader communities.

But even discussing the social sustainability, mostly social aspects are considered inside and outside the companies (e.g., *occupational health and safety*, the *welfare of workers*, and *transparency in labor practices*).

For the particular case of the food supply chain, the United Nations (UN) and its specialized agency, the Food and Agriculture Organization (FAO), offer a holistic approach to food systems' sustainability and lasting development globally. UN Development Program helps countries and territories to achieve the Sustainable Development Goals (SDGs), giving support for reducing poverty, inequalities, and exclusion of various populations, and contributing to build-up sustainable prosperity for people while protecting the planet. The 2030 Agenda for Sustainable Development has foreseen 17 sustainable development goals, which are of critical value for humankind and the earth (i.e., no poverty, zero hunger, good health and well-being, quality education, gender equality, clean water and sanitation, affordable and clean energy, decent work and economic growth; industry, innovation and infrastructure; reduced inequalities, sustainable cities and communities, responsible consumption and production, climate action, life below water, life on land; peace, justice, and strong institutions; partnerships for achieving the goals) (UN, 2015). For example, in June 2022 in Sweden, a high-level meeting under the theme "Stockholm+50: a healthy planet for the prosperity of all – our responsibility, our opportunity" expects to celebrate 50 years of global environmental action and encourage the implementation of the UN Decade of Action for accomplishing the SDGs, and green agreements from last years (UNEP, 2022).

FAO has the main goal of accomplishing *food security* and tries to play a role *against hunger and poverty* to improve the well-being of populations around the globe. FAO has many initiatives regionally aiming to develop resilient food systems worldwide, transforming the existent ones for better nutrition, with a revival of traditions, improving or less affecting the environmental factors (soil, water, etc.). FAO promotes sustainable food systems with a neutral or positive impact on the natural resource environment for the benefit of society in terms of nutrition and welfare. Similarly, at the regional level, European Commission (EC), aiming to achieve SDGs goals and mitigate aftermaths for sustainable Europe's climate by 2050, set the European Green Deal with its core Farm to Fork Strategy. This specific strategy, addressing the sustainable food systems, has been foreseen to boost the economy and improve *people's well-being and health*, with benefits of recovery and the transition, especially after the COVID-19 pandemic (EC, 2020).

5.2.3 Environmental Dimension of Sustainability on the Agri-food Supply Chain

The agriculture and food system have been affected by climate change and unexpected extreme weather events, biodiversity loss, soil erosion, land degradation, water pollution, waste increment, greenhouse gases, etc. All of these proved that economic growth in the food supply chain from farm to fork could not be sustainable without correct addressing of environmental issues and adequate management of the environment. In modern society, to meet the needs of all stakeholders regarding the environment, a lot of policies, strategies, regulations, guidelines, standards, and other initiatives have been developed. Many of these focus on the enterprises or businesses rather than the products. Their main goal is to identify, apply, and evaluate adequate indicators for appraisal sustainability terms in a comprehensive and how much as possible accurate manner. For instance, the Sustainability Assessment of Food and Agriculture (SAFA) systems envisage the evaluation of organizations in the global agri-food supply chain with the purpose of having a holistic framework regarding the sustainability of food and agriculture systems (FAO, 2014). At the regional level, European Union (EU) proposed a management tool, otherwise worldwide applicable, that helps companies to evaluate, report, and improve their *environmental performance* through Eco-Management and Audit Scheme (EMAS) (Regulation (EC) No 1221/2009 revised and consolidated in 2019). Also, European Commission's emphasized the importance of measuring and communicating environmental performance (2013/179/EU) for Organization Environmental Footprint (OEF) studies. In this respect, *environmental footprint impact* categories refer to any resource use or emissions associated with organizational activities.

The family ISO 14000 comprises international standards for planning, implementing, and assessing activities and outputs/inputs related to environmental consequences with both economic and ecological benefits in the long term. This set of standards provides principles, requirements, and guidance for implementation and certification of the *environmental management systems* (EMS), environmental *performance evaluation* at the organizational level, *eco-efficiency assessment,* and *eco-design*. They may assist in external *environmental communication*, eco-labels, environmental declarations and claims. Material flow cost accounting (MFCA) is a management tool for businesses to efficiently use the material resources used in manufacturing and distribution processes for cost savings and better materials valorization. Life-cycle assessment (LCA) is an additional tool to assist an organization in identifying and evaluating environmental aspects (inputs, outputs, and potential environmental impacts system) throughout the life cycle of a product to reduce the overall environmental impact. Other standards mention the requirements for supporting organizations to diminish and calculate the products' greenhouse gas (GHG), water footprint, and carbon footprint.

Besides the instruments offered by international standards, other opportunities exist to evaluate environmental performance using, for instance, environmental key performance indicators- KPIs (e.g., as DEFRA (Department for Environment, Food and Rural Affairs) did in the United Kingdom). Also, for converting inventory data from the life cycle assessments, there is a method named Life Cycle Impact Assessment (LCIA), which encompasses a set of potential impacts.

By environmental evaluation and tracking the impact of business operators on the whole food supply chain, being reactive to technology advancements were developed many *eco-innovative technologies and practices*, such as green technologies, new recycling techniques, emerging technologies and novel products, cleaner production, energy recovery, packaging eco-design, water or waste treatment. As mentioned, innovation is not only a valuable asset for businesses along the food supply chain for returns, profitability, or other economic performance indicators. For example, several food technologies have developed through innovation during the last decades, some of them being potentially green (Barba et al., 2016; López-Pedrouso et al., 2019). They enable food preservation or manufacturing by saving energy or water, higher valorization of food by-products, reducing or recycling waste, and lower nitrogen or carbon footprint along the food supply chain, which finally means an environmental impact downward. Although some of these novel technologies proved to be green, they are still emerging; sometimes are too expensive, and the gained technological and environmental outputs not being economically justified at this very moment. Also, there is another problem regarding the social acceptance, the consumer issues concerning the understanding and acceptance of novel technologies or novel food products in consumption (Bredahl, 2001; Ronteltap et al., 2007; Siegrist & Hartmann, 2020).

5.3 Prospects on Baking Industry Along the Sustainable Agri-food Supply Chain

The baking industry is a part of the food supply chain. As with other food industries, there are similarities and some differences regarding the general sustainability approach of the food system along this specific cereal-to-bread chain (Gava et al., 2014). Considering the baking industry along the food supply chain, as is introduced in Fig. 5.2, there are three main areas, namely agriculture, food processing, and distribution-commerce (incl. food services), to be addressed for sustainability assessment.

So, for an in-depth assessment of the baking sector along the sustainable agri-food supply chain, should perform a multi-criteria analysis (MCA) considering the sustainability dimensions and, on the other hand, all three areas of agriculture, food processing, and trade between which occur numerous interactions (see Figs. 5.1 and 5.2). A very accurate sustainability assessment is challenging to perform, and a successful evaluation depends on how and what indicators are chosen. Nevertheless, there is no generally accepted model or criteria, and scientific research is still ongoing.

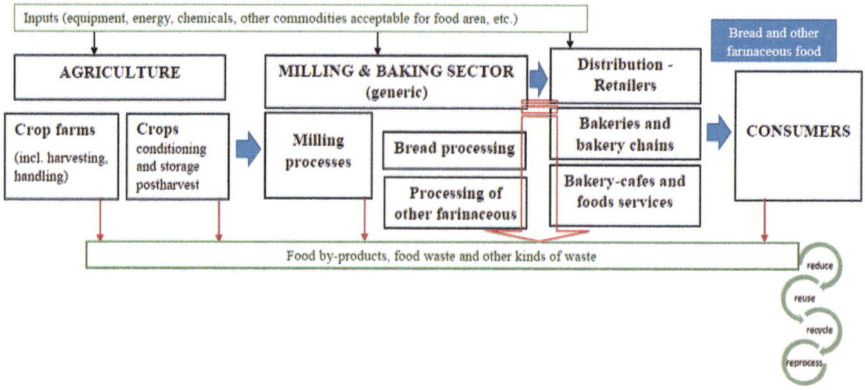

Fig. 5.2 Baking sector along the sustainable agri-food supply chain

Galli et al. (2015) proposed a model using a set of 19 criteria for the sustainability dimensions to explore the performance assessment of this particular wheat-to-bread supply chain. They adopted health as a distinct dimension from the social one. However, FAO assumed nutrition, health (associated with food safety), and well-being within the social dimension of the sustainable food systems. Also, ethics was appreciated by Galli et al. (2015) as an apart dimension of sustainability, but in recent literature, ethical responsibility is comprised of either a social (mainly) or economic dimension. In the model proposed by Galli et al. (2015), the attributes owing to the economic dimension of the wheat-to-bread chain sustainability were Affordability, Farmers' income and value-added received, Governance/even distribution of power, Profitability/competitiveness, Resilience, and local development. Technological innovation was considered by Galli et al. (2015) as an economic attribute but, as mentioned above, believe this criterion addresses both economic and environmental dimensions. Economic and ecological efficiency, and Waste were attributed at the same time to economic and environmental dimensions. Like as environmental criteria, there were taken into account Biodiversity, Resource use, and pollution. The social dimension of sustainability in the model defined by Galli et al. (2015) referred to Connection and labor relations, Consumer behavior, Food security, and Territoriality. Still, Information and Communication are items covered by social dimensions; however, there are essential instruments and processes in management systems (QMS, FSMS, RMS, EMS), which means a holistic approach toward all sustainable dimensions. A similar comment can be made for another criterion, namely Traceability, which cannot be related only to health.

Although the model of Galli et al. (2015) is not complete, being improvable, it should note that some attributes are closely related or overlap with each other to a certain extent because of this particular food commodity – bread, which is a staple food. Reason for which affordability is linked to nutrition value, healthiness, and consumer behavior. At the same time, food security is rather an umbrella for many attributes such as nutrition, biodiversity, food safety, waste, and even technological innovation.

5.3.1 *Agriculture Features for a Sustainable Cereal-to-Bread Chain*

The agriculture segment may comprise crop farms and additional postharvest operations till the grains arrive at food processing units. In this regard, several topics about sustainability in agriculture are discussed further.

Tillage, sowing, and plant protection are only a few works for growing crops. *Technological inputs* in terms of equipment, chemicals, etc., contribute to productivity and increase economic dimension. But some of them can directly or indirectly affect food safety and the environment. For instance, some added chemicals can residually remain in the soil and accumulate further in crops because of improper agricultural practices or using plant protection products (e.g., pesticides). These residual substances become later chemical contaminants for grains and polluters for soil, water streams, or air.

To reduce the environmental impact of having suitable agricultural outputs, various *crop production and protection practices*, such as crop rotation, judicious use of agrochemicals, and pest control, were defined as Good Agricultural Practices (GAP). FAO (2003) presented four GAP pillars: economic viability, environmental stability, food safety and quality, and social acceptability. Case studies performed in different areas or countries showed that improving cultivation technologies and the *correct management* applied for grains' farming meant successful stories for sustainable agriculture. Consequently, complying with GAP is a pathway to contribute to sustainability. Even though sustainable agricultural practices were promoted worldwide, they are not applied as expected because of various economic and social constraints. So, there is a long-distance between theory and practice -promising instruments, such as business models, innovation advancements, and technology, become inconsistent where the financial incentives are lacking, access to information is limited, knowledge transfer is not efficient, or government policies are not strong enough. Worldwide there are huge differences between small and large farms, local cultures and reluctance to change, and specific conditions for agricultural systems. Kheiri (2015) identified some barriers to sustainable agriculture in Takestan county of Iran; the conclusion of his study revealed that the most significant obstacles to sustainable agriculture for wheat farmers consisted of the high cost of sustainable agriculture practices and farmers' attitudes. On the other hand, Houshyar et al. (2017), evaluating wheat production sustainability in Southwest Iran, observed could not use the same set of indicators for sustainability in all farming conditions. So, specific local indicators should be employed for different regions to assess sustainability.

Alternative solutions for conventional farming were considered *eco-friendly and green techniques*, such as replacing chemical fertilizers (nitrogen) with biofertilizers and organic farming. For sure, all these alternatives come with a set of advantages but also limits (Döring & Neuhoff, 2021). For example, nitrogen fertilization significantly affects grain yield and wheat quality (i.e., protein content). Compared with conventional farming, for an adequate driven N-supply from soil organic matter inputs, the production management should be performed by taking into account of maximum quantity and type of organic fertilizers, previous crops and ground residues from crop rotation, and selection of wheat varieties with higher nitrogen uptake efficiency (Bilsborrow et al., 2013).

Generally speaking, *organic farming* is adopted to ensure sustainable crop production. But, comparing the effect of organic and conventional management on the yield and quality of wheat grown in a field trial between 2004 and 2008, Bilsborrow et al. (2013) concluded that wheat yield from the organic production was 40% lower than from the conventional one. Some authors reported other figures in different conditions, but the basic idea was that a lower grain yield was registered for wheat in organic agriculture compared to conventional farming. Besides the different accumulation of nitrogen that affects the protein content of wheat kernels, other research proved that some *wheat genotypes* are more convenient than others to be used either in organic or conventional farming (Al-Ghumaiz, 2019). Bencze et al. (2020) summarized the results of a three-year experiment on hulled wheat varieties and concluded *ancient wheat cultivars* as potential candidates for food system diversification into organic agriculture.

Other interesting results of a comparative study between organic and conventional farming were highlighted by van Stappen et al. (2015). Their conclusions on environmental performance revealed a sensitive distinct between organic and conventional wheat farms in Wallonia-Belgium depending on the working hypothesis, namely due to the functional units considered either as a basis of the mass (kg grain) or land area (ha). The results reported to mass basis and yield showed that environmental impact is less for a conventional farm adequately managed. In contrast, if the land surface is the functional unit, organic wheat farming is more eco-friendly than conventional, at least for some indicators (i.e., global warming, energy demand, and photo-oxidants formation).

Consequently, adequate agricultural management, eventually with sustainable technological *innovations* or innovative practices, using natural fertilizers instead of chemical ones, no intense agronomical treatment with pesticides, fungicides, etc., or partial conversion to organic farming could be a few ways to protect or restore the environment.

As known, wheat is the main crop for making bread, but also pastry, pasta, confectionery, or other mealy products. Different types of wheat are grown depending on pedo-climatic conditions and geographical areas – e.g., soft and hard wheat, winter and spring wheat, red/white wheat, and durum wheat. Also, other cereals, such as rye, rice, corn, oat, millet, barley, pseudocereals or non-grains, are cultivated and used to process an extensive range of farinaceous foodstuffs. According to Skendi et al. (2020), the main cereals produced worldwide are corn, rice, and wheat, with ample food and non-food use applications. Nowadays, wheat, rice, and corn are standardized commodities traded globally. As Bartolini et al. (2014) commented, the process of wheat "domestication" due to intensive breeding programs and decrement in the genetic variability has resulted in the improvement of the yields and agronomic characteristics (with a clear positive effect on economics). But wheat "domestication" also conducted to a genetic erosion (Bonnin et al.,

2014), affecting biodiversity and the other two dimensions of sustainability. Besides standardized crops used worldwide, *local biodiversity* concerning the traditional or national crop varieties is used for domestic markets (Bartolini et al., 2014). Biodiversity, use of locally sourced grains, and *revival of ancient species* and varieties create opportunities for reconnecting staple crop producers, bakers, and consumers for sustainability goals.

In agriculture, a set of processes and activities are performed before and after crop harvest to keep grain quality and safety under control and for agricultural activities resumption within an efficient operational management cycle. Zhang et al. (2018) found out the *technological development* applied to farm equipment and cereal transportation were major driving forces of sustainable grain quality improvement in China. Using a one-step stochastic frontier model for rice, wheat, and corn crops, they observed that grain transportation improves postharvest grain quality. And its impact is more significant as time goes by.

Grain storage is crucial to secure an adequate quantity and quality of harvested grains. Operations of *cereals storage* depend largely on warehouse facilities, regarding the type of storage units, their dimensions or capacities, and technological advancement used for keeping or conditioning the grains during the storage period (if any). The main factors influencing stored grains are temperature and relative humidity, grain moisture, carbon dioxide and oxygen, grain characteristics, and potential food hazards (biological, chemical, and physical). Suitable warehouse conditions ensured during the grain storage may contribute to quality improvement, avoid food safety issues, and reduce storage losses with economic and social effects, in terms of efficiency, food quality, and safety, and not the last the food security. Jayas (2012) pointed out that grain postharvest losses can vary in an extensive range from 1–2% to 20–50%, depending on storage units (barns, storage cells, warehouses, etc.) and how those facilities are managed. These percentages of postharvest losses for grains are counted when the economic efficiency is measured at the organization level or food security is discussed, especially in less developed countries.

In Nepal and India, predominant traditional *postharvest practices*, especially in rural areas, consist of using chemicals for pest control (e.g., insecticides, fungicides) and for grain loss decrement during farm storage (Prakash et al., 2016; Kandel et al., 2021). Noteworthy, these are not considered sustainable in many respects but, at least, can partially solve a problem of regional food security and food supply in countries less developed with agriculture-based economies.

Prakash et al. (2016) acknowledged that postharvest loss in Indian agriculture is higher because often, the storage facilities are not available. Still, as an alternative to the chemical traditional pest control of grain stored, the farmers from various Indian villages and districts followed indigenous eco-friendly practices. In this aim, different methods are applied before grains storage, such as sun drying, ash or a red soil coating, or mixing with various matters (e.g., turmeric powders, lime powder, Margosa leaves, and basil seeds).

Drying is a sustainable option to reduce losses during grain storage and adequate protection against contamination from various biological and chemical hazards. In that way, the quality of harvested crops is preserved better, and grain wastes are reduced with an environmental gain. But implementation of technologies and new sustainable systems for drying and storing grains incurs specific barriers in terms of initial investments, training, maintenance, and others. In this regard, Lermen et al. (2020) performed a survey to identify the perceived value of sustainable product-service systems for drying and storage of grains as potential offers for Brazilian farmers.

Galli et al. (2015) proposed local and global models for the wheat-to-bread supply chain. In their framework, the inputs considered for agricultural production and postharvest grain handling are as follows: land, workers and associative forms (farmer consortia), water, seeds, fertilizers, agricultural machinery, fuel, and energy supply. As mentioned above, the inputs are more. From a farming standpoint, besides the fertilizers, other inputs should be considered as chemicals in general (e.g., substances for crop protection), alternative natural or organic matters for agricultural production, and postharvest operations. Except for material inputs, the financial ones are critical for farmers and business operators within whole agriculture. Financial resources and economic reasons often drive farmers' reluctance to change towards sustainable agriculture.

Knowledge transfer through training and innovation is an important asset not only for human capitalization but for implementation of the new business models, the best agricultural practices, technological advancements and emerging technologies, suitable tools for quality assurance and food safety management systems, digitalization in agriculture 4.0, and so on. And all information introduced must comply with legal requirements and updated provisions put in force. So, information is an essential resource, an input to be exploited in sustainable agriculture.

Based on the assessment of the inputs and outputs of agriculture for supplying grains, various authors concluded that this part has the most significant impact on environmental indicators from the whole grain-to-bread chain. LCA also supports some of these results (e.g., Korsaeth et al., 2012).

Besides the chemicals used in agriculture that can accumulate and damage the soil, water, air, or biotic environment in the long term, the agricultural area envisages losses that can be counted as agri-food wastes and cereal by-products. A problem challenging, especially in large farms and companies, is to reduce the volume of cereal by-products and wastes and, secondly, find alternative solutions more efficiently for higher valorization other than their turn into feed.

Except for environmental performance indicators (e.g., carbon footprint, nitrogen footprint, water footprint), there is a priority to fulfill waste management and reduce the waste at the source, followed by reuse and recycling. Noteworthy, correct management of postharvest operations leads to fewer cereal by-products waste that helps overall food processing and, on the other hand, improves the environmental dimension of sustainability.

Still, a holistic evaluation of agriculture sustainability supposes the coexistence of environmental, economic, and social objectives, which sometimes are synergic conflicting, some partially overlapped or even depending on each other. Iocola et al. (2021) introduced an MCA tool with a hierarchic-structured multi-criteria approach for the sustainability assessment of Italy's organic durum wheat-based farms. This tool consisted of 109 attributes (i.e., 64 indicators and 45 aggregated variables) divided into three groups corresponding to sustainability dimensions.

According to the model proposed by Iocola et al. (2021), the environmental pillar covers three topics: natural resources management, crop practices, and environmental attention. Natural resources management refers to the soil, biodiversity, and water; and comprises 20 indicators with an overall weight of 47%. The second topic, namely the crop practices, consists of fertilization, crop protection management, and energy, with 12 indicators having an overall weight of 38%. The environmental attention focused on climate change management and waste management, setting out three indicators with a weight = 15%.

Three topics define the sustainable economic pillar of Iocola et al. (2021), namely: (1) economic viability (expressed by the financial results or economic efficiency on a yield basis, economic independence from public aid and inputs, and multi-functionality; eight indicators; weight = 50%); (2) product valorization (related to food safety and quality, and certification; three indicators; weight = 25%); (3) markets (trade arrangements/channels, short value chains, and contribution to the development of new supply chain; five indicators; weight = 25%).

Three themes represent the social sustainability pillar: (i) work (as the contribution to employment, work contracts, and workplace safety; four indicators; weight = 24%) (ii) human capital (in terms of cooperation, innovation & training; seven indicators; weight = 47%); (iii) territory development (based on communication and awareness-raising, and the value of the landscape; two indicators; weight = 29%). Data collection for this model encompassed that, in many cases, the tool indicators gave rather estimations than measures related to variables, especially for the environmental pillar along the agri-food supply chain (Iocola et al., 2021).

5.3.2 *Milling and Baking Industries from a Sustainable Perspective of the Food Supply Chain*

Cereals processing in mills supposes cleaning, grading, conditioning, grinding, and sieving of different mealy fractions. Appropriate dry and cleaner seeds with minor foreign matter and homogeneous grains are easier to process in the milling industry. Cleaning is an operation to remove foreign matters and grain impurities in bulk. Magnetic separator, sieve separator, and destoner are only a few types of machinery for eliminating or avoiding physical contamination with metal, glass, stone, and plastic foreign matters. Also, in this stage, cleaning consists of keeping apart grain impurities such as shriveled seeds, other cereals, sprouted or damaged grains, husks, or other biological matters (pest, insect bodies). Other cereal by-products result from grain screening and consist mainly of grain seeds, which do not comply with grading specifications. Regarding grain conditioning, Cappelli et al. (2020) concluded that wheat tempering could be an effective tool for improving the milling process and the breadmaking, with potential benefit on LCA.

Milling is an essential part of the cereal processing along this specific agri-food supply chain that can be performed dry or wet with traditional or newer technologies using various machinery from vintage mills to the innovative milling equipment. Some authors reconsider milling processing as a sustainable prospect considering the *revival of the traditional* stone watermills, for example, which can use green energy and withal promote traditions and craftsmanship and even specific form of tourism. On the contrary, other authors are faithful to technological progress that could solve industrial issues for uniform, safe, and high-quality products and improve some environmental indicators (energy savings, reduced carbon footprint, etc.). *Innovation in the milling industry* has been acknowledged in new or upgraded equipment (such as roller milling and automation systems) and technologies (referring to cleaning, grinding, and sieving systems, de-braning, microfluidization to reduce the bran particle size, and micron technology).

The milling process transforms cereal grains into mealy fractions by grinding, crushing, or cutting to the needed particle size. Appropriate *milling technologies and machinery* are combined to obtain an extensive range of flours and mealy products (e.g., refined flour, wholemeal, etc.) with higher or lower amounts of endosperm particles and specific biochemical composition (referring to damaged starch, gluten, lipids, pentosans, or others).

Through milling, depending on the grinding diagram, variable quantities of the bran or seed coat, aleuronic and sub-aleuronic layers, and germs can be removed from the industrial process. However, when wet milling is discussed, it generates by-products and wastewater. According to Skendi et al. (2020), the significant supplier of cereal by-products is the milling industry along this particular supply chain. Many of these *by-products are valuable* raw materials rich in bioactive compounds and other nutrients, which may be extracted and used as nutraceuticals, pharmaceuticals, additives, and other ingredients (Galanakis, 2022).

Besides milling, other grain malting and pearling technologies also produce various by-products that can valorize further for sustainable applications (Galanakis, 2022).

In general, bakery products are obtained by dough preparation, dough processing, and baking. The dough is prepared by mixing the ingredients according to a chosen recipe and fermentation. The dough processing consists of many operations such as dividing, rounding, shaping, and proofing. Baking is the process through which dough transforms into bread loaves, baking parameters being related to the type of products (incl. Loaf mass and shape) and oven equipment. Mealy products are made from similar raw materials such as flours, water, and other ingredients. But sometimes, they are processed by different technologies (than leavening or baking).

Firstly, *correct management of overall operations* allows better valorization of internal resources of the bakery plant with economic gains (e.g., cut costs by unnecessary dispersal of resources) and reduces environmental impact as carbon, energy, water, and waste footprints. For instance, the control process and proper water management regarding the quality, quantity, and temperature of the water used during the breadmaking process result in water waste decrement. In this regard, Cappelli and Cini (2021) pointed out a few *eco-friendly strategies* for improving dough kneading.

Dough preparation and making-up (e.g., by the humidity level, size, and shape of the dough), proofing, and baking also influence the energy consumption. According to Cappelli and Cini (2021), adequate bread production management regarding the kneading phase and baking is essential to increase the sustainability of bakery products in terms of energy consumption and environmental impact. Still, thermal processes such as baking or drying are the most energy-intensive processes concerning energy flows.

The baking stage is the most energy-consuming of the whole breadmaking process. The energy used in baking, either in terms of heating (burning fuels) or electricity, depends on equipment type, production capacity, type and size of loaves to be baked, and how the process is kept under control. Larger plants and small and medium-bakeries differently approach energy management. For example, Kannan and Boie (2003) and Briceño-León et al. (2021) presented some case studies about the energy consumption of different bakeries in Germany and Ecuador and opportunities for sustainable energy management. Similarly, using LCA, Recchia et al. (2019) commented that the pasta drying stage is the most critical for energy consumption when high-quality pasta Italian production is based traditionally on a low-temperature long-time drying process.

In conclusion, food processing, starting from milling to baking and even reprocessing for reusing or recycling by-products and edible food waste, requires increased attention to manage all operations correctly, with economic and environmental benefits. For example, optimizing the breadmaking process using suitable quantities of water and energy in the best conditions through controlling the process in each stage may reduce the inputs and improve the energy savings and water footprint.

Automation allows a standardized quality and better control of the whole food processing, but it decreases the range of baked goods.

Implementing Lean Manufacturing in baking and other farinaceous industrial sectors has many challenges and difficulties (Jain & Lyons, 2009; Castro & Posada, 2019). Still, it can be a convenient solution for reducing waste and optimizing all resources used in high-automatized plants along this food supply chain.

Analyzing the LCA of bread manufacturing at different scales, various authors showed that environmental impacts differ depending on many factors as following: production capacity (bakeries scale), nature and type of ingredients, breadmaking methods and parameters used, facilities and available equipment, how the process is controlled, and even the kind of packaging applied for commercialization.

Espinoza-Orias et al. (2011) estimated the carbon footprint of sliced white and wholemeal bread produced and consumed in the UK; they concluded that bread manufacturing is the third significant contributor in the PAS 2050-compliant study with approximately 16%. In comparison, the first contributor to the carbon footprint is the agriculture stage (with cc. 32–35%), followed in the second place by bread consumption.

Notarnicola et al. (2017) analyzed the energy flows and GHG emissions related to 21 European bread types using an LCA "from-cradle-to gate" with the following functional units of bread: weight, energy value, and price. They noted the more sustainable bread from environmental and nutritional (energy value) standpoints, are those which contain more ingredients than flour, yeast, water, and salt (e.g., improved with vegetable oils, and milk) and those having a reduced baking time (small loaves or flat shaped bread).

In one way or another, many authors showed *that innovation and technological advancements* in terms of ingredients, techniques, and equipment used for bread-making and farinaceous products could sustain the food supply chain.

As a social demand of consumers, the interest in developing new products increases in the baking sector, and the range of baked commodities are consistently increased. The industry has focused on healthy, nutritional, and functional foodstuffs with this aim. More and more are using alternative raw materials, some new ingredients, and others forgotten recipes. Thus, they try valorizing the biodiversity (e.g., for biofortification) or revival the traditional handcrafts (e.g., ancient sourdough). Similarly, other farinaceous products (pasta, snacks, etc.) and processing routes capture the pathway towards diversification, enhancing quality for pleasure, convenience, and nutrition.

Advancements in technical equipment and innovative baking technology could optimize operation parameters (i.e., kneading speed, dough temperature, mass transfer, heat transfer, processing time, etc.), reducing food waste and energy consumption with obvious environmental impacts.

Cappelli and Cini (2021) pointed out several innovations and improvement strategies to increase the sustainability of pasta and bakery products using LCA analysis. Moreover, LCA is used to improve industrial innovations and new development products, with favorable environmental effects along the whole chain from farm to farinaceous foodstuffs offered to end-consumers.

Skendi et al. (2020) commented on the valorization of cereal by-products in developing new mealy products (e.g., biofortified bread with leaves, minerals, and vitamins, carotene-enriched sorghum porridges, cookies with defatted rice bran, extruded corn snacks with oat), biopolymer-based packaging films and bio-composites for edible coatings. But on the contrary to the innovative scope envisaged towards healthier mealy products, sometimes additional compounds and technical solutions applied in manufacturing have an objectionable effect on dough rheology or end-product quality that make consumers reluctant to accept these products in consumption.

By developing emerging food technologies, a few applications are available as alternative solutions for cereal preservation, making specific farinaceous foodstuffs other than bread, novel packaging formula, and new methods of reprocessing the food by-products for their higher valorization along the food supply chain. The majority of these emerging technologies are under development, and some of them are very promising regarding food quality improvement, food processing, or process control. Despite technological challenges and economic feasibility, the usage of novel technologies must be discussed from case to case because some novel methods often face various environmental and legislative issues.

Galli et al. (2015) appreciated the waste as an attribute of marginal importance for their case studies of wheat-to-bread/pasta sustainability, considering the waste quantity less significant than in other food industries (e.g., dairy). This conclusion is probably related to a higher valorization of food waste and cereal by-products along the particular food supply chain of mealy products. In this regard, Cappelli and Cini (2021) are promoters of circular economy applied in the farinaceous chains, with insights into sustainable innovation.

Innovation in shelf life-prolonging to retard bakery products staling and packaging role for preventing bread waste also influences decrement bread losses during the commercialization until the effective consumption with environmental impacts. Unfortunately, food waste in this sector is large enough, especially for fresh bakery products.

5.3.3 *Sustainable Trade of Farinaceous Commodities Along the Food Chain*

Bakery product distribution and consumption are of paramount importance for a sustainable food supply chain in many respects. Firstly, this area is more connected with the social dimension of sustainability, mainly when it refers to consumers concerning nutrition and well-being. Secondly, as quoted, bread consumption is an important contributor to the carbon footprint (Espinoza-Orias et al., 2011).

Generally speaking, bread became a staple food with an affordable price and accessibility for the consumers. But the bakery market is very dynamic, competitive, and still increasing. For sure, this farinaceous market sometimes endures syncope, faces bottlenecks, and is shaken by external causes. For instance, the global food crisis and the war in Ukraine already affect the whole cereal-to-bread supply chain. Galli et al. (2015) clearly distinguished between global and local wheat-to-bread chains taking into account the geographical distance between wheat production and food consumption and the organization level of the supply chain exerted by local vs. global "actors", specific features of milling and baking technologies used, ingredients and recipes, product identity and quality perception as regards of the industrial and artisanal way of processing bread, traditions, and local culture. Analyzing the wheat-to-bread supply chain models, Galli et al. (2015) concluded that the local food supply chain should be approached differently from the global one, referring to the material inputs and technological flows and the supplying distribution channels farinaceous commodities.

The main distribution channels of bakery commodities are regular retailers like grocery stores and other specialty foods stores, supermarkets, or hypermarkets, besides the bakeries chains and traditional neighborhood bakeries. Especially during the Covid-19 pandemic period, online shops and home delivery systems developed for this kind of baked foodstuffs. But bakery-cafes and other food services (restaurants, pubs, fast foods, etc.) will keep their important role for people eating and socializing. Any market segment should be tackled differently, from mass distribution and traditional food to higher value-added commodities.

Based on the literature review, Galli et al. (2015) discussed several attributes for sustainability assessment of wheat-to-bread chains, comparing two Italian case studies as examples of Italy's local and global bread system. They observed the relevance of some attributes rather than others along this food supply chain from grain to bread. For instance, without generalizing their results, Galli et al. (2015) pointed out the importance of nutrition and consumer behavior as attributes that help differentiate local and global sustainable mealy food systems. Also, Bartolini et al. (2014) and Gava et al. (2014) highlighted a variety of local/global contrasts along the wheat-to-bread supply chain. According to Galli et al. (2015)'s findings, the performance of the local Italian chain is better and complementary in relation to those attributes where the global chain is more constrained. Still, the global chain has advantages in resource use and pollution, with environmental benefits (based on emissions and energy flow per kg).

Environmental impact may differ depending on product type or how mealy food is consumed. For instance, Espinoza-Orias et al. (2011) noticed that the second important contributor to the overall carbon footprint along the wheat-to-bread chain in the UK is bread consumption (with a share of 25%) due to the refrigerated storage of end-product and bread toasting. Also, bread waste cannot be neglected at least in the European Union, with a negative impact on the environmental footprint.

5.3.4 A Sustainable Approach to Waste and By-products Along the Cereal-to-Farinaceous/Bread Chain

Along the cereal-to-bread/farinaceous chain, agriculture is the first stage for developing a circular economy. A circular economy envisages the valorization of agricultural wastes, by-products, and co-products by employing innovative technologies and new business practices for profit-making and positive environmental impact (Toop et al., 2017; Barros et al., 2020).

Agri-food wastes and cereal by-products are agricultural losses resulting from the farm area's harvesting and postharvest processing. For instance, the cereal by-products resulting from conditioning (incl. cleaning) before storage and grain screening may partially be recovered within food, feed, or non-food value chains. Some other recovered waste can also be reprocessed for different industrial applications within the circular economy. As Skendi et al. (2020) reviewed, the agricultural residues and cereal by-products can be valorized further through alternative processes to obtain an extensive range of products, such as bioplastics, biofertilizers, biofuels (i.e., biogas, biodiesel, bioethanol), various fermentation metabolites and nutrients (phytochemicals, proteins, fibers and saccharides, lipids, other plant-based compounds). The literature proposes numerous applications to valorize by-products and waste cereal fractions into chemicals, microbiological compounds, enzymes, etc. For instance, wheat bran may be converted to biobased products and fits well with the biorefinery concept, but meantime is widespread use in baked products for fiber content increment (Galanakis, 2022).

Skendi et al. (2020) reviewed cereal by-products recovered from the milling industry and valorize as enriched flours, cereal gum, cereal starch, cereal gluten, cereal germ oil, bran oil, bran gum, phenolic compounds, polysaccharides or dietary fibers, essential fatty acids. High-added value can be obtained through fermentative processes in which cereal by-products become microbial substrates and result in antimicrobials or microbial metabolites, enzymes, antioxidants, and other compounds (Galanakis, 2022).

Sustainable valorization of by-products derived from different cereal processing stages can be seen in targeted applications (as product types or quality for compounds obtained) as well as in green or eco-friendly technologies used for these valorization processes (e.g., ultrafiltration or membrane systems), with overall benefits in economic outcomes and environmental impact. Besides the cost-efficient reuse of food wastes and cereal by-products with economic outputs, their valorization through reprocessing allows enlargement of raw materials base, food diversification (e.g., gluten-free products, bakery foodstuffs enriched or source of vitamins, fibers, etc.), and sustainable circularity (based on cradle-to-cradle principle).

Also, food waste from this supply chain, such as bread waste, can be partially valorized through various technologies, e.g., anaerobic digestion and animal feed production. For example, Vandermeersch et al. (2014) investigated the opportunities to valorize food waste resulting from the retail sector in Belgium based on environmental performance and two models of LCA analysis.

5.4 Conclusion

Sustainability is a complex concept that envisages many interactions among its dimensions, i.e., social, economic, and environmental, and various items or issues that demand a contextualized, well-balanced, and holistic approach.

The baking sector is part of the food supply chain, particularly cereal-to-bread and other farinaceous food commodities chains. Baking sustainability should be considered for all three areas of agriculture, food processing, and trade. Different models and methods were applied in order to perform a multi-criteria analysis of baking sustainability. The peculiar features of the food supply chain under investigation, working hypothesis, functional units, chosen criteria and sub-criteria for appraisal are only a few considerations that influence the sustainability assessment. Although the number of studies on LCA and sustainability evaluation of the cereal-to-bread chain has consistently multiplied in the last 20 years, the research is still in progress. New rooms for the performance of the sustainable baking industry seem to open from scientific, technical, and policy-making standpoints.

References

2013/179/EU Commission Recommendation of 9 April 2013 on the use of common methods to measure and communicate the life cycle environmental performance of products and organisations Text with EEA relevance, *OJ L 124, 4.5.2013*, pp. 1–210.

Abdelkafi, N., & Tauscher, K. (2016). Business models for sustainability from a system dynamics perspective. *Organization and Environment, 29*, 74–96. https://doi.org/10.1177/1086026615592930

Al-Ghumaiz, N. S. (2019). Sustainable agriculture in organic wheat (Triticum Aestivum L.) growing in arid region. *International Journal of Design & Nature and Ecodynamics, 14*(1), 1–6. https://doi.org/10.2495/DNE-V14-N1-1-6

Barba, F. J., Orlien, V., Mota, M. J., Lopes, R. P., Pereira, S. A., & Saraiva, J. A. (2016). Chapter 7 – Implementation of emerging technologies. In C. M. Galanakis (Ed.), *Innovation strategies in the food industry* (pp. 117–148). Academic Press. https://doi.org/10.1016/B978-0-12-803751-5.00007-6

Barros, M. V., Salvador, R., de Francisco, A. C., & Piekarski, C. M. (2020). Mapping of research lines on circular economy practices in agriculture: From waste to energy. *Renewable and Sustainable Energy Reviews, 131*, 109958. https://doi.org/10.1016/j.rser.2020.109958

Bartolini F, Brunori G, Marescotti A, Gava O, & Galli F (2014). Sustainability performance of local vs global food supply chains: The case of bread chains in italy. In: *Proceedings of the 4th World Sustainability Forum*, 1–30 November 2014, MDPI: Basel, Switzerland. https://doi.org/10.3390/wsf-4-g008.

Başaran, B. (2018). 1 – Integrated management systems and sustainable development. In: Kounis, L. D. (ed.) *Quality management systems – A selective presentation of case-studies showcasing its evolution*, 20p. https://doi.org/10.5772/intechopen.71468.

Bencze, S., Makádi, M., Aranyos, T. J., et al. (2020). Re-introduction of ancient wheat cultivars into organic agriculture—Emmer and einkorn cultivation experiences under marginal conditions. *Sustainability, 12*(4), 1584. https://doi.org/10.3390/su12041584

Bigliardi, B., & Galanakis. (2020). 10 – Innovation management and sustainability in the food industry: Concepts and models. In C. Galanakis (Ed.), *The interaction of food industry and environment* (pp. 315–340). Academic Press. https://doi.org/10.1016/B978-0-12-816449-5.00010-2

Bilsborrow, P., Cooper, J., Tétard-Jones, C., et al. (2013). The effect of organic and conventional management on the yield and quality of wheat grown in a long-term field trial. *European Journal of Agronomy, 51*, 71–80. https://doi.org/10.1016/j.eja.2013.06.003

Bocken, N. M. P., Short, S., Rana, P., & Evans, S. (2013). A value mapping tool for sustainable business modelling. *Corporate Governance, 13*(5), 482–497. https://doi.org/10.1108/CG-06-2013-0078

Bonnin, I., Bonneuil, C., Goffaux, R., Montalent, P., & Goldringer, I. (2014). Explaining the decrease in the genetic diversity of wheat in France over the 20th century. *Agriculture, Ecosystems and Environment, 195*, 183–192. https://doi.org/10.1016/j.agee.2014.06.003

Bredahl, L. (2001). Determinants of consumer attitudes and purchase intentions with regard to genetically modified food – Results of a cross-national survey. *Journal of Consumer Policy, 24*, 23–61. https://doi.org/10.1023/A:1010950406128

Briceño-León, M., Pazmiño-Quishpe, D., Clairand, J. M., & Escrivá-Escrivá, G. (2021). Energy efficiency measures in bakeries toward competitiveness and sustainability – Case studies in Quito, Ecuadorr. *Sustainability, 13*, 5209. https://doi.org/10.3390/su13095209

Cappelli, A., & Cini, E. (2021). Challenges and opportunities in wheat flour, pasta, bread, and bakery product production chains: A systematic review of innovations and improvement strategies to increase sustainability, productivity, and product quality. *Sustainability, 13*, 2608. https://doi.org/10.3390/su13052608

Cappelli, A., Guerrini, L., Parenti, A., Palladino, G., & Cini, E. (2020). Effects of wheat tempering and stone rotational speed on particle size, dough rheology and bread characteristics for a stone-milled weak flour. *Journal of Cereal Science, 91*, 102879. https://doi.org/10.1016/j.jcs.2019.102879

Castro, M. R. Q., & Posada, L. G. A. (2019). Implementation of lean manufacturing techniques in the bakery industry in Medellin. *Gestão & Produção, 26*(2), e2505. https://doi.org/10.1590/0104-530X-2505-19

Döring, T. F., & Neuhoff, D. (2021). Upper limits to sustainable organic wheat yields. *Scientific Reports, 11*, 12729. https://doi.org/10.1038/s41598-021-91940-7

EC. (2020). *A farm to fork strategy for a fair, healthy and environmentally-friendly food system*. Communication from the Commission to the European Parliament, the Council, the European Economic and Social Committee and the Committee of the Regions. European Commission, Brussels. COM. 381 final.

Eizenberg, E., & Jabareen, Y. (2017). Social sustainability: A new conceptual framework. *Sustainability, 9*(1), 68. https://doi.org/10.3390/su9010068

Espinoza-Orias, N., Stichnothe, H., & Azapagic, A. (2011). The carbon footprint of bread. *International Journal of Life Cycle Assessment, 16*, 351–365. https://doi.org/10.1007/s11367-011-0271-0

FAO. (2003). Development of a framework for good agricultural practices, 17th Session, Rome, 31 March–4 April 2003. COAG/2003/6. Available at: https://www.fao.org/3/y8704e/y8704e.htm#P86_24256. Accessed 17.02.2022.

FAO. (2014). SAFA – Sustainability assessment of food and agriculture systems guidelines (version 3.0), E-ISBN 978-92-5-108486-1. Available at: https://www.fao.org/3/i3957e/i3957e.pdf. Accessed 17.02.2022.

Galanakis, C. M. (2022). Sustainable applications for the valorization of cereal processing by-products. *Foods, 11*(2), 241. https://doi.org/10.3390/foods11020241

Galli, F., Bartolini, F., Brunori, G., et al. (2015). Sustainability assessment of food supply chains: An application to local and global bread in Italy. *Agricultural and Food Economics, 3*, 21. https://doi.org/10.1186/s40100-015-0039-0

Gava, O., Bartolini, F., Brunori, G., & Galli, F. (2014). Sustainability of local versus global bread supply chains: A literature review. In: 3rd AIEAA (Italian Association of Agricultural and Applied Economics) conference "feeding the planet and greening agriculture: Challenges and opportunities for the bio-economy", June 25–27, Alghero, Italy (No. 173096). https://doi.org/10.22004/ag.econ.173096.

Gimenez-Escalante, P., & Rahimifard, S. (2019). A methodology to assess the suitability of food processing technologies for distributed localised manufacturing. *Sustainability, 11*(12), 3383. https://doi.org/10.3390/su11123383

Goni, F. A., Gholamzadeh Chofreh, A., Estaki Orakani, Z., et al. (2021). Sustainable business model: A review and framework development. *Clean Technologies and Environmental Policy, 23*, 889–897. https://doi.org/10.1007/s10098-020-01886-z

Hopwood, B., Mellor, M., & O'Brien, G. (2005). Sustainable development: Mapping different approaches. *Sustainable Development, 13*(1), 38–52. https://doi.org/10.1002/sd.244

Houshyar, E., Smith, P., Mahmoodi-Eshkaftaki, M., Azadi, H., & Tejada Moral, M. (2017). Sustainability of wheat production in Southwest Iran: A fuzzy-GIS based evaluation by ANFIS. *Cogent Food & Agriculture, 3*(1), 1327682. https://doi.org/10.1080/23311932.2017.1327682

Iocola, I., Colombo, L., Guccione, G. D., et al. (2021). A multi-criteria qualitative tool for the sustainability assessment of organic durum wheat-based farming systems designed through a participative process. *Italian Journal of Agronomy, 16*(1), 1785. https://doi.org/10.4081/ija.2021.1785

Jain, R., & Lyons, A. C. (2009). The implementation of lean manufacturing in the UK food and drink industry. *International Journal of Services and Operations Management., 5*(4), 548–573. https://doi.org/10.1504/IJSOM.2009.024584

Jayas, D. S. (2012). Storing grains for food security and sustainability. *Agricultural Research, 1*, 21–24. https://doi.org/10.1007/s40003-011-0004-4

Kandel, P., Kharel, K., Njoroge, A., et al. (2021). On-farm grain storage and challenges in Bagmati Province, Nepal. *Sustainability, 13*, 7959. https://doi.org/10.3390/su13147959

Kannan, R., & Boie, W. (2003). Energy management practices in SME-case study of a bakery in Germany. *Energy Conversion and Management, 44*, 945–959. https://doi.org/10.1016/S0196-8904(02)00079-1

Kheiri, S. (2015). Identifying the barriers of sustainable agriculture adoption by wheat farmers in Takestan, Iran. *International Journal of Agricultural Management and Development, 5*(3), 159–168. https://doi.org/10.5455/ijamd.175275

Korhonen, J., Honkasalo, A., & Seppälä, J. (2018). Circular economy: The concept and its limitations. *Ecological Economics, 143*, 37–46. https://doi.org/10.1016/j.ecolecon.2017.06.041

Korsaeth, A., Jacobsen, A. Z., Roer, A. G., et al. (2012). Environmental life cycle assessment of cereal and bread production in Norway. *Acta Agriculturae Scandinavica, Section A – Animal Science, 62*(4), 242–253. https://doi.org/10.1080/09064702.2013.783619

Krechovská, M., & Procházková, P. T. (2014). Sustainability and its integration into corporate governance focusing on corporate performance management and reporting. *Procedia Engineering, 69*, 1144–1151. https://doi.org/10.1016/j.proeng.2014.03.103

Kuhlman, T., & Farrington, J. (2010). What is sustainability? *Sustainability, 2*(11), 3436–3448. https://doi.org/10.3390/su2113436

Lermen, F. H., Ribeiro, J. L. D., Echeveste, M. E., Milani Martins, V. L., & Tinoco, M. A. C. (2020). Sustainable offers for drying and storage of grains: Identifying perceived value for Brazilian farmers. *Journal of Stored Products Research, 87*, 101579. https://doi.org/10.1016/j.jspr.2020.101579

López-Pedrouso, M., Díaz-Reinoso, B., Lorenzo, J. M., et al. (2019). 3 – Green technologies for food processing: Principal considerations. In F. J. Barba, J. M. A. Saraiva, G. Cravotto, & J. M. Lorenzo (Eds.), *Innovative thermal and non-thermal processing, bioaccessibility and bioavailability of nutrients and bioactive compounds* (Woodhead Publishing series in food science, technology and nutrition) (pp. 55–103). Woodhead Publishing. https://doi.org/10.1016/B978-0-12-814174-8.00003-2

McMurrian, R. C., & Matulich, E. (2006). Building customer value and profitability with business ethics. *Journal of Business and Economics Research, 4*(11), 11–18. https://doi.org/10.19030/jber.v4i11.2710

Nosratabadi, S., Mosavi, A., Shamshirband, S., Kazimieras Zavadskas, E., Rakotonirainy, A., & Chau, K. W. (2019). Sustainable business models: A review. *Sustainability, 11*(6), 1663. https://doi.org/10.3390/su11061663

Notarnicola, B., Tassielli, G., Renzulli, P. A., & Monforti, F. (2017). Energy flows and greenhouses gases of EU (European Union) national breads using an LCA (Life Cycle Assessment) approach. *Journal of Cleaner Production, 140*, 455–469. https://doi.org/10.1016/j.jclepro.2016.05.150

Perrings, C., & Ansuategi, A. (2000). Sustainability, growth and development. *Journal of Economic Studies, 27*(1/2), 19–54. ISSN 0144-3585.

Pipatprapa, A., Huang, H. H., & Huang, C. H. (2017). The role of quality management & innovativeness on green performance. *Corporate Social Responsibility and Environmental Management, 24*(3), 249–260. https://doi.org/10.1002/csr.1416

Prakash, B. G., Raghavendra, K. V., Gowthami, R., et al. (2016). Indigenous practices for eco-friendly storage of food grains and seeds. *Advances in Plants & Agriculture Research, 3*(4), 101–107. https://doi.org/10.15406/apar.2016.03.00101

Recchia, L., Cappelli, A., Cini, E., Pegna, F. G., & Boncinelli, P. (2019). Environmental sustainability of pasta production chains: An integrated approach for comparing local and global chains. *Resources, 8*, 56. https://doi.org/10.3390/resources8010056

Regulation (EC) No 1221/2009 of the European Parliament and of the Council of 25 November 2009 on the voluntary participation by organisations in a Community eco-management and audit scheme (EMAS), repealing Regulation (EC) No 761/2001 and Commission Decisions 2001/681/EC and 2006/193/EC, *OJ L 342, 22.12.2009, pp. 1–45.*

Ronteltap, A., van Trijp, J. C. M., Renes, R. J., & Frewer, L. J. (2007). Consumer acceptance of technology-based food innovations: Lessons for the future of nutrigenomics. *Appetite, 49*(1), 1–17. https://doi.org/10.1016/j.appet.2007.02.002

Schaltegger, S., Hansen, E. G., & Ludeke-Freund, F. (2016). Business models for sustainability: Origins, present research, and future avenues. *Organization & Environment, 29*(1), 3–10. https://doi.org/10.1177/1086026615599806

Siegrist, M., & Hartmann, C. (2020). Consumer acceptance of novel food technologies. *Nature Food, 1*, 343–350. https://doi.org/10.1038/s43016-020-0094-x

Skendi, A., Zinoviadou, K. G., Papageorgiou, M., & Rocha, J. M. (2020). Advances on the valorisation and functionalization of by-products and wastes from cereal-based processing industry. *Foods, 9*(9), 1243. https://doi.org/10.3390/foods9091243

Toop, T. A., Ward, S., Oldfield, T., Hull, M., Kirby, M. E., & Theodorou, M. K. (2017). AgroCycle – developing a circular economy in agriculture. *Energy Procedia, 123*, 76–80.

Torelli, R. (2021). Sustainability, responsibility and ethics: Different concepts for a single path. *Social Responsibility Journal, 17*(5), 719–739. https://doi.org/10.1108/SRJ-03-2020-0081

UN. (2015). *Transforming our world: The 2030 agenda for sustainable development*. A/RES/70/1. New York, USA.

UNEP (2022) Stockholm+50: A healthy planet for the prosperity of all – Our responsibility, our opportunity. Available at: https://www.unep.org/news-and-stories/press-release/stockholm50-international-meeting-accelerate-action-towards-healthy. Accessed 27.02.2022

van Stappen, F., Loriers, A., Mathot, M., Planchon, V., Stilmant, D., & Debode, F. (2015). Organic versus conventional farming: The case of wheat production in Wallonia (Belgium). *Agriculture and Agricultural Science Procedia, 7*, 272–279. https://doi.org/10.1016/j.aaspro.2015.12.047

Vandermeersch, T., Alvarenga, R. A. F., Ragaert, P., & Dewulf, J. (2014). Environmental sustainability assessment of food waste valorization options. *Resources, Conservation and Recycling, 87*, 57–64. https://doi.org/10.1016/j.resconrec.2014.03.008

Zhang, M., Duan, F., & Mao, Z. (2018). Empirical study on the sustainability of China's grain quality improvement: The role of transportation, labor, and agricultural machinery. *International Journal of Environmental Research and Public Health, 15*(2), 271. https://doi.org/10.3390/ijerph15020271

Chapter 6
Sustainability Assessment of the Baking Industry Complying with Standards Requirements: A Case of Romania

Alexandrina Sîrbu

6.1 Introduction

Bread and other farinaceous foodstuffs are seen as staple food worldwide, this specific supply chain from farm to consumers being an essential part of the entire food supply chain globally. As for Europe, the bakery industry is in the top five of the food industry. According to Fortune Business Insights (2022), only the world bakery market sized 397.90 billion USD in 2020, of which USD 137.05 billion counted for Europe. They reported a bakery market segmentation for 2020, where bread represented the significant segment (almost half), followed by a second segment of cake and pastries, 14.61% biscuits and cookies, and the last narrowed part of other bakery products.

As Briceño-León et al. (2021) reviewed, the distribution of the industry according to the bakeries'size (as production scale) and segmentation by distribution channels between industrial bakery commodities, traditional bakeries, and food services differ from region to region and from a country to another. As for Europe, between 2010 and 2020, the industrial sector of 45% was covered by approximately 1000 plant bakeries, while 190,000 small and medium bakeries ensured the European market segment of 55% bakery products (Briceño-León et al., 2021; Federation of Bakers, 2019). The industrial baking sector in Europe is better represented in countries such as the United Kingdom, Netherlands, and Bulgaria. In contrast, Greece and Spain are areas for traditional and smaller craft bakeries (Federation of Bakers, 2019).

Also, the European consumption patterns vary widely in both quantity and range of bakery products. With a slightly decreasing trend during the last decade, it is estimated to have an average yearly consumption of 45–50 kg of bread per

A. Sîrbu (✉)
Constantin Brancoveanu University of Pitesti, FMMAE, Ramnicu Valcea, Romania

J. M. Ferreira da Rocha et al. (eds.), *Baking Business Sustainability Through Life Cycle Management*, https://doi.org/10.1007/978-3-031-25027-9_6

inhabitant, having fresh bread as the star of best-sold bakery food commodities. However, it registered growth for pastries and biscuits.

Because of regional and global trends and policies in force, the baking sector along its food supply chain experienced different paradigms for lasting development, circular economy, or sustainable growth. Thus, the shape of the baking industry changed at national and local levels in many respects. The cereal-to-bread supply chain is essential among other food supply chains. A sustainable approach is necessary for all dimensions, i.e., social, economic, and environmental. Noteworthy, along the cereal-to-bread chain in Europe, there are solid links between agriculture, milling, food processing, and distribution. Some of the larger or multinational companies in the baking industry also own farms and other agricultural or milling assets (inclusive grain storages).

On the other hand, as a movement against the fast-food lifestyle and in order to promote local food traditions and cultures, since the 1980s, an organization has grown internationally, i.e., Slow Food, that envisages global sustainability along the food supply chain from farm to fork (Slow Food, 2004). They addressed the variety of the food supply equally, including the cereal-to-bread chain (Slow Food, 2017). Its members, with people from over 160 countries, have been actively involved in combating the disappearance of biodiversity, local food, and sustainable archaic practices in agriculture and food production. They attract attention to the food we eat and our food choices, their philosophy pledging for good, clean, and fair agro-food products worldwide (Slow Food, 2021).

From a social standpoint, the baking sector employs approximately a third of the labor force involved in the European Food and Drink industries sector, but with lower labor productivity than other food industries. Automated machinery, new equipment, and legislation contributed to better working conditions for operators in the baking industry. And human resources were differentially capitalized (in terms of workers' protection, occupational safety and health, labor practices, professional training, added value through innovation and knowledge transfer, etc.) within this supply chain.

In comparison with other food chains, the bakery sector is more focused on consumers. Although the main offer consists of bread as a staple food, nowadays, there is a considerable diversification of baked and other farinaceous foodstuffs to meet the consumers' demand in relation to customers' choices and preferences. For instance, affordable, nutrition, convenience, and healthier alternative are the main attributes considered by consumers when they refer to bakeries and other mealy foodstuffs. Noteworthy, no matter the range of mealy products, food safety along the whole cereal-to-bread/mealy products chain must be an implicit pre-requisite (in many world regions and countries, food safety is foreseen in food laws).

As mentioned, the baking market is very fragmented between small and medium enterprises (SMEs) and large companies. Also, the main baked foodstuffs offered on this market are fresh, which means the period for their commercialization is smaller, and an intense focalization on clients is foreseen. These are the reasons why the bakery market is intensely competitive, and the economic criterion is compulsory for survival. Consequently, to meet the requirements of consumers, many companies invest in innovation and technology for financial performance increasing and not at last environmental protection or at least compliance with national or regional environmental laws and policies. For economic sustainability, organizations that act along the cereal-to-bakery and farinaceous products supply chain work to implement sustainable business models, management systems, and even corporate practices to increase their profitability, continual improvement, and sustainability in the long term. Firms try to valorize better business opportunities for a competitive advantage in the market and sustainable growth at each level. Obviously, the tools used for economic gain differ from SMEs and more prominent companies, but the main purpose is the same, namely to capture, keep and even increase a particular market share.

The concerns towards environmental aspects are considered different depending on the country, plant or bakery scales, regional policies, and national environmental laws and policies. Many focus on environmental processes, eco-technologies and eco-practices, some for environmental performance and ecological impact assessment, and others for eco-design and environmental claims.

The Federation of European Manufacturers and Suppliers of Ingredients to the Bakery, Confectionery and Patisserie Industries (Fedima) acknowledge the initiatives and supporting actions of its members to contribute to a more sustainable European bakery industry with the purpose of achieving at least six (Fedima, 2021) of the UN Sustainable Development Goals, i.e., SDGs 3, 8, 12, 13, 15 and 17 (UN, 2015). In this aim, some of Fedima's members found new ways to reduce food waste and packaging through responsible consumption and production (SDG 12). They acted to sustainable packaging, responsible sourcing, and reduced energy emissions, aimed at achieving SDG 13 (climate action) and SDG 15 (life on land). Others promoted good health, nutrition, and well-being through an adequate, improved diet (SDG 3). By building partnerships for achieving the goals (SDG 17), some offered decent work and ensured economic growth (SDG 8) (Fedima, 2021).

6.2 Research Methodology

This paper aims to describe a synopsis of the sustainability features of the Romanian baking industry. In this context, a summary-up of the Romanian milling & bakery industry has been made within the food supply chain. And a qualitative research was performed for sustainability assessment of the Romanian baking industry complying with some standards requirements.

Firstly, extensive documentary research was performed. The data was collected from trade and media press (newsletters, leaflets, online information), professional associations, industrial business operators' websites, legislation, scientific studies, and other technical papers (inclusive standards). The information covers aspects of:

- the Romanian bakery market, with outcomes and challenges facing the milling & bakery industry along the food supply chain;

– sustainable bread supply, as a general approach, with the specific case of Romania;
– the importance of standards for sustainability implementation along the food supply chain and impact on Romanian bakery businesses.

The study introduced further is not extensive enough to have a very in-depth blueprint of the baking sector nor outline a SWOT (Strengths, Weaknesses, Opportunities, and Threats) analysis. Still, a SWOT analysis is not the purpose but to identify the adequate context of the research. So, an overview of the bread industry in Romania is introduced to have the big picture of its features along the cereal-to-bakery products supply chain and identify the context of organizations. Generally speaking, the baking industry ensures adequate, safe, nutritionally, and affordable baking goods for the Romanians. And an up-dating on the sustainability in the Romanian baking industry is proposed.

Secondly, in the period July 2020 to August 2021, a research was conducted based on a survey with representatives of several business operators. As information sources for defining the premises and scope of the research, it took into account the literature review on business models, organizational culture, management systems, standard requirements, Life cycle assessment (LCA), environmental issues, and the multi-criteria approach sustainability concepts.

Standardization is a factor in promoting technical progress, achieving and improving the quality of food commodities (incl. Food safety), economic rationalization, as well as facilitating trade in domestic and international markets. Based on an empirical approach, at this stage is proposed a brief assessment of how the Romanian baking manufacturers comply with standards requirements towards sustainability. For the social dimension of sustainability was considered food safety with an impact on the health and well-being of the consumers. At the same time, quality management was foreseen as a tool for economic performance. Concerning environmental aspects should be noted that neither official food waste, carbon footprint, nor energy consumption statistics are available at the national level, the reason for which the study does not focus on environmental performance assessment but on how the environmental processes are kept under control within the Romanian baking industry. Assumed research criteria were leadership, resources allocation, management review, personnel, and environmental issues. As depicted in Table 6.1,

Table 6.1 Research model proposed for sustainability assessment according to some standard requirements

Sustainability dimension	Management system (MS)	Research criteria (standard requirements)
Social	Food safety	People (personnel)[a]
Economic	Quality	Leadership[a]
		Resources allocation[a]
		Management review[a]
Environmental	Environmental	Environmental issues

[a]Criterion is general for all considered MSs

following the standard requirements, these five sections address many sustainability dimensions than one.

So, the experimental data have been collected from 40 companies by proposing a model about environmental issues, food safety, and quality using management systems for a more sustainable supply chain following specific standard requirements (of ISO 14001:2015, ISO 9001:2015, and ISO 22000:2018).

The methodology used in this case study has been based on qualitative research. A survey was conducted based on a questionnaire, having closed or open questions, and completed with the additional direct interviews to clarify or understand specific issues related.

6.3 An Overview on Romanian Bakery Industry

The Romanian food market is placed on second top of the largest markets in Central Eastern Europe after Poland, being very dynamic, even in specific and challenging conditions of the COVID-19 pandemic period of 2021 (FRD, 2021).

The latest data shows that Romanians are ranked the first in the EU-27 in terms of baked goods consumption. Romanian bread consumption is placed second after Turkey as part of the Balkan region (comprising areas of South-Eastern Europe). Although the total consumption of bread in Romania has continued to decrease in the last three decades, it remains up to 60% more than the European average.

Traditional commerce ensures approximately 50% of the total sales, and the preferences for fresh white bread are still on top of the Romanians' consumption habits. The preferred places for bread purchase are traditional neighborhood markets and grocery stores, followed by bakeries shops and super- and hypermarkets (Scurtu Cristu et al., 2016). Approximately a decade ago, a growing sector became In-store bakeries. Also, bread and other bakery products are sold within the food service sector for consumption outside the home, mainly as part of the menu. Bakery-cafes developed mostly in cities. Especially since the end of March 2020, during the Covid-19 pandemic, online shopping for food flourished on e-commerce platforms (with 42–48% compared to 2019), and home delivery systems developed for various baked foodstuffs and mealy products.

Among the factors that influence consumption patterns and determine trends in the bakery market, it mentions fast food, consumption outside the home, globalization of markets, and health concerns. The consumption patterns for bakery products are influenced by age, gender, education, geographical area, environment (rural-urban), and individual preferences (Sîrbu, 2009b). Scurtu Cristu et al. (2016) analyzed the average bread consumption by regions in Romania; they observed the highest consumption of bread was recorded in the South-Western Region of Oltenia (with more than 10 kg/inhabitant/month), while the lowest was found in the North-East Region (with 6.5 kg/inhabitant/month). Also, consumers' types of bread and desired quality slightly varied from one region to another (Sîrbu, 2009). Regarding bread, there are also differences between regions of Romania in terms of preferences for packed vs. unpacked bread and the loaf size, respectively, in the sold portions (Scurtu Cristu et al., 2016).

Regarding the reasons driving the consumers' preferences, the results seem to be different between packed and unpacked bread. For instance, Scurtu Cristu et al. (2016) highlighted that among the first criteria for packed bread assessment, the taste, price, and brand seem to be the most important. On the contrary, other sources showed that the main attributes of fresh unpacked bread are quality (in general), freshness, and ingredients (recipes) (Arta alba, 2019). Still, sliced bread is considered more convenient to buy and consume, becoming of interest to consumers as well as the bakers.

However, it is interesting that the same consumption trends quoted by Sîrbu (2009) have been kept in the last two decades concerning bakery products, as follows:

- increasing demand for packed "long-life" products to the detriment of fresh bread;
- decreasing bread consumption in favor of other mealy products (biscuits, industrial pastries, cookies, pasta, etc.);
- increasing demand for ethnic (e.g., pita) and traditional bakery products;
- an increasing trend for convenient ready-to-bake products, such as frozen dough and part-baked products;
- increased interest in healthy bakery products, dietetic and functional foods, and even organic products.

Obviously, the tendencies changed, as a rate, in different periods, depending on other macroeconomic factors and the political context. For instance, during the COVID-19 pandemic in the period 2021–2022, the sales of packed bread vertiginously increased, while the demand for expensive functional or organic food had a slower rate, also related to the decrement of purchasing power of the consumers.

Most Romanians' consumption trends can be correlated with social changes in recent decades and, in context, are somewhat similar to those of other European consumers. However, there are certain peculiarities in quantity demanded on the market, the accepted price, products' quality appreciation by consumers, and preferred assortments.

Even if the bread market is consistent and competition in the market is intense, that does not mean irregularities appear. Besides very active marketing, most issues related to non-compliant with the food laws appeared in bread labeling with non-declared ingredients or incomplete mentions with a potential to mislead the consumers (Scurtu Cristu et al., 2016).

Different reports showed that European average yearly bread consumption decreased in the last three decades from approximately 75–80 kg/inhabitant to cc. 45–50 kg/inhabitant nowadays; compared to this, the changes in the bakery products market in Romania were more significant. Firstly, in Romania, the bread consumption decreased from about 121.87 kg/inhabitant in 1990 to cc. 108–110 kg/inhabitant in 2007–2008 (Sirbu, 2009) and below 100 kg/inhabitant after the 2010s. An estimated yearly consumption of bread among Romanians from 2015–2016 is around 80–85 kg/inhabitant (Deselnicu et al., 2020; Scurtu Cristu et al., 2016).

Besides the decrease in individual bread consumption (as average figures), another factor affecting the total production volumes and sales was the population decrement from 23.2 million inhabitants in 1990 to cc. 19.3 million residents in 2020. In that way, a reduction in total production happened with cc. 70% for the period mentioned above. Even so, the baking industry is an important branch of the Romanian food industry in terms of goods offered and production volumes, revenues, labor force involvement, and other economic indicators.

According to the Classification of Activities in the National Economy (NACE codes) 2022 in Romania, the Manufacture of bakery and farinaceous products is included in section C – Manufacturing, Class 107 (Codes from 1071 to 1073), as part of the Manufacture of food products (CAEN, 2022). This class refers to the manufacture of bread, fresh and preserved pastry, cakes, rusks and biscuits, pasta (macaroni, noodles, etc.), couscous, and similar farinaceous products. Bakery and farinaceous products consist of an extensive range of leavened and unleavened bread, pastry, pies, cakes, pizza, and other flour-based products. The assortment of baked foodstuffs continues to be enriched, being difficult sometimes to distinguish between various ranges of farinaceous products.

Firme.Info (2022) reported that, in Romania, there are registered 12,948 companies with a NACE code 107, with different distribution among 1071 NACE (11,804 firms), 1072 NACE (912 firms), and 1073 NACE (232 firms). The best represented area is bakery activity for manufacturing leaven bread (cc. 90% as a number of companies, but not in production volume). Although the number of enterprises in the baking industry considerably increased in the last decade (e.g., from 8449 companies in 2014 – source: Scurtu Cristu et al., 2016), of these registered companies, only cc. 70% are still running bakery and farinaceous production business in 2022. That means a reconfiguration of the baking sector in terms of producers and not only in distribution channels. Like the European baking market, the Romanian one is fragmented between smaller bakeries and large plants.

After the declared turnover, the leading players in the baking industry are Vel Pitar (bread and pastry) followed by CHIPITA ROMANIA Ltd. (biscuits, cookies, preserved pastry), CROCO Ltd. (biscuits, cookies, preserved pastry), FORNETTI ROMANIA Ltd. (bread and fresh pastry), PHOENIXY Ltd. (biscuits and preserved pastry), and LA LORRAINE Ltd. (bread and fresh pastry) (Firme.Info, 2022).

However, do not forget, when bread production discusses, that the food supply chain involves milling and baking. In this respect, the companies, especially the larger ones with activities in NACE code 107, also have registered other NACE codes such as milling (NACE code 106 – Manufacture of grain mill products, starches and starch products). In this context, the shape of the milling & baking industry changes in-depth. Consequently, the main actors in the top five companies with a turnover higher than 50 million euros are Vel Pitar SA. (which is part of the same group with Sapte Spice, having primary the milling activities), Oltina Impex Prod Com Ltd., Boromir Ind Ltd., and Dobrogea Group SA. Except for the financial outcomes, they differ in shareholding type, national dispersion, and production scale. Some companies are joint-stock, and others are limited companies. Some are concentrated in a specific county, and others have manufacturing distributed in many branches in different regions in Romania.

The first eleven companies, which produced with cc. 30 large industrial plants, made a turnover of 6.5 billion RON in 2019, representing approximately 25% of this industry sales (Deselnicu et al., 2020). These data prove that the industrial milling and baking sector is highly concentrated.

As Deselnicu et al. (2020) pointed out, the milling and baking industry is one of the challenging Fast Moving Consumer Goods (FMCG) industries. And this statement is supported by the collection of firms with activities engaged in the bakeries sector as well as how changed the top-ranking of companies from one period to another (for comparison, see Sîrbu (2009) and Deselnicu et al. (2020). That means it is essential to use all resources (human, financial, material) of the organization adequately to produce and sell bakery products and find the best strategies to keep a competitive place in this market. For this purpose, at least the larger companies engaged in the milling and baking industry are applying a differentiation strategy.

Besides the harsh competition between the more prominent players, there is also a certain dynamic of the competition for SMEs, who either wish to grow or survive. Competition in this specific market for distribution channels and increasing sales is between producers no matter the manufacturer scale without any evident partition between smaller and larger firms. For example, larger companies have their own retail chains as neighborhood shops. At the same time, even smaller bakers wish to jump with their products from the local markets and corner bakeries into the hypermarkets.

Deselnicu et al. (2020) commented on the forces driving the Romanian baking industry competition, analyzing four strategic groups existent in this food sector. These strategic groups were distinguished by considering the geographical coverage and range of mealy products marketed by the leading players as strategic dimensions.

However, the rules of the "game" change all the time within this specific food market because, firstly, bread is a staple food; and secondly, all socio-political and economic pressure is apparent on bread commodities and end-consumers.

6.4 Assessing the Sustainability in Baking Industry Complying with Standards Requirements

6.4.1 Sustainability on Baking Sector in Romania - Premises

Sustainability can be conceptually approached along the food supply chain considering all its dimensions, i.e., economic, social, and environmental. Meantime, the baking industry as a part of the cereal-to-bread supply chain should be proceeded in a distinct manner, having particular features when sustainability discusses.

Many authors analyzed sustainability along the whole cereal-to-bakery and farinaceous products supply chain (e.g., Galli et al., 2015; Gava et al., 2014). In order to perform a sustainability assessment of the cereal-to-bakery and farinaceous products supply chain should be addressed three main areas, i.e., agriculture, food processing, and trade (wholesale distribution, retails, and other forms of commerce, including the food services). These areas are strongly interconnected and sometimes dependent. Although various models for sustainability evaluation were proposed considering multi-criteria analysis and multiple indicators, an accurate assessment is still difficult to conduct, and this topic for scientific research is still open.

Next, it refers to the Romanian baking sector considering the bread manufacturing and processing of other farinaceous, as well as trade towards the end-consumers. For sure, the baking industry is not out of the general context of the whole supply chain from farm to fork. But following, the attention focuses on the baking industry, which comprises farinaceous food processing and commercialization activities.

Sustainable development of this sector became evident in the 2000s when the bakery and other farinaceous products' market became a more competitive market (Sîrbu, 2009). The reasons that drove these changes were due to the significant growth of the leading players and market reshaping, adoption of Community *acquis* and various European policies and regulations (related to food, environment, etc.), and entering international markets.

A Few Aspects of the Social Dimension of the Romanian Baking Industry Since the pre-accession to the EU, Romanian legislation was harmonized with the "Community *acquis*". In this respect, starting in 2000–2006, food and environmental laws intensively changed and updated toward a sustainable approach. Government strategies aligned with those of the European Union for public health and food safety had an indirect impact on sustainability, even if, at that very moment, sustainability was not a primary goal envisaged.

This market is consumer-oriented, and consumer behavior continues to be an essential variable that challenges the Romanians and business operators in the baking industry. For example, demand for convenient and ready-to-eat products increased the percentage of sliced and packed bread commodities, which can be easier divided into small portions, have prolonged shelf life, and, not at last, appear safer (due to avoiding additional hands on the foodstuffs but no otherwise).

Even if bread remains an essential food in the Romanian daily diet, the assortment of other baked commodities changed visibly in the domestic bakery market. The industry has started to supply healthier and functional bakery foodstuffs using new or many ingredients or trying to revive already forgotten traditional methods or handcrafts.

It is well that health concerns have grown among Romanians in recent decades, but sometimes social media manipulates consumers' decision-making. For instance, an intensive "information" against white bread has been launched in the last years; regular white bread was considered unhealthy and responsible for obesity and cardiovascular diseases. For sure, it is excellent to increase the number or change some ingredients for variety and additional functional characteristics in the diet. However, bread is still a staple food for Romanians. Besides that, in Romania, this kind of regular white bread is made without adding sugar (as in other international recipes), and they use wheat flour 650 type (with 0.65% ash) but no refined white flours as 450 or 480 types. Regarding salt content, according to the Romanian standards and laboratory determinations, in most bread, sodium chloride content varies in percentage from 1 to 1.3 or 1.5 maximum, a level lower than other European figures (e.g., see Joossens et al., 1994; Pérez Farinós et al., 2018).

Indeed, bread consumption is higher than in other European countries. This food pattern can be explained in Romanian traditions and how Romanians associate bread with other food in their daily meals. No less true that, for Romanians with lower purchasing power, bread often appears in food consumption patterns by replacing more expensive foodstuffs as a substitution effect.

On the other hand, communication and social media influence bread consumption and food patterns (Lădaru et al., 2021).

If considering human capital and labor force as a component of social sustainability, it is necessary to highlight that in 2019 the Romanian milling and bakery industry engaged almost 45,000 employees, having still registered a labor shortage of nearly 10% (Barbu, 2019).

Towards a Sustainable Economic Baking Industry Businesses run in the bread supply chain cover the Romanian market, being restrictive neither national nor local. As a whole, different companies use inputs or supply goods globally. So, beyond the national competition in the baking market, significantly larger enterprises also face competitive rules abroad. That is more challenging because, compared with other food sectors, most of the leading players in this milling and bakery market are companies with Romanian capital investments (e.g., Oltina Impex Prod Com, Boromir Ind), part of them starting as small family-type businesses in the 1990s.

In the pre-accession period of the EU, the efforts to attract investment funds intensified, aiming at re-tech and automation, increasing production capacities or building new plants or facilities, and innovating new products, processes, and adaptive marketing and management systems.

During the last decades, many outcomes in product innovation were registered for diversification and adapting to consumers' requirements, considering the global or regional trends in nutrition and well-being (inclusive cereal by-products high valorized through bio-fortification of the baked foodstuffs).

Food manufacturing occurs in artisan bakeries, small-scale, or large industrial plants in the milling and baking industries. Production scale, type of raw materials handled, technological inputs, and how the processes are operated all these influence financial performances that ensure the economic sustainability of the companies. Knowledge transfer, access to an experience economic environment, scientific advancement and theories, and self-learning lessons urged Romanian bakers to apply business models based on managerial principles and the Deming cycle. Thus, it became compulsory to correctly manage overall operations within the baking industry to improve financial outcomes and economic efficiency. No matter the bakery scale, all wish to gain money at the end of the day.

Environmental Issues Addressing the Baking Sector Using different tools, such as LCA and various environmental indicators (e.g., carbon footprint, energy consumption, greenhouse gas (GHG) emissions, water footprint), in literature, different authors try to estimate at least partially the environmental impact of bread processing. For example, Andersson and Ohlsson (1999) discussed the relation of the production scale with environmental issues, such as energy use, emissions, and waste management. Espinoza-Orias et al. (2011) evaluated the carbon footprint of bread produced and consumed in the United Kingdom. Kannan and Boie (2003) studied energy management aspects for bakeries'sustainability in Germany, while Briceño-León et al. (2021) discussed ecological issues of energy consumption in bakeries of Ecuador. But, in-depth studies on environmental performance evaluation and the environmental effects generated by bread production and consumption in Romania did not find.

Although environmental performance assessment is not available, many bakeries, especially the industrial plants, control their processes with ecological impact for energy savings, reducing electricity consumption, and recovering heat resulting from the baking process in industrial ovens. Generally speaking, through automation, the breadmaking process is better kept under control with optimized quantification of all inputs and cut costs in terms of energy, electricity, and water.

The interest in applying the best manufacturing practices increased simultaneously with the reinforcement of the regulatory field of food safety in Romania and when bakers observed that adequate production management could save their money in many respects. In this case, the potential environmental impact of water, energy, or waste management was appreciated as a second goal after the direct economic benefit. Also, technical equipment advancements and innovations applied in the baking industry proved to be sustainable in both economic and environmental means.

Representatives of Romanian patronages of the milling and baking industry – ROMPAN and ANAMOB raised the issue of food waste within this specific food supply chain, too.

However, it is worth noting that many authors that analyzed the environmental aspects related to the bread manufacturing identified various factors which influence ecological impacts, such as manufacturing scale, recipe used (as regards the ingredients, breadmaking methods, and processing parameters), manufacturing practices, equipment and available infrastructure (warehouses, facilities, etc.), packaging (Andersson & Ohlsson, 1999; Kannan & Boie, 2003; Espinoza-Orias et al., 2011; Galli et al., 2015; Notarnicola et al., 2017).

6.4.2 Standards, As Tool for Management Systems' Implementation

As known, the International Organization for Standardization (ISO) offers various standards applicable to different kinds of organizations in many areas of interest. Most of them are voluntary, but they are widely used because they foresee general or specific requirements that support the implementation and operation of various management systems. Sets of standards dealing with several management systems can help organizations to improve their objectives related to many requirements in fields such as quality management, food safety, environmental processes, risk management, conformity, occupational health and safety, information security, accounting aspects and better usage of resources, business continuity, etc. (ISO, 2022). As introduced in Table 6.2, many types of management systems or parts thereof can be implemented in organizations along the food supply chain, singularly or as integrated systems. There are different opinions about successive implementation or many integrated management systems in a whole one, depending on organizations' needs, resources, and benefits.

This study further considers a few specific requirements of standard families of ISO 9000, ISO 22000, and ISO 14000 for particular management systems implemented in bakeries in Romania. At the organizational level, mostly larger bakeries implemented various management systems with more international standard requirements and later private-labeled certification.

Quality management system implementation (based on ISO 9000 family) envisages quality assurance at an organizational level. It promotes economic performance by reasonably providing all resources, controlling processes, and improving efficiency (cutting costs). Also, it can positively influence the food industry on green performance (Pipatprapa et al., 2017).

Food safety management based on HACCP (Hazard Analysis Critical Control Points) can be systemically applied in accordance with requirements stipulated in ISO 22000 family. In this way, food safety addresses the social dimension of health and well-being by supplying as safe products for consumption as possible on the food market. But FSMS is managed within economic means by food business operators. Additionally, some private certification schemes were developed under the Global Food Safety Initiative (GFSI) umbrella.

The family ISO 14000 comprises a set of standards that express principles, specific requirements, and guidelines either for implementation or certification of the EMS, environmental performance evaluation, eco-efficiency, or eco-design at the organizational level.

By evaluating the activities and inputs/outputs to environmental issues, these consequences are discussed from both economic and ecological standpoints. For instance, ISO 14001 is considered rather a process standard in so far as companies flexibly choose environmental objectives (Mosgaard et al., 2022).

Noteworthy that between management systems, such as QMS, FSMS, and EMS, there are a few similarities concerning managerial principles, resources allocation, communication, and a few other processes. But, they are distinct in scope, eligibility, objectives, and systemic approach. Also, quality and food safety management systems focus on their customers, such as end-consumers.

As shown in Table 6.2, the applicable standards can address one or many dimensions of sustainability along the bakery supply chain. The sustainability approach is not restrictive to ISO 9001, ISO 22000, and ISO 14001, but those standards were more known among all when asked Romanian bakers. Only a few larger bakers have

Table 6.2 International standards applicable to different management systems

Type of systems	Applicable standards
Quality management system (QMS)	ISO 9001:2015 Quality management systems — Requirements, ISO/TS 9002:2016 Quality management systems — Guidelines for the application of ISO 9001:2015, ISO 9004:2018 Quality management -Quality of an organization – Guidance to achieve sustained success ISO 22006:2009 Quality management systems — Guidelines for the application of ISO 9001:2008 to crop production
Food safety management system (FSMS)	ISO 22000:2018 Food safety management systems — Requirements for any organization in the food chain ISO/TS 22002–1:2009 Prerequisite programmes on food safety — Part 1: Food manufacturing ISO/TS 22002–2:2013 Prerequisite programmes on food safety — Part 2: Catering ISO/TS 22002–3:2011 Prerequisite programmes on food safety — Part 3: Farming ISO/TS 22002–4:2013 Prerequisite programmes on food safety — Part 4: Food packaging manufacturing ISO/TS 22002–5:2019 Prerequisite programmes on food safety — Part 5: Transport and storage ISO 22005:2007 Traceability in the feed and food chain — General principles and basic requirements for system design and implementation
Environmental management systems (EMS)	ISO 14001:2015 Environmental management systems — Requirements with guidance for use, ISO 14002-1:2019 Environmental management systems — Guidelines for using ISO 14001 to address environmental aspects and conditions within an environmental topic area — Part 1: General ISO/DIS 14002–2 Environmental management systems — Guidelines for using ISO 14001 to address environmental aspects and conditions within an environmental topic area — Part 2: Water ISO 14004:2016 Environmental management systems — General guidelines on implementation ISO 14006:2020 Environmental management systems — Guidelines for incorporating ecodesign
Business continuity management systems (BCMS)	ISO 22301:2019 Security and resilience Business continuity management systems — Requirements ISO 22313:2020 Security and resilience — Business continuity management systems — Guidance on the use of ISO 22301

(continued)

Table 6.2 (continued)

Type of systems	Applicable standards
Information security management system (ISMS)	ISO/IEC 27001:2013/COR 2:2015 Information technology — Security techniques — Information security management systems — Requirements — Technical Corrigendum 2 ISO/IEC 27003:2017 Information technology — Security techniques — Information security management systems — Guidance ISO/IEC 27013:2021 Information security, cybersecurity and privacy protection — Guidance on the integrated implementation of ISO/IEC 27001 and ISO/IEC 20000–1 ISO/IEC TS 27022:2021 Information technology — Guidance on information security management system processes
Occupational health and safety management system (OH&SMS)	ISO 45001:2018 Occupational health and safety management systems — Requirements with guidance for use ISO/DIS 45002 Occupational health and safety management systems — General guidelines for the implementation of ISO 45001:2018
Innovation management system (IMS)	ISO 56002:2019 Innovation management — Innovation management system — Guidance
Social responsibility	ISO 26000:2010 Guidance on social responsibility ISO/TS 26030:2019 Social responsibility and sustainable development – Guidance on using ISO 26000:2010 in the food chain

implemented occupational health and safety management system. Of interviewed respondents, a single person acknowledged the business continuity standard, and another said that their company started to implement standards for information security.

There were no concerns about the innovation management system, even though many business operators stated that they are interested in the innovation process as a business opportunity for competitive advantage in the market.

6.4.3 *Assessing the Sustainability in Baking Industry Complying with Standards Requirements – A Case Study*

The study aimed to provide an insight into the Romanian baking industry's response to sustainability, considering some management systems' peculiarities, which operate in this sector at the organizational level. The first stage introduced external issues relevant to understanding the organizational context for baking manufacturers and outlined premises for the sustainability of this specific sector in Romania. The questionnaire consisted of a set of questions that could analyze a few attributes associated with food safety, food quality, and environmental management systems. For this, five sections were developed.

The survey was initially applied to 40 bakeries 'representatives. The respondents of the baking business operators were from at least one category, i.e., food technologist, baker producer, manager, or team leader for quality/food safety/environment management.

From the initial number of companies participating in the survey, answers of four small bakers from the countryside were considered apart because they did not implement any management system. However, based on additional interviews, those four bakers admitted they comply with food safety requirements based on the HACCP method because they follow the food law. Representatives of Romanian official control at county level, i.e., the Sanitary-Veterinary and Food Safety Directorate (DSVSA), have registered their activities as business food operators. DSVSA guide them and periodically perform control. But above-mentioned bakers did not introduce any particular management system. They learn from experience or knowledge acquisition from various sources (other professionals, fairs, news, etc.). They consider their bakery scale small enough and do not need a whole system to control the breadmaking process and costs associated with inputs (flour, other raw materials, energy, or electricity). They carefully analyze the inputs and outputs to avoid wasting money and making a profit. To the extent that revenues are satisfactory, they do not seem too concerned about increasing profitability on a scientific basis. Regarding the environmental aspects, they consider their production scale too small to have an ecological impact. But food waste and other edibles are valued for animal feed. Instead, they were very proud of the bakery products they provided to customers and the satisfaction of faithful consumers, believing that real sustainability means continuing to run their business in a friendly environment for the next generation.

The 36 complete questionnaires showed that respondents came mainly from SMEs but balanced between small, medium, and large companies. All implemented FSMS until this very moment. Also, some introduced progressively many management systems, successively carried out apart or as an integrate management system. Various management systems applied at the organizational level were chosen as a result of how the external environment of firms changed and internal needs appeared. For instance, representatives of average-scale bakeries pointed out they shifted from one MS to another depending on specific conditions. But only a few large companies integrated more management systems into one.

Management systems' implementation seems to be conditioned by market reasons. All business operators handled FSMS because food safety requirements are mandatory by food law; standard requirements are easier to follow for any bakery along the food supply chain, no matter the production scale. If a quality management system was more convenient for larger companies, the smaller ones considered they must involve too many resources. If it is not compulsory, for them is more trouble (e.g., additional tasks in the job description, more procedures, and more recordings to keep, allocation of additional funds for a holistic approach). Similarly, many SMEs commented regarding ISO 14001 adoption. So, they supported compliance with standardized management systems in how their efforts to provide resources are justified by gaining visible competitive advantages in a short time. In other words, when asked why they should implement any management system, the main reasons that drove almost producers to act for MS were socio-political factors and economic constraints. In this respect, they enumerated legislation in force, legitimation, customer expectations, and market requirements.

Although the research intended to focus only on management systems' implementation initially, the certification process seems crucial for Romanian manufacturers because it legitimates their efforts and outcomes toward the market and their direct customers. So, when discussing the management systems implemented within organizations, all respondents referred to the certification of the systems they have actively giving effect at the firm level. Consequently, many systems are also certified by independent third-party audit bodies.

In analyzing the dynamics of MSs within the baking industry, many baker manufacturers first started implementing a quality management system, afterward FSMS with or without an additional EMS. Some bakers stated that they introduced one specific or a couple of management systems at the beginning and later gave up a particular MS; either they adopted different standards, the organizational culture changed, or the whole process of compliance with standard provisions went beyond their interest.

Still, QMS certification for SMEs has dropped recently in favor of food safety, the FSMS certificate being considered a helpful tool in trade relations between partners. Also, food safety certification under private label schemes sometimes becomes compulsory, especially when using larger distribution channels for domestic hypermarkets and international trade.

To and Lee (2014) ranked Romania in seventh place in the top-ten countries in terms of EMS certificated according to ISO 14001 in 2010. They indeed referred to all industrial production but not only the food sector. However, in the last decade, similar to QMS, the interest of Romanian firms in EMS implementation in full or certification has slowed down along the cereal-to-bread supply chain after an increasing trend.

Regarding food safety, from respondents of this survey, cc. 90% of firms had certified FSMS. The other 10 percent represent small bakers who implemented food safety management based on the HACCP method complying with the national food legislation in force.

Of interviewed companies, 61% had International Featured Standards (IFS) certificates. The others had food safety management systems certified according to ISO 22000 and FSSC 22000. Surprisingly, certification granted according to ISO 22000 requirements was available for only 28% of the companies involved in the survey. A few medium and large companies had FSMS ongoing with simultaneous certification labels such as IFS and ISO 22000, IFS and FSSC 22000, IFS and BRC, ISO 22000 and FSSC 22000.

Additionally, related to the social dimension of the sustainable Romanian baking industry, it should emphasize that at least 8% of analyzed firms were concerned with social compliance. They implemented an occupational health and safety management system. Of these, two large companies addressed improving working conditions considering ethical and responsible business practices. In this case, a leading company performed social audits according to Sedex Members Ethical Trade Audit (SMETA) protocol. Another company used Business Social Compliance Initiative (BSCI) to monitor and assess workplace standards.

There is proof that governance models are applied, at least for large companies, to steer the Romanian baking industry towards sustainability and business competitiveness.

Regarding quality management, QMS was certified for 33% of the analyzed companies. Some respondents concluded that, to a certain extent, the top management decided if they continue or not to ask for a third-party audit aiming at certification. But that does not mean that percentage is representative of quality management system implementation. Medium and large companies have combined various management systems with a common structure to control their economic performance, assure conformity to clients, and continually improve (incl. Reducing predictable losses and costs). The last versions of ISO standards promote this kind of managing manifold processes in ways to consider making it easier on simultaneous integration in whole or parts thereof with other management systems and facilitating management system implementation in any organization along the food supply chain, no distinguishing scale-size.

Of respondents, one person highlighted that quality management focused on sustainability is generated internally by the principal owner's beliefs.

Regarding the environment, there were voluntary initiatives of companies to integrate the ecological dimension of sustainability based on standards. Of interviewed people, only 22% of firms are implemented and certified EMS; and 17% are large companies. Respondents from companies with EMS considered that standards offered them specific guidelines in addressing environmental issues at the company level and helped solve problems in bakeries. The larger companies used ISO 14001 as a tool for improving processes and operational outcomes with economic performance and environmental benefits. Beyond their ecological objectives and QMS, standards help large firms to a strategic approach for efficient management of all resources (inclusive energy-saving, proper water management, waste disposal or recycling, etc.).

But in some cases, respondents admitted that even if EMS is not implemented in full, parts thereof are integrated into another management system. For instance, they introduced waste prevention programs and other specifically ecological measures for operating an EMS into the FSMS or QMS's documentation.

All interviewed people recognized that ISO 14001 supports regulatory compliance. Some of them assimilated between environmental performance and the regulatory provisions. Many SMEs have considered they lack appropriate resources to improve the environmental outcomes.

Noteworthy, Johnstone and Hallberg (2020) explored the contextual factors supporting EMS with ISO 14001 adoption in SMEs for improved sustainability. Comparable comments with Johnstone and Hallberg's (2020) results were observed for SMEs under discussion. SMEs with EMS ongoing acknowledged they thought it was better to adopt ISO 14001 for following environmental laws with benefits for legitimizing their environmental actions with control bodies or an improving image in the market, facing potential competitors and end-consumers.

Even if the ISO 14001: 2015 considered EMS to be easier enforced in SMEs, the interest manifested by bakers for EMS implementation fell. Apart from setting environmental goals, the organization must assess its environmental impacts that have implications for the organization and should establish key performance indicators (KPIs) for each environmental objective. When respondents were asked about environmental KPIs, all indicated waste management.

According to the literature review, Romania has a lower recycling rate in the UE and risks infringement or penalties. Before the pandemic period, it was quoted that cc. 20% of food waste is lost before consumption.

Based on the survey, most baking firms follow the environmental requirements, mainly pushed by environmental legislation and less induced in the spirit of sustainable principles. Convincing in this respect is the argument that "polluter pays".

However, some similitudes between FSMS, QMS, and EMS complying with international standards occur concerning managerial principles, resource allocation, management review, a few processes, and even communication. So, a standard backbone structure is available for all these management systems. The methodology applied for the effectiveness of these MSs, follows the recognition of the common requirements (e.g., risk-based thinking, improvement promotion).

Leadership and top management commitment shall demonstrate for the management systems. All respondents acknowledged that policies and objectives are assumed and internally communicated by top management. For those that implemented many systems, they are recognized distinctly between quality, food safety, and environmental goals for the effectiveness of each MS. A more vital link with the company's strategic approach is also related to its efficiency and is perceived for quality objectives rather than food safety or environmental ones. Except for the top management, for different employees at an intermediate level, the specific objectives for food safety and environmental goals and policies seem to support rather the applicable statutory and regulatory provisions.

Managerial principles based on the Deming cycle, namely PDCA (Plan-Do-Check-Act), are applied in each MS according to the latest versions of ISO 9001, ISO 22000, and ISO 14001 have spawned a basic architecture for joint requirements and processes, such as resource allocation and management review.

A set of questions was asked about resources that support management systems implementation. 80% of respondents were directly or indirectly involved in planning the resources, especially the financial ones. A few were in charge of raw materials. Human resources (HR) are issued mainly by top management and a dedicated HR department. Even though they almost agreed with the statement that the company should provide adequate resources, many bakers pointed to financial funds as threats or constraints for business operating, mainly when they invest in infrastructure, equipment, maintenance, and re-tech discussions. However, contrary to first look and declared statements, the bakery companies focus on suitable allocation and optimization of all resources, and in support of this assertion can observe how flourishing and dynamic the baking market with indubitable economic outcomes is.

Concerning management review, it is foreseen as part of performance evaluation. But top management in larger companies dealt differently with this process than smaller bakers. If leadership in large firms used different tools (e.g., Pareto chart, Gantt diagram) and statistics for reporting tasks, small companies insufficiently developed documentation or did not pay too much attention to keeping minutes. Still, periodically they conducted a fundamental management analysis and monitoring.

Some topics related to the personnel are still sensitive, such as employees' shortages, gross salaries, or competencies. The opinions are divided between employers and employees, top management and workers at other levels. However, most average-to-large firms agreed with the specific standard requirement, which stipulates that they should provide necessary human resources for an adequate operating of any MS. And acquiring, monitoring, and improving the competence of people are highlighted in terms of training. Cc. 75% acknowledged they performed training by following the PDCA cycle in the company, and topics planned for training are yearly adapted for standard requirements for specific MS. Based on answers given in this respect, the frequency, training level, and allocated resources consistently differ between smaller bakers and larger companies.

6.5 Conclusion

The Romanian bakery market is economically attractive and dynamic, even if its business operators face many challenges and constraints in terms of decreasing consumption, intensive competition, and others. Depending on firm size and own management, Romanian bakery companies try to increase or consolidate their position on the domestic market, gain revenues and improve their image to consumers, official control, and other stakeholders. Complying with the EU and national regulations regarding quality assurance, food safety, environmental aspects, and others, the bread industry in Romania addresses sustainability, and many firms implement some management systems.

The standardization contributes to economic development by doing business correctly and facilitating commercial activities on the market. This study considered a few standard requirements applying quality, food safety, and environmental aspects for the sustainability assessment in the Romanian baking industry. Still, the interest of almost all interviewed enterprises focused more on food safety than quality management or environmental performance. Especially for smaller businesses, the socio-political factors seem to be the main drivers for any management system implementation, besides the economic objectives and improved satisfaction of their customers.

References

Andersson, K., & Ohlsson, T. (1999). Life cycle assessment of bread produced on different scales. *Int J Life Cycle Assess, 4*(1), 25–40. https://doi.org/10.1007/BF02979392

Arta albă. (2019). Piaţa pâinii, între tradiţie şi inovaţie (Bread market between tradition and innovation. Arta alba, 27th March 2019 (in Romanian), (Last Accessed on February 27, 2022; Available at: https://artaalba.ro/piata-painii-intre-traditie-si-inovatie/).

Barbu, P. (2019). Romanian milling and bakery industry needs 5,000 employees. Business Review, Agriculture 12th June (Last Accessed on September 27, 2021; Available at: https://business-review.eu/business/agriculture/romania-lacks-nearly-5000-employees-202135).

Briceño-León, M., Pazmiño-Quishpe, D., Clairand, J. M., & Escrivá-Escrivá, G. (2021). Energy efficiency measures in bakeries toward competitiveness and sustainability —case studies in Quito. *Ecuadorr Sustainability, 13*, 5209. https://doi.org/10.3390/su13095209

CAEN. (2022). Lista completă şi actualizată a codurilor CAEN 2022 (complete and updated list of NACE codes 2022) (in Romanian) (last accessed on March 27, 2022; Available at: https://caen.ro/).

Deselnicu, D. C., Bulboacă, M. R., Dumitriu, D., & Alexandrescu, L. (2020). Analysis of the bakery industry strategic groups in Romania. *ICAMS 2020 – 8th International Conference on Advanced Materials and Systems III, 4*, 271–276. https://doi.org/10.24264/icams-2020.III.4

Espinoza-Orias, N., Stichnothe, H., & Azapagic, A. (2011). The carbon footprint of bread. *International Journal of Life Cycle Assessment, 16*, 351–365. https://doi.org/10.1007/s11367-011-0271-0

Federation of Bakers. (2019). European Bread Market (last accessed on April 13, 2022; Available at: https://www.fob.uk.com/about-the-bread-industry/industryfacts/european-bread-market/).

Fedima. (2021). Position Paper for a Sustainable European Bakery Industry (Last accessed on February 27, 2022; Available at: https://www.fedima.org/images/210811_Sustainability_Position_Paper_Final_Version.pdf).

Firme.Info. (2022). List of firms within NACE categories and 107 codes (Last accessed on February 27, 2022; Available at: http://www.firme.info/industria-alimentara-COD-CAEN-10/fabricarea-produselor-brutarie-produselor-fainoase-COD-CAEN-107.html)

Fortune Business Insights. (2022). Bakery Products – Market Research Report (Last accessed on April 26, 2022; Available at: https://www.fortunebusinessinsights.com/industry-reports/bakery-products-market-101472).

FRD. (2021). Romanian Food Market 2021 (Last accessed on April 26, 2022; Asvailable at: https://www.frdcenter.ro/wp-content/uploads/2021/07/Romanian-Food-Market-2021-by-FRD-Center-2.pdf).

Galli, F., Bartolini, F., Brunori, G., et al. (2015). Sustainability assessment of food supply chains: An application to local and global bread in Italy. *Agricultural and Food Economics, 3*, 21. https://doi.org/10.1186/s40100-015-0039-0

Gava, O., Bartolini, F., Brunori, G., & Galli, F. (2014). Sustainability of local versus global bread supply chains: a literature review. In *3rd AIEAA (Italian Association of Agricultural and Applied Economics) conference "feeding the planet and greening agriculture: challenges and opportunities for the bio-economy", June 25–27* (Vol. No. 173096). https://doi.org/10.22004/ag.econ.173096

ISO. (2022). ISO standards (Last accessed on March 27, 2022; Available at: https://www.iso.org/standards.html).

ISO 14001:2015 Environmental management systems — Requirements with guidance for use

ISO 22000:2018 Food safety management systems — Requirements for any organization in the food chain

ISO 9001:2015 Quality management systems — Requirements

Johnstone, L., & Hallberg, P. (2020). ISO 14001 adoption and environmental performance in small to medium sized enterprises. *J Environ Manag, 266*, 110592. https://doi.org/10.1016/j.jenvman.2020.110592. PMID: 32310124.

Joossens, J. V., Sasaki, S., & Kesteloot, H. (1994). Bread as a source of salt: an international comparison. *J Am Coll Nutr, 13*(2), 179–183. PMID: 8006300. https://doi.org/10.1080/07315724.1994.10718392

Kannan, R., & Boie, W. (2003). Energy management practices in SME-case study of a bakery in Germany. *Energy Convers Manag, 44*, 945–959. https://doi.org/10.1016/S0196-8904(02)00079-1

Lădaru, G. R., Siminică, M., Diaconeasa, M. C., Ilie, D. M., Dobrotă, C. E., & Motofeanu, M. (2021). Influencing factors and social media reflections of bakery products consumption in Romania. *Sustainability, 13*, 3411. https://doi.org/10.3390/su13063411

Mosgaard, M. A., Bundgaard, A. M., & Kristensen, H. S. (2022). ISO 14001 practices – a study of environmental objectives in Danish organizations. *Journal of Cleaner Production, 331*, 129799. https://doi.org/10.1016/j.jclepro.2021.129799

Notarnicola, B., Tassielli, G., Renzulli, P. A., & Monforti, F. (2017). Energy flows and greenhouses gases of EU (European Union) national breads using an LCA (life cycle assessment) approach. *Journal of Cleaner Production, 140*, 455–469. https://doi.org/10.1016/j.jclepro.2016.05.150

Pérez Farinós, N., Santos Sanz, S., MªÁ, D. R., et al. (2018). Salt content in bread in Spain, 2014. *Nutr Hosp, 35*(3), 650–654. https://doi.org/10.20960/nh.1339. PMID: 29974775.

Pipatprapa, A., Huang, H. H., & Huang, C. H. (2017). The role of Quality Management & Innovativeness on green performance. *Corporate Social Responsibility and Environmental Management, 24*(3), 249–260. https://doi.org/10.1002/csr.1416

Scurtu Cristu, M., Cristu, C., & Stanciu, S. (2016). Analysis of the bakery industry sector in Romania. In *The 28th international business information management association conference: vision 2020 innovation management, development sustainability, and competitive economic growth, Seville, Spain, 9-10 November 2016, I – VII:1939-1947.*

Sîrbu, A. (2009). *Merceologie alimentară – Pâinea şi alte produse de panificaţie (Science of Food commodities – Bread and other bakery products).* AGIR Publisher, Bucharest (in Romanian).

Sîrbu, A. (2009b). Preferinţele de consum ale românilor pentru produsele de panificaţie. Studiu de caz efectuat în Râmnicu Vâlcea (Romanians' consumption preferences for bakery products. A case study conducted in Râmnicu Vâlcea), ROMPAN-Actualităţi în industria de morărit – panificaţie (ROMPAN-Newsletter in the milling & bakery industry, Bucharest) I, *2009*, 41–47. (ISSN 1584-7888) (Last accessed on November 26, 2021; available at: http://www.rompan.ro/uploaded_files/file/2009_1_revista.pdf)

Slow Food. (2004). *Terra Madre.* Slow Food Publisher.

Slow Food. (2017). What's the price of a loaf of bread? (Last accessed on Feburary 27, 2022; Available at: https://www.slowfood.com/whats-price-loaf-bread/).

Slow Food. (2021). About us (Last accessed on December 27, 2021; available at: https://www.slowfood.com/about-us/).

To WM, Lee PKC. (2014). Diffusion of ISO 14001 environmental management system: global, regional and country-level analyses. *Journal of Cleaner Production, 66*, 489–498. https://doi.org/10.1016/j.jclepro.2013.11.076

UN. (2015). Transforming our world: the 2030 agenda for sustainable development. A/RES/70/1. New York, USA

Chapter 7
Across American Overview on Sustainability Approach Throughout Baking Industry: An Analytical-Descriptive Approach

José G. Vargas-Hernández and Muhammad Mahboob Ali

7.1 Introduction

Sustainable baked food options for choices and their impact that these have on food preferences determined by the consumer behavior which underlie the needs on achieving sustainability in the baking industry across America. This theoretical study is more explicitly considered as food safety and security based on food culture and heritage linking policy making and implementation at transnational, national and local levels with sustainable development of natural and environmental resources, water, renewable energies, etc., which contribute to mitigate emission of gases and climate changes while attending the potential to offer baked goods that deliver healthier and more nutritious lifestyle. While there is an overwhelming number of studies and research examining the issues related to sustainable development, it requires substantially more attention the analysis of food in the baking industry at global, regional, national, and grass root levels across America.

Specifically, the analysis of this research study was focuses on the potential, challenges, and opportunities that the across American baking industry has in relation to its sustainable development, social and economic concerns, and their impact and effects on nutrition, health, and lifestyle. This study on the sustainability of across American baking industry is divided to facilitate the analysis in three main topics. The study also begins with conceptualizing the sustainability the second

J. G. Vargas-Hernández (✉)
Postgraduate and Research Department, Tecnológico Mario Molina Unidad Zapopan, Zapopan, Mexico
e-mail: jose.vargas@zapopan.tecmm.edu.mx

M. M. Ali
Dhaka School of Economics, Constituent Institution of the University of Dhaka, Dhaka, Bangladesh

J. M. Ferreira da Rocha et al. (eds.), *Baking Business Sustainability Through Life Cycle Management*, https://doi.org/10.1007/978-3-031-25027-9_7

topic, the impact and implications in across American baking industry which includes some details related to their reality, challenges, and opportunities.

The Sustainable Development Goals evolved from the former Millennium Development Goals have prompted ferocious debates at all levels of governments and sectors on how to address the multiple issues and problems leading to policy making.SDG-2 relates to zero hnger,SDG-3 depicts good health and well-being,SDG-6 opines that clean water and sanitation and SDG-17 refers to the partnership for the goals. Alternative concern is keeping low food prices and accessible to low income consumers to avoid the risk of food bio insecurity. Consumer behavior refers to characteristic purchasing behavior, variety-seeking purchasing behavior, dissonance-reducing purchasing behavior, multifaceted purchasing behavior. In case of bakery industry, consumer behavior may change due to rational and also sometimes irrational behavior. Rationality does not accept consumers are aware of their favorites, causes and choice procedures while irrational behavior of the consumer takes as decision creation which founded on spirits somewhat than evidences by the social order through moral hazard investigation.

The across American market for bakery products is fragmented due to national, regional, and local competitive bakery companies with their own strategic approaches to implement, boost and maintain their competitive brand presence in a market share among the consumers. Some of these strategies are the market expansion, partnerships, mergers, strategic alliances, joint ventures, acquisitions, etc. The across American countries have witnessed a steady consumption growth of bakery and bread products due to an increasing demand in a market driven by healthy nutrition, accessibility, instant products, convenience. The trend towards consumption of nutritious and healthy bakery products is increasing the demand of organic and ancient grains, custom blends, wheat flours, rich in proteins and nutrients, and the minimum processing with lesser preservatives and trans fats. Health-consciousness orientation is an innovation trend dominating the market of bakery products catered with the launching of new products containing wholegrain, fibers gluten- free and other healthy ingredients, associated with high value and price. Packaging innovation tend for more portable and single-self-serving increasing the foodservice outlets. Mexico is the second largest baking market in Latin-America and the eight largest world-wide. Grupo Bimbo, a Mexican producer and commercial bakery company is the world's largest. However, there are many medium, small, and micro business called "panaderías" that have altogether the largest share of the Mexican market. Food safety and food biosecurity are complementing aims for achieving better quality while reducing scarcities and enhance freedom from hunger. Food is best before dates of increasing waste and losses and becoming poisonous threatening food biosecurity, as well massive and not targeted recalls to safety concerns lead to food waste. Due to the enhancing healthy lifestyle concept in North American countries, there is an increasing demand for healthy bread containing whole grain, high in fiber, and gluten-free. In Latin American countries, competition among bakery manufacturers is uneven due to the informal distribution channels and diversity of bread products. The ecological, financial, and public influence on bakery products to create value chain which has involved by the courtesy of an extensive variety of backers across America. Meanwhile, American Bakers Association (2022) describes that supply chain disturbances and exaggerated food prices in the economy was started recuperating from the COVID-19 epidemic's effects which lead to food supply and food security might be at jeopardy.

Research question of the study is whether across America to be able to distribute better nutrition, health, and lifestyle to the consumers?

The study besides including introduction, in section:2 discussed literature review, while in section: 3, narrates methodology of the study. Present scenario in the baking industry across America was described in section:4 and section:5 opines Discussion. Conclusion, Implication and Future Research work will be described in section:6.

7.2 Literature Review

Food is a form of cultural heritage that have an impact on food preferences and choices supported by outcomes of environmental concerns, social context, economic growth, and food culture. Cultural acceptability of baked food is an element of food security concept contributing to the basic needs and wellbeing of consumers beyond the nutritional adequacy (Maxwell & Smith, 1992). Baking industry's food production systems must be supported by better biosecurity and inputs from healthy animals and welfare. Biosecurity and safety food risks are traced back as landfills where food ingredients are cultivated (Oivanen et al., 2000). A Food security has been a common challenge throughout human history, although the concept has evolved with the industrial food systems (Carney, 2012; Maxwell, 1996; Maxwell & Devereux, 2001). Moldy and spoiled grains already lost should not enter the food, feed chains and green energy and instead should be incinerated to recover the energy (SOU, 2007) to cut carbon footprints. The food bio-waste residues can be used in composting for organic fertilizers, recovering nutrients and energy leading to lowering the resource footprint of production (Albihn & Vinnerås, 2007), It has been underestimated the total energy consumption in the baking industry supply chains, due to the lack of data, despite the possibility to determine the contribution of baked cakes through emissions on a life cycle basis from food and drugs sector (Druckman & Jackson, 2009). Flour often undergoes a kill step through baking during production, although salmonella was in food-poisoning outbreaks (Eglezos, 2010; Neil et al., 2012). The different domains of nature are interrelated, interdependent and interconnected among the energy, water, and food, etc. and must be addressed like that (Hoff, 2011). Social value is a component of food utilization that provides socio cultural and religious benefits (Ingram, 2011). Water has been identified as the most crucial domain that influences all other domains (World Economic Forum, 2011). Sustainable food biosecurity requires availability or production of food, access, nutritional sufficiency, and safety, and stable conditions (Helland & Sörbö, 2014).

Inevitably, food production and consumption preferences and choices, access, sovereignty, and safety are linked to environmental problems and concerns, have an impact on the dynamics derived from climate change concerns across a wide range of areas including price stability and utilization (Food and Agriculture Organization (FAO) of the United Nations, 2009; Carlsson-Kanyama, 1998; Vermeulen et al., 2012). According to the Agriculture Organization (FAO), food security exists when people have physical, social, and economic access to sufficient, safe, and nutritious food that allow them to meet their dietary needs and food preferences for a more active and healthy life. Food securities are supported by availability, access, utilization, and stability (Food and Agriculture Organization (FAO) of the United Nations, 2009). Recycling of materials as a waste management practice, green water and green energy consumption create savings in water footprint. The primary material packaging for cupcakes is made from virgin fibers due to food regulations (Ecoinvent, 2010). As an umbrella concept, food security is context-specific nutritional dimension (Gibson, 2012; Pottier et al., 2016).

The link tactic refers to the connections between climate, food, energy, and water. These connections between the domains of energy, water, food, and climate, etc., have originated a nexus approach reflecting the growing concerns on examining the links between resource and food biosecurity, energy policies, etc., among others (Allouche et al., 2015; Hellegers et al., 2008). The nexus approach between water, energy food, energy, and climate domains focus on the identification, demonstration and modelling the interrelationships and connections supported by the argument these domains are under pressure and demands by the growing trade-offs among the sustainable development goals, the economic growth and population growth leading to an acceleration of the ecosystem degradation (Bazilian et al., 2011; Rothausen & Conway, 2011; Scott et al., 2011; Hoff, 2011; Hermann et al., 2012; Bizikova et al., 2013; Howells et al., 2013; Howells & Rogner, 2014).

The different domains are intrinsically interconnected leading to growing concerns on resource biosecurity that results from food and energy crisis and must be governed by linking energy and food biosecurity policies (Allouche et al., 2015; Hellegers et al., 2008). Nutrition science offers insights that benefits the perspective of food as socio cultural construction (Paxson et al., 2016). A sustainable feeding people requires disruptive changes in supply chains of food able to reduce losses and waste (Kearney, 2019; Sundström et al., 2014; United Nations, 2019). Most of proteins and edible energy are lost in the conversion from plant to animal-based food. The wasted and lost foodstuffs represent around one third of food produced and around 8% of global greenhouse emissions. Reducing food losses and waste is a challenge and opportunity for business (Unilever, 2019). The U.S. Food and Drug Administration (FDA) has issued several warnings about the potential dangers of eating raw flour in bakery products such as cakes and cookie dough (U.S. Food and Drug Administration, 2016).

Limited supplies and lack of access to nutritious and safe foods are threats to food biosecurity affecting more the low-income consumers who may trigger social unrest (Arezki & Brückner, 2011; Bazerghi et al., 2016; Helland & Sörbö, 2014; Johnstone & Mazo, 2011). Some bakery products such as the cheesecake requires storage refrigeration that must be assumed in an energy saver refrigerator (Siemens, 2015).

Intelligent packaging using sensors to monitor the characteristics of bakery products through the supply chain improves food safety and biosecurity while reduces food losses and waste (Newsome et al., 2014; Poyatos-Racinero et al., 2018). The use of block-chains trace-back outbreaks verifying the origin and fate of the bakery products, facilitates to speed up and trust of procedures and allows the consumers to check the bakery products in real time on the spot (Ahmed & Broek, 2017).

Besides other family sensors, the most relevant are for identification tags, baking food package integrity, freshness and temperature Poyatos-Racinero et al., 2018). Intelligent tags trace and track a baked product in real time information to identifying on the food supply chain, analyzing the causes of food losses and waste, controlling inputs in the production process and food frauds (Manning & Soon, 2016).

The nexus approach focuses on the interconnectivity between different domains, as in education, health, and food that are linked on the community-based cooking learning programs on nutritional food (Iguchi et al., 2014). The Food Consumption Scores methods used to measure food security have been criticized because it does not consider the historical context and its complexities on local specific parameters (Pottier et al., 2016).

Consumers adopt eating behaviors based on a sociocultural, economic, and environmental background and supported by food preferences and choices and food security. Local baking industry must attend to local sociocultural contexts (Pottier, 1999). Consumers want to purchase their baked cakes food whenever they believe is convenient creating a challenge to food retailers of a dynamic resource allocation in maintaining their inventories at appropriate stock levels and avoiding food waste, losses, and stock-outs (Arunraj & Ahrens, 2015). Implementing food baked cake waste and losses prevention strategies aimed to achieve long term food safety and sustainability such as changing criteria for fresh produce, is a solution that requires a trade-off based on economic, social, and environmental justifications (Mourad, 2016). Studies on baking industry have focused on issues related to social and environmental consequences of genetically modified organisms, the threat of globalization consumption tendencies on traditional food cultures and practices, negative impacts of meat production and consumption, organic food, consumer practices (Hull et al., 2016; Klein, 2009; Pottier et al., 2016; Wilk, 2016).

Reduction of post-consumer baked cake waste is liked to reduction of freshwater and marine ecotoxicity. Post-consumer baked cake waste is underestimated (DEFRA, 2015) despite that any potential reduction would translate into potential improvements of the environment, freshwater ecotoxicity and savings. Circular food systems applied to baking industry meaning that most of recycled nutrients are efficient in biosecurity, environmental sustainability and resource footprints, although the cycle of nutrients have the risk to become a cycle of pathogens and chemical hazards (Monsees et al., 2017). A strategic hierarchy for reducing baked cakes' losses and waste are in order of source reduction, reusing and reprocessing, recycle as feed for other animals, recover the energy, compost for nutrients, raw materials for other processes, recover the energy by incineration and dumping the garbage in landfills (Vågsholm & Boqvist, 2020). The analytical metadata collected with big data instruments and sensors provide a holistic perspective of processes of the various bakery products across diverse transportation and storage operations and conditions of temperature and packaging leading to better predicting baked cake supply (Vågsholm & Boqvist, 2020).

Manufacturing, distribution and transportation use packaging materials and shopping bags to facilitate distribution and marketing operations of the products to be contained, handled, and transferred to the final consumer (Konstantas et al., 2019b). Environmental and health impacts related to decrease the content of sugar in baked cakes must be mitigated by product reformulation knowing that constitutes a substantial part of the weight (Hashem et al., 2018). The best option for environmental impacts including global warming potential are the pies, although they are the worst option on ecotoxicity. Cheesecake followed by the pie have the higher human toxicity impact while the best alternative are the whole cakes (Konstantas et al., 2019b). The mineral depletion (MD) of cheesecake is higher than any other cake caused by the impact of raw materials production and transport due to the use of iron, chromium, copper, and nickel in infrastructure and equipment (Konstantas et al., 2019b).

The environmental impact should have the lowest level of products such as the whole cakes ranking top with cake slices that have the lowest freshwater eutrophication despite of some toxicity potential. Freshwater ecotoxicity (FET) and marine ecotoxicity (MET) show similar trend across the different bakery products with the lowest levels in whole cakes and the highest levels in cheesecake The hotspots of marine ecotoxicity are the raw materials with the major sources of the impact are the releases of copper, nickel, and zinc as well as the end of life stages (Konstantas et al., 2019b). Foods from producers and retailers that might be a risk for the population as for example, nuts, soy, wheat, eggs, etc., containing allergens, borne pathogens, foreign plastic and metal materials, etc. (Maberry, 2019). The environmental impacts of baked cakes at the sectoral level are estimated based on the consumption data of the estimated market value shares and the market prices on product impacts (Keynote, 2015a, 2015b; Konstantas et al., 2019a, 2019b). The cheesecake has as primary sources of FFD impact the milk and soft cheese and the whole cakes have the lowest fossil fuels depletion (Konstantas et al., 2019b).

COVID-19 has created some opportunities and challenges although the growth for Canadian bakeries depends on food service rebounds, exports, and labor. One challenge is the call for shifting towards the usage of eco-friendly green packaging with less plastic to continue post-pandemic, although the costs will increase the price and not all consumers are willing to pay more (Burak, 2020). Food preferences and decisions of consumers of bakery products are framed by forms of cultural heritage and cultural knowledge linked to environmental variables such as the diversity of natural resources, food security and climate change (Kapelari et al., 2020). Bakery products are available across all distribution's channels but is growing fast in modern retail chains as consumers are attracted, in part, as the result of demand for a healthier food, the influence of culture and the marketing power, which affect the consumption habits to demand pastries, cakes, sweet biscuits and cookies, sweet pies, et. Healthy food concerns for busy consumers are a growing adoption of ready-made pastries and cakes (Report Linker, 2022).

The market of across America bakery products is segmented by types of products: cakes and pastries, bread, biscuits & Cookies, and Morning Goods. Regarding the distribution channel of these products the main are the hypermarkets, supermarkets, specialty stores, convenience stores, online retailing, and others (Mordor intelligence, 2022).

Demand for bakery, bread and tortilla products in México has been strong and will continue and industry revenue has increased but is being pushed down due to the wheat supply and the dynamics of the exchange rates (Research & market, 2022).In México, the bakery industry comprises the manufacture of bread, biscuits, pastas for soup, frozen bakery products, corn tortillas, flour tortillas and other premixed flours including mill nixtamal (IbisWorld, 2022).

The Mexican market of bread and bakery food is served by more than 45,000 bakeries out of which 97% is micro-businesses. The manufacturing and commercial operations of Grupo Bimbo encompasses South American, central American, Caribbean, and South and southern west of United Sates (Baking Business, 2022).

The Latin American market of baking industry is dynamic, diverse, and complex that requires good knowledge economic, social, cultural, geographic, and political experience to understand the consumer behavior patterns shopping and spending in specific regions, countries, and localities. Baking industry in Latin America is highly segmented growing rapidly in a very competitive market going through long-term transformation, although the pace of change varies across the different countries. The Latin American market of baked goods has been growing and expanding at an average annual rate of 7.09% from 2018–2023 (Baking, 2022).

Additional organic and natural ingredients of baked goods based on different natural and organic grains and seeds to provide nutrition and texture is a trend that continues to grow. Some examples of this diversification of baked goods are on the shelfs, such as cakes and pastries having a mixture of salty and sweet ingredients, bred with chocolate, butter popcorn and caramel, trail seed mix with salty nuts and chocolate (Baking, 2022).

The baking industry in Brazil in 2018 had more than 63,000 traditional bakeries with sales approaching US $30 billion (Baking, 2022).

7.3 Methdology of the Study

The study was based on conceptual view and descriptive analysis. As such the study used published materials to analyze the research question which was given in the section:1 and also literature review. Exact sources of literature reviews were citing and also given in the references. Time period of the study was between January, 2022 and March, 2022. However, the study did not consider any quantitative analysis. The study did only subjective judgement based on literature reviews in the area of our theoretical research and to get answer of the research question.in the backing industry across America to get a sense of the natures of ideas which researchers work to analyze and descriptions narrated in this article.

7.4 Present Scenario in the Baking Industry Across America

An estimated one in six North Americans succumb to a foodborne bacterial disease in one year, according to DeVault (2018) with estimated data from Centers for Disease Control (2011). Out of that estimation, 130,000 people require hospitalization and 3000 dye (Eglezos, 2010). There are no clear figures for Central American, South American, and Caribbean countries, but it is suspected that the number of deaths caused by foodborne bacterial diseases are very high. Recent regulations on labeling prevent marketing of baked goods, snacks, cookies, and sweet goods to target an obesity epidemic that affects more than half the Latin American population, and to caution consumers about products that have saturated fats, high in sugar and calories. This regulation has prompted Grupo Bimbo to innovate in healthier products. Similar initiatives are being developed by bakery companies in Chile, Uruguay, Perú, etc. Chile in 20,016 and México in 2020 are the leading countries in Latin America on health initiatives in regulating wellness and health with mandatory warning labels on packaged foods and beverages aimed to caution on the consumption of baked goods that are high in sugar, calories sodium and sutured fats. These health regulation on baked food enable consumers to opt for healthier consumption. A rapid urbanization process poses a challenge to food biosecurity for urban consumers to being able to have access to food in places where the socioeconomic safety nets are incomplete. Baking industry is food manufacturing. Food baking industry and manufacturers try to provide safe products to their customers employing safety and quality processes taking a proactive approach to food safety by avoiding financial risks out of recalls from potential contamination sources. The growing population, the improving economic conditions, an emerging market and the rising living standards in México is leading towards growth in bakery production, bread, and rolls despite the rising prices that are damaging the purchasing power of middle classes. In Mexican food culture, consumers are looking for more convenient, easy, and quick meals. The Latin American market dynamics of bakery products are fueling growth in high density urban areas. In Latin American Countries, there is a tendency to local innovation, although industry must be aware of the consumers price sensitivity. Consumers prefer artisanal bakery cakes, pastries, and breads to packaged baked goods produced by commercial bakeries who are garnering their market share and compete through diversification, expansion, reformulation, and repackaging. However, traditional small bakeries are forced by the market competition to survive. Latin American consumers of frozen baked goods are preferring quick-service of frozen and convenient bakery foods among artisans, catering, supermarkets, and hypermarkets. The packaging losses have an impact on the increase of fossil fuel depletion and the water footprint. The biggest companies sharing the market include Grupo Bimbo SAB de CV, Flowers Foods, El Paso Baking Co., Campbell Soup Company and Panaderia Rosetta. The most active companies dominating the bakery market are the Kellogg, Hostess Brands, Bimbo Bakeries, Canyon Bakehouse, Allison's Bread, Mondelēz, Dave's Killer Bread, Boudin Bakery, among others in the market of United states. Bakeries, including tortilla manufacturing companies in Canada is the second largest food manufacturing sector and it is the fastest growing food sector with a growth rate of 5.4% as an average since 2015.

Proximity to the product and the potential risk to contamination are considered to classify in zones the production facilities in relation to the product. Zone 1 refers direct contact to product. Zone 2 includes non–food-contact in the processing area and the pathways. Zone 3 is the area of immediate contacting leading to contamination by means of accidental human traffic and machines and zone 4 are the remaining areas including storage. Character and belongings of mental, demonstrative, societal, cultural and reasoning inspirations on choice creation for purchasing of bakery products across America by the consumer behavior.

Potential contaminants in the baking-handling industry environments are ever-present being introduced through raw materials, energy, water, the physical environment including heating, ventilation, and air conditioning systems know as HVAC, pests, employees, and other food products, etc. Economic allocation of resources has been observed in reductions of cheesecake environmental impacts in areas such as natural land transformation, agricultural land occupation, terrestrial acidification, marine eutrophication. The cheesecake effects have an impact on the amount of raw milk to produce milk powder used. Whole cakes observed reduction in ozone depletion and pies with urban land occupation.

Sensitive analysis considers the packaging effects on the results in manufacturing and refrigerated storage processes though the distribution chain of bakery products. Cheesecake product needs of frozen storage subject to the uncertainty of electricity consumption. Uncertainty analysis of the impacts of the ambient-storage baked cakes change less than the impacts of cheesecakes. The large amount of packaging cupcakes during post-consumption waste management benefits the recycling. Cost- benefit analysis of bakery products needs to be assessed of across America so that consumer durables can be purchase without losing marginal utility of the bakery product from the bakery is positive.

The long-term food biosecurity has become a matter of national and local policy making concern driving to the development of novel sources of food safety and resilient supply chains. Food resilience ca be enhanced by eating the cereals mixed in the baked cakes.

Sustainability in the baking industry must reduce resources, environmental footprints and eliminate food waste and losses. The environmental, economic, and social impacts from production, distribution and consumption of baked cakes may vary from product to product.

Cheesecakes and pies are the higher contributors to the impacts at the sectoral level due to the environmental burdens and can be improved by targeting the raw materials and energy. The raw materials and energy stage are the main hotspots which should be targeted for improvement opportunities to reduce the impacts in the baked cakes supply chains. The production, distribution and consumption of baked cakes are linked to the life cycle environmental impacts which results that the type of cheesecake is the worst option.

Baked cakes increase food waste in situations when are beyond the best-before dates to be discarded by retailers and consumers who often left over and discard from catering establishments and households. Allergens introduce into the products and processes from the unlikeliest of sources. Proactive programming must facilitate the detection and monitoring of pathogens and allergens posing risks to their products and customers before they enter the process of food products as the responsibility. If the manufacturing facilities switching back and forth between production of different products that contain specific allergens and allergen free. For example, switching production of snack cakes between cakes with almonds and cakes free of tree nuts.

Pathogens can enter along the way on the food manufacturing process, from the field to the product at the final consumer. Environmental testing used to detect specific pathogens identifies food pathogens introduced into a food-handling environment not eliminated by sanitation practices that could be passed on to the food ingredients being processed. The factors of demand for baked cakes are volatile and correlated with product characteristics, consumer behavior, nutritional ingredients, healthy food with low calories, customer visit for shopping, price, promotions, discounts, events or festivals, weather, season, food safety, quality, etc.

Improvements on post-consumption of food waste of baked cakes lead to impact reductions on sugar as raw material, energy, agricultural land occupation, terrestrial ecotoxicity and photochemical oxidants, which increase the savings. The sugar content of baked cakes has been identified as a relevant policy making concern aligned with policy developments. Temperature sensors detect abuses along the supply chain using dynamic dating of baked cakes for shelf life. The indicators of the sensor turn red if the quality of baked cake declines and is not suitable for human consumption. Packaging changes in the baked cake industry are difficult to implement due to the need to maintain environmental sustainability, food safety and biosecurity and shelf life, despite those reductions in packaging result in savings.

Intensification of circular food economy systems including baked cakes production is part of the solutions to future food biosecurity, considering resistance to potential public health risks from the possibilities of trade-offs between environmental sustainability, food safety and food biosecurity to intensify food production aided by antimicrobials. Cupcakes has the lowest contribution due to the low level of consumption relative to other bakery products. Food access is affected by preferences and choices encompassing economic, sociocultural, environmental, and religious values which have an influence on consumer demand of certain types of baked food.

Circular baking production systems implies more efficiency on environmental and resource footprints; recycling of nutrients, by-products and food losses and waste; improves transparency of supply chains and the balance food supply and demand; and reduces transportation and storage. There are some risks when implementing a circular and recycling system of baking production. The design of circular food production systems must avoid cycles of biological and chemical hazards, persistence of pathogens in the feed and food chains. Applications of forecasting of baking industry supply and demand, monitoring the supply chain based on big data strategies and artificial intelligence through the sensors to include safety considerations, lead to more efficient control of processes, reduce the food losses and waste, and reassures trust between the baking industry and final consumers, but ignoring food safety concerns is a recipe for disaster.

7.5 Discussion

In Latin America, local fresh bakery products, from the farm-to-fork and artisan, it's a way of life to enjoy in demographically, economically, and culturally diverse market where international tendencies of food movements coincide with the local deep-rooted food traditions and heritage. The bakery industry in Latin America elaborates their baking and bread products for households based in artisan bakeries with traditional knowledge and skills and only a minimum percentage of bakeries is implementing new technologies. So far, traditional bakeries resist and refuse to disappear. The analysis of the consumer behavior of bakery products is important to determine the market demand of the baking industry. There is a wide range of motives and preferences that consumers of bakery products have when making choices. The proponents of the nexus approach sustain that the energy, water, food, and climate sectors are interconnected in such a way that actions in one sector may have an impact on the other two. The interconnections among these three domains are crucial and not limited only to these sectors. Other Latin American nations are following these initiatives with their own to focus on health and nutritious food. The Pan American Health organization (PAHO) has launched the Nutrient Model to discourage consumption of identified unhealthy products. Brazil, Ecuador, and Costa Rica have issue regulations on front-of-packaging labeling of bakery goods and food marketing to school children. These regulations are shifting consumption from processed baked foods such as biscuits and bread toward organic, nutritious, and more fresh food.

Logistics and distribution are the bakery operations that have an impact in urban and rural areas where there are diverse point of sales offering the wide array of baked goods that must be fresh by the time consumers needed. Basic infrastructure for the network of distribution represents a challenge for any wholesale baker to launch low-expenditure and longer shelf-life goods to overcome the complexities of the regional logistics and distribution network. The Mexican Bimbo bakery company is famous in reaching every morning the localities more remote of the Mexican territory.

Brand-building initiatives of bakery companies develop trust which pay off in the long run by meeting and exceeding the expectations of consumers who usually become loyal, despite the price sensitivity. Latin American consumers are willing to trade up the best choices when they get more value with a limited budget, which in turn leads to rely on known brands and not accept to go wrong on private labels. The nature of baked cake manufacturing requires permanent cleaning of production facilities leading to loss of ingredients and some finished products. Baking manufacturing operations vary depending on the type of product, but the core include mixing, forming, baking, filling, finishing, and packing processes before the production goes into the distribution and marketing. Some of the products are the cakes, pies, cheesecake biscuit, etc. After the manufacturing process, the bakery products are shipped to the distribution centers to be temporarily stored and then subsequently sent to the different sales points such as the retailers to be displayed and available for purchasing. Pesticides used in agriculture, chlorine and phosphorus are the major cause of terrestrial ecotoxicity (TET). The cheesecake has also the highest impact on primary energy demand (PED) per kg. Whole cakes have the lowest primary energy demand. The pie has the highest global warming potential (GWP) becoming the best alternative. Manufacturing has a high impact of global warming potential. Cheesecake requires more urban land occupation (ULO) and agricultural land occupation (ALO) than the other cakes and do not vary in other products. Raw materials, packaging and transport requires urban land occupation in industrial buildings and roads. Whole cakes and cupcakes reduce urban land occupation due to recovery of energy and materials. Cheesecake requires more natural land than pie and whole option. The difference between frozen and ambient-storage cakes is made of production and leakage of refrigerants. Composting is created and developed out of the apple peel and other organic ingredient losses and waste in manufacturing process as well as for post-consumer baked food such as cake waste. This system of composting can be credited as an organic fertilizer for displacing equivalent and similar number of chemical fertilizers. The Sustainable Development Goals (SDGs) has addressed the nexus between concerns and policies on water, energy, and food as interconnected, interactive, and interdependent concerns. Transnational, regional, national, and local problems of availability of fresh water affect food, energy, health, and other areas of a resilient sustainable development. A regional focus of the Latin America water is concerned and connected to the bakery production companies to implement practices of sustainable environmental development. The cheesecake has the higher ozone depletion (OD), and it has also the highest fossil fuel depletion (FFD) among all other bakery products followed by cupcakes. Contributions across all products to fossil fuel depletion (FFD) with high percentages are the raw materials and manufacturing process. The production, distribution and consumption of baked cakes, cheesecakes and pies contribute to sectorial impacts, despite the low impact at product level. Reducing food waste and losses and waste is a political objective. These wasted and lost foodstuffs amount around one fourth of the energy content of food production representing a potential for improvement of food biosecurity and the opportunity to feed more hungry people. The baking industry across America is aware of its contributions to sustainable development framed by the green economy and based on the interdependence of the use of renewable natural resources, water, and energy. Ensuring the sustainable management and availability of water connected to food and climate change, contribute to economic development and a more sustainable environment. Bakery companies must ensure safe water consumption as an ingredient in their production processes, as well as in protecting human health, for instance, by spreading knowledge about water usage and handling wastewater The estimated volumetric blue and green freshwater consumption in production and distribution processes in the bakery industry may lead to reduce the water footprint. The water and energy consumption at the across America bakery companies used for production and distribution operations should be frame by the sustainable development goals. Regarding sustainability concerns, most of the large companies in the baking industry across America are conscious about the use of alternative energies ensuring access to more reliable, affordable, sustainable, and clean energy, such as fuel cells for onsite electricity generation to mitigate or reduce carbon emissions. Renewable energy technologies drive sustainable development of the baking industry leading to adapt new and transfer innovative technologies in collaborations between the baking companies and across countries. However, national policies have an influence on the development of more efficient and sufficient energy production demanded by the bakery industry. The Brazilian baking industry is intending to offer a diversification of baked organic and artisanal goods to achieve a better market positioning in a competitive market of local manufacturers that offer many options appealing for consumers of baked goods based on whole grain, added nutrients, gluten-free.

Several global and local commercial bakery industry, food and beverages brands are investing in advanced technology to be identified, applied, and adapted in production and distribution processes. Cheesecake is the least environmentally sustainable impact with the higher level of ecotoxicity than any other bakery products. The cheesecake has the higher levels of terrestrial acidification than pies and other type of cakes. The shelf life of baked cakes labeled as best before dates is a quality management control to guarantee that consumption before the date provided that the storage instructions is safe and healthy but poisonous thereafter dates.

Socio-economic, education and cultural background have a role in supporting consumers to make informed decisions when it comes to choosing bakery products. Cultural knowledge is relevant for the consumers to make food choices and preferences. Implementing intelligent packaging for baked cakes supported by artificial intelligence adds the benefits of better diet control for nutrients and reduction of wasted or lost food. The baking industry is going through significant growth and transformation in Latin American countries through different pace of changes subject to the economic development, increasing urbanization, food patterns, use of new technology in producing and processing, etc. The emerging consumer sectors in Latin America is gradually changing habits of accessibility and consumption based on pre prepared and packaged foods. Because many of the Latin American urban dwellers in emerging middle and upper income of the market segments of consumers live similar lifestyles, convenient access to baked goods on-the-go breakfast and smaller-portion snacks bringing benefits of time saving and enhanced taste, leading to bakery companies to experiment with diversity of new branded goods. Multidisciplinary approaches based on complementary and competing visions should give the baking industry better tools for ensuring quality and safety (Nychas ct al., 2016; Ropodi et al., 2016). The use of IT technology offers an efficient source reduction, reprocessing, and recycling of food. Big data analyses and artificial intelligence instruments provides continuous benefits for the baking industry in achieving the better sustainability by ensuring food quality, safety, and biosecurity while reducing food waste and losses. Consumer behavior are culture-bound, though the idea that all consumers involve in purchasing with precise vital policymaking modes or elegances appears to be workable solutions.

Global, regional and local sustainability concerns are taking greater relevance in Latin American such as losses waste food that have effects on food supply and prices due to the lack of infrastructure and physical capital, the conduct of the distribution channels, sales points, and consumers, etc. The impact on the sustainability of food systems leads to increasing prices for consumers, lower profits for producers and distributors, reduced food availability, safety, and security food, etc. Across America need to allocate healthier nutrition, health, and lifestyle to the consumers by purchasing hygienic, healthy, evading weighty gravies, and not fad diet so that adding value to maintain good health.

7.6 Conclusion, Implication and Future Research Work

Assumptions about the relationship between food policies and sustainable development in across American baking industry led to conclude that is weak, if not lacking, to be able to address all the interconnected and interrelated social, environmental, and economic issues and concerns. Regionalization of markets and transnational rivalry strength professional to operate in a diverse baking industry's atmosphere In fact, the research findings are directed towards explaining this lack of engagement between policy making and implementation in the baking industry across America to be able to deliver more sustainable food and contributing to better nutrition, health, and lifestyle of consumers. Food waste refers to the discarding of safe and nutrition's food for human consumption. Food waste is discarding the safe and nutritious foods, and food loss is the lost supplies along the food supply chain between the producer, the market, and consumers. Food losses and waste occurs due to problems in processing, handling, packing, refrigerating, transportation, storage and retail because of infrastructure inadequacies, inefficient supply chains, improper packaging, inappropriate legal and incentive system frameworks, custom clearance, etc. Regarding the highest level of photochemical oxidants formation (POF) is incurred by the cheesecake and lowest by whole cakes. The main hotspot across the cakes, are some specific raw materials such as butter, milk powder, eggs, soft cheese, and sugar. Butter accounts for the higher percentage of global warming potential of whole cakes. Raw materials of baked cakes such as butter, sugar and palm oil are the most critical hotspot that have an impact on environment and contribute to the different life cycles stages. Phosphate releases is a hotspot for milk powder, cheesecake, and soft cheese production are high at the farming stage and flour for whole cakes, cake slices and cupcakes.

Previous research findings on sustainable baking and food industry do not have a simple, and direct impact on policy making leading to discussions without an agreement on the relationship between policy making and the development of a sustainable baking industry. The lack of consensus on discussions about the implications and impact of food in the baking industry on sustainable development led to conclude that the researchers should get more involved to work with manufacturers and commercial partners of baking goods to focus on the needs and interests of consumers and producers while focusing on the sustainability concerns.

Source reduction and reprocessing of baking production is an option to sustainable intensification by eliminating food waste or loss and increasing the output with the same footprint, which requires the baking industry and consumers to adapt their quality requirements and specifications (Johnson et al., 2018). The recommendation is to double the production and consumption of a diet rich in plant-based fruits, vegetables, legumes, and nuts and to halve or reduce the consumption of sugar mixed in the baked cakes confers both human health and environmental benefits. If the global baking industry system change in this direction, the food security improves and becomes more resilient.

The trade-offs between food safety and security, economic, social, and environmental sustainability are vital and should be based on evidence and risk. Monitoring the production environment of the baking and milling industry for bacteria and pathogens in a proactive approach can guarantee a safe product. However, developing circular food production systems to recycle nutrients is a challenge for trade-offs between food safety, food biosecurity and sustainability. Consumer behavior to purchase bakery products may play a rudimentary part for venders to understand market behavior and then suggest tailor-made plans in bakery industry for long run sustainability through creating competency among the American countries. An evaluation of social and environmentally sustainable policies and their related practices applied to the baking industry across America, may influence policy making and activism towards the design and implementation of more sustainable practices of natural resources, water, and energy in relation to the food safety and development, although this may entail some ethical dilemmas to deliver better nutrition, health, and lifestyle to the consumers. However, the connections between some of the sustainable development goals are very weak, lacking structure and transparency, as for example the goal on hunger and food biosecurity is connected to equality, health, infrastructure ecosystem protection, climate change, disasters, etc., yet there is not any reference to any potential interconnection among water, energy, food, etc. Food biosecurity resilience is the ability to deal with shocks, risks including food safety, stress in production, distribution, and consumption of food without increasing the risks of hunger, malnutrition and food borne diseases. Food in bio insecurity is a threat to public health, social sustainability, and political stability.

In future across America, cost-benefit analysis on sustainability approach throughout baking industry needs to be done as a research work. How does integrating framework for sustainability tactic through baking industry in cross-America may be developed by the researchers. In-depth study needs to be considered in terms of country wise participative, suitable, and supportable implementation and extenuation in terms of components of the bakery products for which qualitative and quantitative analysis through mixed research work may be done. Sustainability with economic viability of the baking industry al through America may be done by analysis cost-benefit analysis and maintenance of excellence of food security and availably for the masses for which policy level research work may be done.

References

Ahmed, S., & Broek, N. T. (2017). Food supply: block-chain could boost food biosecurity. *Nature, 550*, 43. https://doi.org/10.1038/550043e

Albihn, A., & Vinnerås, B. (2007). Biobiosecurity and arable use of manure and biowaste - treatment alternatives. *Livestock Science, 112*, 232–239. https://doi.org/10.1016/j.livsci.2007.09.015

Allouche, J., Middleton, C., & Gyawali, D. (2015). Technical veil, hidden politics: interrogating the power linkages behind the nexus. *Water Alternatives, 8*(1), 610–626.

Arezki, R., & Brückner, M. (2011). *Food prices and political instability*. IMF working paper. WP/11/62. IMF (Washington, DC). Available online at: https://www.imf.org/en/Publications/WP/Issues/2016/12/31/Food-Prices-and-Political-Instability-24716 (accessed March 1, 2022).

American Bakers Association (2022). Bakery supply chain challenges infographic, https://americanbakers.org/news/bakery-supply-chain-challenges-infographic, (accessed March 4, 2022).

Arunraj, N. S., & Ahrens, D. (2015). A hybrid seasonal autoregressive integrated moving average and quantile regression for daily food sales forecasting. *International Journal of Production Economics, 170*, 321–335. https://doi.org/10.1016/j.ijpe.2015.09.039

Baking Business. Com (2022). Latin American baking industry undergoing significant transformation obtained from https://www.bakingbusiness.com/articles/48886-latin-american-baking-industry-undergoing-significant-transformation(accessed March 1, 2022)

Bazerghi, C., McKay, F. H., & Dunn, M. (2016). The role of food banks in addressing food inbiosecurity: a systematic review. *Journal of Community Health, 41*, 732–740. https://doi.org/10.1007/s10900-015-0147-5

Bazilian, M., Rogner, H., Howells, M., Hermann, S., Arent, D., Gielen, D., et al. (2011). Considering the energy, water, and food nexus: towards an integrated modelling approach. *Energy Policy, 39*(12), 7896–7906.

Bizikova, L., Roy, D., Swanson, D., Venema, H. D., & McCandless, M. (2013). *The water–energy–food biosecurity nexus: towards a practical planning and decision-support framework for landscape investment and risk management*. IISD.

Burak, K. (2020). Fall 2020 bakery outlook – moving back to normal https://www.fcc-fac.ca/en/knowledge/economics/fall-2020-bakery-outlook-moving-back-tonormal.html (accessed March 1, 2022).

Carlsson-Kanyama, A. (1998). Climate change and dietary choices — how can emissions of greenhouse gases from food consumption be reduced? *Food Policy, 1998*(23), 277–293.

Carney, M. (2012). Food security and food sovereignty: what frameworks are best suited for social equity in food systems? *Journal of Agriculture, Food Systems, and Community Development, 2012*(2), 71–87.

Centers for Disease Control. (2011). *Foodborne disease outbreak surveillance*. Published online at www.cdc.gov. Centers for Disease Control, Atlanta, GA.

DEFRA. (2015). *Food statistics pocketbook*. DEFRA.

DeVault, J. D. (2018). Environmental monitoring in the milling and baking industry. *Great Plains Analytical Laboratory, Kansas City, M. Cereal foods world, 63*(1). https://doi.org/10.1094/CFW-63-1-0032

Druckman, A., & Jackson, T. (2009). The carbon footprint of UK households 1990–2004: a socio-economically disaggregated, quasi-multi-regional input–output model. *Ecological Economics, 68*, 2066–2077.

Eglezos, S. (2010). Microbiological quality of wheat grain and flour from two mills in Queensland. *Journal of Food Protection, 73*, 1533–1537.

Ecoinvent. (2010). Ecoinvent V2.2 database. Swiss Centre for life cycle inventories,

Food and Agriculture Organization (FAO) of the United Nations. (2009). *Declaration of the World Summit on Food Security. 2009*. Available online: http://www.fao.org/tempref/docrep/fao/Meeting/018/k6050e.pdf

Gibson, M. (2012). Food security—a commentary: what is it and why is it so complicated? *Food, 2012*(1), 18–27.

Hashem, K. M., He, F. J., Alderton, S. A., & MacGregor, G. A. (2018). Cross-sectional survey of the amount of sugar and energy in cakes and biscuits on sale in the UK for the evaluation of the sugar-reduction programme. *BMJ Open, 8*(2018), e019075.

Helland, J., & Sörbö, G. M. (2014). *Food biosecurity and social conflict. CMI Report 2014* (Vol. 1). Christian Michelssen Institute.

Hellegers, P., Zilberman, D., Steduto, P., & McCornick, P. (2008). Interactions between water, energy, food, and environment: evolving perspectives and policy issues. *Water Policy, 10*(S1), 1–10.

Hermann, S., Welsch, M., Segerstrom, R. E., Howells, M. I., Young, C., Alfstad, T., et al. (2012). Climate, land, energy, and water (CLEW) interlinkages in Burkina Faso: an analysis of agricultural intensification and bioenergy production. *Natural Resources Forum, 36*(4), 245–262.

Iguchi, M., Ehara, T., Yamazaki, E., Tasaki, T., Abe, N., Hasimoto, S., & Yamamoto, Y. (2014). Ending the double burden of malnutrition. In *Addressing the food and health nexus in the sustainable development goals (POST2015/UNU-IAS policy brief)* (Vol. 6). UNU-IAS.

Ingram, J. A. (2011). Food systems approach to researching food security and its interactions with global environmental change. *Food Security, 2011*(3), 417–431.

Hoff, H. (2011). Understanding the nexus. In *Background paper for the Bonn 2011 nexus conference*. Stockholm Environment Institute.

Howells, M., Hermann, S., Welsch, M., Bazilian, M., Segerstro, M. R., Alfstad, T., et al. (2013). Integrated analysis of climate change, land-use, energy, and water strategies. *Nature Climate Change, 3*(7), 621–626.

Howells, M., & Rogner, H. (2014). Assessing integrated systems. *Nature Climate Change, 4*(7), 246–247.

Hull, E., Klein, J. A., & Watson, J. L. (2016). Supermarket expansion, informal retail and food acquisition strategies: an example from rural South Africa. In *The handbook of food and anthropology* (pp. 370–386). Bloomsbury Academic.

IbisWorld. (2022). What are Bakery Products & Tortilla Manufacturing industry in Mexico? Bakery Products & Tortilla Manufacturing in Mexico - Market Research Report https://www.ibisworld.com/mx/industry/bakery-products-tortilla-manufacturing/95/

Johnson, L. K., Dunning, R. K., Bloom, J. D., Gunter, C. C., Boyette, M. D., & Creamer, N. G. (2018). Estimating on-farm food loss at the field level: a methodology and applied case study on a North Carolina farm. *Resources, Conservation and Recycling, 37*, 243–250. https://doi.org/10.1016/j.resconrec.2018.05.017

Johnstone, S., & Mazo, J. (2011). Global warming and the Arab spring. *Journal of Surviv, 53*, 11–17. https://doi.org/10.1080/00396338.2011.571006

Kapelari, S., Alexopoulos, G., Moussouri, T., Sagmeister, K. J., & Stampfer, F. (2020). Food heritage makes a difference: the importance of cultural knowledge for improving education for sustainable food choices. *Sustainability, 12*(4), 1509. https://doi.org/10.3390/su12041509

Kearney, A.T. (2019). *How Will Cultured Meat and Meat Alternatives Disrupt the Agricultural and Food Industry? Dusseldorf; AT Kearney Studie zur Zukunft des Fleischmarkts bis 2040*. Available online at: https://www.atkearney.com/retail/article/?/a/how-will-cultured-meat-and-meat-alternatives-disrupt-the-agricultural-and-food-industry (Accessed July 5, 2021).

Keynote. (2015a). *Ice creams & frozen desserts*. Keynote.

Keynote. (2015b). *Biscuits and cakes market report 2015*. Keynote.

Klein, J. (2009). Creating ethical food consumers? Promoting organic foods in urban southwest China1. *Socio-Anthropologie, 17*, 74–89.

Konstantas, A., Stamford, L., & Azapagic, A. (2019a). Economic sustainability of food supply chains: life cycle costs and value added in the confectionary and frozen desserts sectors. *Science of the Total Environment, 670*, 902–914.

Konstantas, A., Stamford, L., & Azapagic, A. (2019b). Evaluating the environmental sustainability of cakes. *Sustainable Production and Consumption, 19*, 169–180. https://doi.org/10.1016/j.spc.2019.04.001

Maberry, T. (2019). *A look Back at 2018 food recalls. Food safety magazine, E-newsletter, February 19.* Available online at: https://www.foodsafetymagazine.com/enewsletter/a-look-back-at-2018-food-recalls-outbreaks/ (accessed July 4, 2019).

Manning, L., & Soon, J. M. (2016). Food safety, food fraud, and food defense: a fast evolving literature. *Journal of Food Science, 81*, 823–834. https://doi.org/10.1111/1750-3841.13256

Maxwell, S. (1996). Food security: a post-modern perspective. *Food Policy, 1996*(21), 155–170.

Maxwell, S., & Devereux, S. (2001). *The evolution of thinking about food security. In food security in sub-Saharan Africa* (Vol. 2001, pp. 13–31). ITDG Publishing.

Maxwell, S., & Smith, M. (1992). Household food security: a conceptual review. In S. Maxwell & T. Frankenberger (Eds.), *Household food security: concepts, indicators, measurements: a technical review* (pp. 1–72). UNICEF and IFAD.

Monsees, H., Kloas, W., & Wuertz, S. (2017). Decoupled systems on trial: eliminating bottlenecks to improve aquaponic processes. *PLoS One, 12*, e0183056. https://doi.org/10.1371/journal.pone.0183056

Mordor intelligence. (2022). *North America Bakery Products Market - Growth, Trends, COVID-19 Impact, and Forecasts (2022–2027).* Mordor intelligence, obtained from https://www.mordorintelligence.com/industry-reports/north-america-bakery-products-market(accessed March 1, 2022).

Mourad, M. (2016). Recycling, recovering, and preventing "food waste": competing solutions for food systems sustainability in the United States and France. *Journal of Cleaner Production, 126*, 461–477. https://doi.org/10.1016/j.jclepro.2016.03.084

Neil, K. P., Biggerstaff, G., MacDonald, J. K., Trees, E., Medus, C., Musser, K. A., Stroika, S. G., Zink, D., & Sotir, M. J. (2012). A novel vehicle for transmission of Escherichia coli O157:H7 to humans: Multistate outbreak of E. coli O157:H7 infections associated with consumption of ready-to-bake commercial prepackaged cookie dough—United States, 2009, Clinical Infectious Diseases 54:511, 2012

Newsome, R., Balestrini, C. G., Baum, M. B., Corby, J., Fisher, W., Goodburn, K., et al. (2014). Applications and perceptions of date labeling of food. *Comprehensive Reviews in Food Science and Food Safety, 13*, 744–769. https://doi.org/10.1111/1541-4337.12086

Nychas, G. J. E., Panagou, E., & Mohareb, F. R. (2016). Novel approaches for food safety management and communication. *Current Opinion in Food Science, 12*, 13–20. https://doi.org/10.1016/j.cofs.2016.06.005

Paxson, H., Klein, J. A., & Watson, J. L. (2016). Rethinking food and its eaters: opening the black boxes of safety and nutrition. In *The handbook of food and anthropology* (pp. 268–288). Bloomsbury Academic.

Pottier, J. (1999). *Anthropology of food: the social dynamics of food security* (p. 1999). Polity Press.

Pottier, J., Klein, J. A., & Watson, J. L. (2016). Observer, critic, activist: anthropological encounters with food insecurity. In *The handbook of food and anthropology* (pp. 151–172). *Bloomsbury Academic.*

Poyatos-Racinero, E., Ros-Lis, J. V., Vivancos, J. L., & Martinez-Manes, R. (2018). Recent advances on intelligent packaging as tools to reduce food waste. *Journal of Cleaner Production, 172*, 3398–3409. https://doi.org/10.1016/j.jclepro.2017.11.075

Report Linker. (2022). Market Focus: Trends and Developments in the Bakery and Cereals Sector in Mexico. http://www.reportlinker.com/p01497760/Market-Focus-Trends-and-Developments-in-the-Bakery-and-Cereals-Sector-in.Mexico.html#utm_source=prnewswire&utm_medium=pr&utm_campaign=Breakfast_and_Cereal. (accessed March 1, 2022).

Research and market. (2022). Bakery Products & Tortilla Manufacturing in Mexico - Industry Market Research Report https://www.researchandmarkets.com/reports/5027770/bakery-products-and-tortilla-manufacturing-in (accessed March 1, 2022).

Ropodi, A. I., Panagou, E. Z., & Nychas, G. J. E. (2016). Data mining derived from food analyses using non-invasive/non-destructive analytical techniques, determination of food authenticity, quality & safety in tandem with computer science disciplines. *Trends in Food Science and Technology, 50*, 11–25. https://doi.org/10.1016/j.tifs.2016.01.011

Rothausen, S. G. S. A., & Conway, D. (2011). Greenhouse-gas emissions from energy use in the water sector. *Nature Climate Change, 1*(4), 210–219.

Scott, C., Pierce, S., Pasqualetti, M. J., Jones, A. L., Montz, B. E., & Hoover, J. H. (2011). Policy and institutional dimensions of the water–energy nexus. *Energy Policy, 39*(10), 6622–6630.

SOU. (2007). *Betankandet om Jordbruket som Bioenergiproducent (Report on Agriculture as Supplier of Bioenergy). (In Swedish).* Available online at: http://www.regeringen.se/sb/d/8963/a/81974 (Accessed January 16, 2020).

Siemens. (2015). *Dishwasher specifications model* SN26T597GB. www.

Sundström, J. F., Albihn, A., Boqvist, S., Ljungvall, K., Marstorp, H., Martiin, C., et al. (2014). Future threats to agricultural food production posed by environmental degradation, climate change, and animal and plant diseases – a risk analysis in three economic and climate settings. *Food Security, 6*, 201–215. https://doi.org/10.1007/s12571-014-0331-y

Unilever. (2019). Reducing food loss waste Available online at: https://www.unilever.com/sustainable-living/reducing-environmental-impact/waste-and-packaging/reducing-food-loss-and-waste/ (accessed March 1, 2022).

United Nations. (2019). *Revision of World Population Prospects*. Available online at: https://population.un.org/wpp/ (accessed June 26, 2019).

U.S. Food and Drug Administration. (2016). *Raw dough's a raw deal and could make you sick.* Published online at www.fda.gov/ForConsumers/ConsumerUpdates/ucm508450.htm. FDA, Silver Spring, MD, 2016

Vågsholm I, Arzoomand N. S. ,& Boqvist, S. (2020) Food biosecurity, safety, and sustainability—getting the trade-offs right. Frontiers in Sustainable Food Systems 4:16. https://doi.org/10.3389/fsufs.2020.00016

Vermeulen, S. J., Campbell, B. M., & Ingram, J. S. (2012). Climate change and food systems. *Annual Review of Environment and Resources, 2012*(37), 195–222.

Wilk, R. (2016). Is a sustainable consumer culture possible? In S. A. Crate & M. Nuttall (Eds.), *Anthropology and climate change: from encounters to actions* (Vol. 2016, pp. 301–318). Routledge.

World Economic Forum. (2011). *Water biosecurity. The water-food-energy-climate nexus.* Island Press. http://www3.weforum.org/docs/WEF_WI_WaterBiosecurity_WaterFoodEnergyClimateNexus (accessed March 1, 2022)

Chapter 8
Asian Overview on Sustainability Approach in Baking Industry

Sharad Kumar Kulshreshtha and Ashok Kumar

8.1 Introduction

In the context of global food security, reducing food consumption and wastage is like a hot cake. Multi-million tonnes of food is wasted annually in the European Union, and out of a total, around 40% happens in the retail sector (Sanco, , 2014). Reducing food waste is one of the giant future challenges to meet the global food demand & food security (OECD, 2011). Among the bakery products, varieties of Bread have accounted for the highest level of wastes at the household level in European Union (Katajajuuri et al., 2014; Prieffer et al., 2013). According to (CVVM, 2014), in the Czech Republic, it was found that bakery products were the most frequently wasted food among all the food categories (Ratinger et al., 2016).

Baking is the science of preparing food using dry heat normally in an oven, or on hot stones. The dry heats of baking change the form of starches in the food and cause its outer surfaces to brown, giving it an attractive appearance and taste. The process of baking is used for the preparation of Rusk, varieties of bread, cakes, pastries and pies, tarts, quiches, cookies, and crackers. The most common baked item in a bakery is bread. Variations in the ovens, ingredients, and recipes used in the baking result in a wide variety of baking products. The two major trends that are being followed in the baking industry are Processed Baking and Artisan Baking. Processed baking involves chemical compounds, processes, shelf life, etc. Processed bread which is wheat bread might have nearly 20 ingredients; artisan bread will

S. K. Kulshreshtha
Department of Tourism & Hotel Management, North-Eastern Hill University, Shillong, India

A. Kumar (✉)
Department of Tourism & Hotel Management, North-Eastern Hill University, Shillong, India

Institute of Hotel Management Catering Technology & Applied Nutrition, Shillong, India

J. M. Ferreira da Rocha et al. (eds.), *Baking Business Sustainability Through Life Cycle Management*, https://doi.org/10.1007/978-3-031-25027-9_8

have around five ingredients. Artisan baking does not involve chemical compounds; it is organic and requires a lot of skill in preparation.

8.1.1 Consumption Culture of Baked Products in Asia

As Asian lifestyles are changing day by day, there are better opportunities for the baking industry to grow and develop its business. Bakery products, due to affordability and high nutritional value are in huge demand. Bakery being a traditional activity holds an important place in the food processing industry. The escalating presence of the small, medium, and large bakery chains has further elicited growth in the bakery & confectionary sector. Organic bakery products as well as those catering to health-conscious consumers can gain ground in sales, particularly those focused on gluten-free, sugar-free, and keto-friendly. In the Asian market, white bread still dominates in the list of bakery items. The bakers should incorporate local ingredients, cooking styles and adopt local diets for stimulation, such as rendang in Indonesia/Malaysia, mala from China, tom yum in Thailand, etc. As the spending power of Asians is growing, younger generations are craving local baked products and as a result, the baking industry is gaining more momentum. The bakery industry in Asia is promoting its own traditional Asian bakery products or a fusion of Asian and western cuisine (Fig. 8.1).

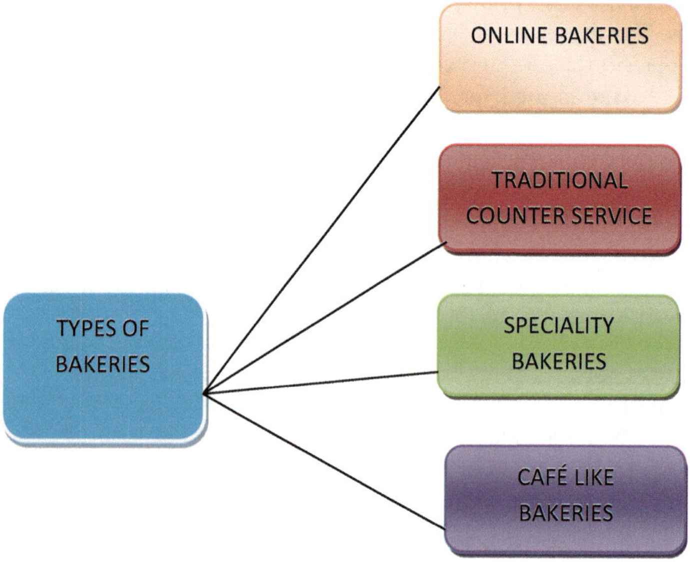

Fig. 8.1 Types of Bakeries

8.1.2 *Asian Baking Industry*

Asian baking Industry is investing in technology to automate their production lines to meet rising demand. Global market research company, Euromonitor International offers a forecast regarding the Asian baked goods market, which is expected to grow from a value of USD74bn this year, to almost 4% up to 2024. Asian Baking industry embraces technology for conversion, and to expand their business. As per Euromonitor, the growth is due to the expansion of marketing distribution channels, manufacturing facilities, and product innovations. The growth is also pushed by the rising artisanal baked goods production as consumers are demanding fresher and healthier products, including products that have no preservatives, low sugar, or lower trans-fat.

China and Japan's top baked goods sales have held 75% of regional sales for more than a decade, and these two high rankers are expected to achieve further growth in the region. Bakery and pastry industries in Asia-Pacific are gearing up to offer more innovative treats with the help of the latest technology. In a recent report, a Market and Market research firm estimated the global bakery processing equipment market at USDD8bn in 2018 and projected to reach a value of nearly USD10.3bn by 2023 with a compound annual growth rate (CAGR) of 5.2% from 2018. The research firm added that Asia-Pacific is estimated to account for the largest market share in 2018 — driven by the growing demand for baked products in developing countries such as India, China, and Australia (Tables 8.1, 8.2, and 8.3).

8.2 Review of Literature

Global sustainability can achieve by sustainable consumption and production. The concept of sustainable development in the bakery industry can be conquered by including three areas of changes: i) reducing overconsumption, ii) Shifting consumption patterns to a more sustainable one, iii) Reducing waste and recycling for future use. Among the developed and western countries, reducing food consumption and wastages are often associated with health concerns and recommendations (Duchin, 2005). The Government policies and research concentrates more on sustainable food product development and minimize food waste. Sustainable food product development may include reducing meat & meat products consumption or increasing the consumption of locally produced food and organic products. Due to many reasons about 33 percent of the food produced for human consumption is either lost or wasted every year (Gustavsson et al., 2011). The four major reasons for food wastages are: 'leftover on the plate', 'expired date of manufacture', 'looked, smelt or tasted bad', and 'leftover from cooking'(Prieffer et al., 2013; Ventour, 2008). According to (Gusia, 2012) the main reasons for food wastages at the household level are long storage in the fridge, faulty storage, and excess cooking. To

Table 8.1 Major bakery players in Asia

S.No.	Name of the Bakery	Year of establishment	Bakery products
1.	*Parle*	1929 (Mumbai, India)	Biscuits, Cake, Rusk,
2.	*Britannia*	1892 (Kolkata, India)	Biscuits, Bread, Cake, Rusk
3.	*Modern*	1965 (Chennai, India)	Biscuits, Bread, Cake, Rusk, Muffin, Indian Bread, Crème Pie, Sweet fills, Pizza Base,
4.	*ITC sun feast*	1910 (Kolkata, India)	Biscuits, Bread, Cake, Rusk
5.	*Goodman Fielder*	1986 (Hongkong)	Bread, Dairy Products
6.	*Bakers Delight Holdings*	1980 (Victoria, Australia)	Bread and its varieties
7.	*Mondelez International*	2012 (Chicago)	Confectionery
8.	*Monginis*	2019 (Mumbai, India)	Cakes, Pastries, Chocolates & Confectioneries
9.	*Harvest Gold India*	1993 (New Delhi, India)	Bread, Buns, and Pizzas
10.	*Surya Food & Agro Ltd.*	1992 (Noida, India)	Biscuit & Cookies, Cakes & pastries, Chocolate, Toffee & candies
11.	*Toly Bread Co. Ltd*	1997 (Shenyang, China)	Bread
12.	*Orion China*	1995 (Langfang, China)	Cakes, Pastries, Chocolates & Confectioneries
13.	*Yamazaki Baking*	1948 (Chiyoda, Japan)	Biscuit & Cookies, Cakes & pastries, Chocolate, Toffee & candies
14.	*Goldilocks Bakeshop Inc.*	1959 (Manila, Philippines)	Bread and bread-type rolls, cakes, pies, chocolate popcorn, and other bakery products
15.	*Pasco Shikishima Corp.*	1920 (Aichi, Japan)	Bread, Japanese Sweets, and Confectionery)

Source: Authors compilation

curtail food wastages (Mont, 2002) proposes a concept of a sustainable product-service system (SPSS). According to Mont (2002), the advantage of adopting a sustainable product-service system approach is that it focuses on various consumption options, which contribute, to environmental protection without reducing consumer welfare. Thus, Warde (2005) and Mylan (2014) argued that various patterns of food consumption cannot be transformed only by educating or persuading individuals to make different decisions. Garonne et al. (2014) introduced a conceptual model of surplus food generation and management. According to the concept, food availability includes three food categories i.e. consumed food, surplus food, and food scrap.

Table 8.2 Facts &figures of baking industry in Asia

Facts & figures of baking industry	Organization
The Asia Pacific was the largest region in the global bread and bakery product market, accounting for 31% of the market in 2019	Business wire
Asia and Africa will dominate the global bakery market in terms of growth in the coming years, characterizing the Asia region as "a dynamic and fast-growing region for bakery"	Mintel's Global Annual Review of "Bread, Bakery and Cakes 2018
The baking industry is seeing 7% to 8% growth in Asia as a whole and as much as 12% to 13% growth in India	Indian Bakers Federation (I.B.F.)
Asia is following the United States in its ban on trans fats. The Food Safety and Standards Authority of India (F.S.S.A.I.) is proposing to limit the maximum amount of trans fat content in vegetable oils, vegetable fats, and hydrogenated vegetable oil to 2% in its goal to eliminate trans fat in India by 2022	The Food Safety and Standards Authority of India
Asia-Pacific bakeries are gearing up to offer more innovative treats with the help of technology	Food & Hotel Asia, Euromonitor
The $74bn Asian market is set to expand and innovate over the next five years, opening the door for new technology, according to a report from Food and Hotel Asia (FHA)	Food & Hotel Asia (FHA)
The Indian bakery market is valued at INR 850 crores, growing at a phenomenal rate of 20–25% annually, becoming one of the top growing businesses in India	TechnoPack, in Economic Times
The U.S. Accounts for Over 27% of Global Market Size in 2020, While China is Forecast to Grow at a 7.1% CAGR for the Period of 2020–2027	Global News Wire

Source: Authors

8.3 Research Questions

1. *RQ 1 What is the history of consumption and production of bakery items in Asian Countries?*
2. *RQ 2 What are the bakery processes & sustainable working practices adopted for Sustainable approaches in the bakery industry.*
3. *RQ 3 What are the latest trends, Ideas & Innovations practiced making bakery 'greener' and more sustainable?*

8.4 Research Objectives

The objectives of our research are:

(i) *To Study, the bakery processes & sustainable working practices adopted for Sustainable approaches in the bakery Industry.*
(ii) *To highlight the latest trends, Ideas & Innovations practiced making bakery 'greener' and more sustainable.*

Table 8.3 Bakery business development & strategies

Bakery product development	Newer products are developed & launched as per the need, demand, and choices of the consumers
Bakery product innovations	Frozen bakery, In-store bakeries, Online bakeries
Bakery processing & equipment	Techniques & Method Used, latest technologies, mechanized equipment
Bakery product packaging	Different Types of Packaging, development in packaging technology and equipment
Bakery product quality	HACCAP guidelines, Stringent quality control guidelines
Bakery marketing and distribution channels	Artisan Baker, Retail, Catering, Online channel Supermarkets/ Hypermarkets, Convenience Stores, E-Commerce, Other
Bakery branding	World trade is shrinking, global brands are available worldwide
Bakery sustainability	Sustainability through reducing wastages, reducing the cost of operation, usage of local & organic ingredients, sustainable equipment
Global bakery market	The Asia Pacific was the largest region in the global bread and bakery product market, accounting for 31% of the market in 2019
Global bakery competitors	The countries covered in the global bakery & confectionary market are Australia, Brazil, China, France, Germany, India, Indonesia, Japan, Russia, South Korea, UK, USA
Future of bakery industry in Asia and world	The bakery sector in the Asian continent is growing rapidly, the baking industry looks to gain more momentum in the days to come. The future of Asian bakery may be in the form of traditional Asian bakery products, more Western foods, or fusions of both

Source: Authors

8.5 Research Methodology

The present study is explorative research which includes information, facts & figures on the bakery processes & sustainable working practices adopted for Sustainable approaches in the bakery Industry. The paper structures are based on the review of related literature on sustainable food consumption and food & bakery waste. With the recent development in terms of the consumption of bakery products, different approaches adopted at the macro and micro level to attain sustainability have been highlighted. This study involves personal experience, close observation, and some secondary sources of information such as research papers, food blogs, books, and some useful websites.

8.6 Asian Bakery on Sustainable Approach

The global baking industry is targeting the Asian market. The North American and European bakery markets have reached the maturity phase of the product life cycle(PLC). Mordor Intelligence has found that quick-service bakeries(QSB's),

Table 8.4 Emerging trends in Asian baking industry

S. No.	Emerging trends	Attributes
1.	E–retailing of bakery products	Due to e-retailing, a range of bakery products has evolved and made available for consumers with a click of a mouse
2.	Technological advancements	Due to the rising demand for premium and healthy products, the diversification and up gradation of technology has been a necessity in the bakery industry. The technological advancements have also introduced fuel-efficient equipment that causes less pollution along with more durability and hygienic food processing
3.	Innovation in ingredients	Along with the technological advancements, Asian bakeries are also using innovative ingredients such as Lame Quick and Spongolit, produced from healthier mono and poly-saturated fatty acids (PFSA)
4.	Improved packaging solutions	Packaging techniques play a momentous role in increasing shelf life, nutritive value, distribution of goods, and marketing. Several bakers are using vertical pouches or sachets to retain the quality and augment the appearance of their baked products
5.	Market diversification	New bakers and entrepreneurs are entering the market, resulting in cut-throat market competition. The concept of cloud baking and the emergence of home-based bakeries have given a boost to the bakery industry. MNC's and other established companies have ventured and diversified into the bakery and related products

Source: Authors

making pizzas, puff products, pastries, cupcakes, and similar ready-to-eat items, have tripled over the past 10–15 years. As per the recent market trends, frozen/chilled bakery products are expected to rise by 10% in the next five years. More and more baked products are displayed and made available in supermarkets across Asian countries (Tables 8.4 and 8.5).

8.6.1 Five Essences for 'Greener' and More Sustainable Bakery

8.6.1.1 Sustainable Resources

To attain sustainability we can invest in energy-efficient tools and equipment which will help to reduce the cost of operations and create an environmentally friendly bakery. Some energy-efficient appliances may be installed to minimize the overhead cost.

8.6.1.2 Use Local and Organic Ingredients

For Greener and sustainable practices we should buy local produce. We should go for local organic ingredients such as local farm eggs, organic butter, milk, and flour from a local farmer or mill instead of relying on deliveries from far distances. This

Table 8.5 Small equipment for sustainable baking

S. No	Products	Features to attain sustainability
1	Silicone cupcake cases	Silicone cupcakes are an ideal individual treat. As with most individual portion sizes, this often means more packaging and reduces food waste and consequently food cost
2	Silicone cupcakes and muffin cases	These can be reused time and time again. They are pretty, and the perfect size, they save money and reduce waste. Reusing property can help in sustainability
3	Glass jars	Jars are pretty and very handy reusable containers for leftovers. Jars can be an ideal way to pick up dry ingredients. Nuts, seeds, oats, and even dried pulses can be bought from big packets and stored in smaller jars for easy usage
4	Bee's wraps	Beeswax-coated cotton sheets can be used to keep your food fresh and protect them from drying out or soaking up other odors in the fridge. These are a much better investment than rolls of plastic wraps and can be washed to use over and over again
5	Reusable baking mats	They are becoming increasingly popular, and are ideal for lining baking trays and cookie sheets before baking. They can be washed and used again. This will avoid paper flapping around in the oven, rolls that are too short for the baking tray, and most importantly, no more waste
6	Reusable piping bags	Piping bags are used to creating those luxurious swirls on top of cakes & Pastries. If we have to cut down on our baking-related waste, a reusable piping bag is something one should consider. They can come in either silicone or cotton fabrics and can be filled and squeezed in the same way as a regular piping bag, but can also be rinsed and saved to use again another time
7	Glass/S S containers	Glass and stainless steel containers look good, they don't stain as plastic tubs do, and they can help in avoiding the use of cling film. Furthermore, when they are not used in baking they are great to use as lunchboxes or to store the ingredients in
8	Baking spray	While reusable baking mats are ideal for wide baking trays, round cake tins aren't quite the right fit. To get rid of that greaseproof paper, a baking spray may be a good choice. Apart from saving on paper waste, it will also stop you from greasing every square centimeter of tin with butter or oil. Instead, you just spray all over the inside of the round cake tin and, once baked, your cake should pop out automatically
9	Bulk local ingredients	Buying larger quantities of items or ingredients at once is a simple way to consume less packaging & reduce carrying cost. Dry ingredients like chocolate last ages once opened, and it saves on countless little individual packets or wrappers. Purchasing local products will directly help the farmers and local distributors
10	Frozen ingredients	A very simple way to prevent waste and reduce carbon footprint can be to buy frozen ingredients. Berries are the best example. Buying frozen ingredients also means they will last longer if you don't need a full quantity of the ingredient to be used at once

Source: Authors compilation

will contribute to lowering our 'Carbon Footprint'. We could also consider making some of our ingredients on the premises. It's more eco-friendly by making its buttercream, chocolate sauces, and chocolate chips. As far as possible, choose "green" suppliers or those who are also committed to sustainability.

8.6.1.3 Diminution in Packaging Waste

The ultimate sustainability goal is to make any bakery a plastic-free bakery. This can be achieved by eliminating the use of single-use plastics, and by reducing the choice of ingredients used in packaging and selling. In the case of bakery cum café and quick service bakeries, paper or metal straws in place of plastic one, eco-friendly disposable is a good choice. We can also encourage the consumers regarding reusable cups to cut down the expenditure on takeaway coffee cups.

8.6.1.4 Water Sustainability

The catering establishments are one of the largest consumers of water. With the utilization of low flow spray valves while cleaning the baking utensils or with water-saving sinks and toilets we can minimize the water usage up to a larger extent. Recycling the used water for washroom and gardening can also be done to decrease the water usage.

8.6.1.5 Donate or Repurpose Unsold Bakery Products

Minimising food waste is good for the sustainable environment. The unsold bakery products can be donated to welfare homes, charitable trusts, and societies, health care centers rather than throwing them away. The leftovers can be repurposed such as leftover bread can be turned into bread crumbs or toasts. The excess croissants can be transformed into delicious puddings such as trifles, or surplus loaves into bread croutons. Consumers nowadays are changing their preferences and using their buying power to go for fresh and organic food. The present generation is moving towards farm-to-table concepts. Sustainable practices may not only cut down on waste and food costs, however, but they will also make bakery more marketable.

8.6.2 Sustainability Vis-à-Vis Bakery Industry

Bakeries, snack producers, and supply-chain partners are taking strong steps toward minimizing their eco-footprint while still maximizing growth. Sustainability means different things to different people. For some, it's all about minimizing emissions or reducing the carbon footprint. For others, it's strengthening local economies and communities. Still, others emphasize organic or natural ingredients. But whatever the definition, sustainability is no longer a "bonus" for any food manufacturer. It's a mandate and a mandate that's coming directly from consumers. According to Nielsen's "Global Corporate Sustainability Report," up-and-coming consumers—namely, Millennials and Generation Z—are the most willing to pay extra for goods they perceive to be sustainable. Large bakery chains are investing in waste management programs designed to convert non-recycled plastics into energy, which helps to meet the energy need of the plant as well as in fueling nearby areas. For the large bakery houses, sustainability is all about a local focus. They procure their ingredients from local suppliers and maximize the usage of local ingredients as well to attain food security (Tables 8.6 and 8.7).

Challenges in Bakery Industry in Asia (a) *Food Safety Standards and Regulations*

(b) *Global Bakery Product Competition*
(c) *Availability of the Latest baking equipment and technologies*
(d) *Demand and supply of Ingredients*
(e) *Bakery consumer behavior research*
(f) *High cost of bakery production*
(g) *High cost of Technology*
(h) *High Rate of Taxation*
(i) *Bakery Business Financial Support*
(j) *Various level of Industry forms-Micro, Small Medium, Large Baking Enterprises*
(k) *Product innovation and quality improvement*
(l) *Environment issues and Business sustainability*
(m) *Marketing Issues*
(n) *Quality human resource and Professional bakers*
(o) *Training and development*
(p) *Traditional and online marketing*
(q) *Supply chain management and logistics*

8.7 Conclusion

Millennials and Generation Z are the most willing to pay extra for goods they perceive to be sustainable. It has been found that Asians are looking for bakery items that are made using organic ingredients that contain no pesticides or chemical fertilizers. Large bakery chains are investing in waste management programs designed to convert non-recycled plastics into energy, which helps to meet the energy need of the plant as well as in fueling nearby areas. For the large bakery houses, sustainability is all about a local focus. They procure their ingredients from local suppliers and maximize the usage of local ingredients as well to attain food security. The Food Safety and Standards Authority of India (F.S.S.A.I.) under the Ministry of Health & Family Welfare has proposed to limit the maximum amount of trans fat

Table 8.6 Emerging Baking institutions in Asia

S. No.	Name of the Institution	Courses	Country
1	International Culinary Institute	Higher Diploma in Baking & Pastry Arts	Hong Kong
2	UNITAR, Selangor	Diploma in Culinary Arts	Malaysia
3	Sunway University, Shah Alam	Diploma in Culinary Arts	Malaysia
4	Dubai College of Tourism	Diploma in Culinary Arts	Dubai, UAE
5	HTMI	Higher Diploma in International Culinary Arts	Dubai, UAE Switzerland
6	ATI College, Kota Kinabalu	Diploma in Pastry Arts	Malaysia
7	Institute of Technical Education	Technical Diploma in Culinary Arts	Singapore
8	TEG International College	Diploma in Culinary & Catering Operations	Singapore
9	International College of Yayasan Melaka(ICYM), Malacca	Diploma in Culinary Arts	Malaysia
10	Thai Chef School, Bangkok	Pastry & bakery Course with Internship	Thailand
11	Tsuji Institute of patisserie, Osaka	Confectionery Arts & Management Course	Japan
12	Le Cordon Bleu, Kobe	Pastry & Bakery Course	Japan
13	Tokyo Culinary & Confectionery Arts Academy, Shinjuku	Confectionery Arts & Management Course	Japan
14	Prima Baking Training Centre	Pastry & bakery course	Sri Lanka
15	Le Cordon Bleu, Kaohsiung City	Diploma in Pastry Arts	Taiwan
16	Bhutan Institute of Tourism & Hospitality	Confectionery Arts & Management Course	Bhutan
17	First Gourmet Academy, Manila	International Diploma in Professional Culinary Arts and Pastry	Philippines
18	Rahnema e Danesh Culinary School, Tehran	Diploma in Bakery & Confectionery	Iran
19	Le Cordon Bleu, Shanghai	Diploma de Patisssserie	China
20	International Institute of Culinary Arts, IICA, New Delhi	Advance Diploma in Culinary arts	India
21	Culinary Academy of India, CAI, Hyderabad	Diploma in Culinary Arts	India
22	Academy of Pastry & Culinary Arts (APCA)	Diploma in Pastry Arts	Singapore Malaysia Indonesia India
23	Institute of Hotel Management Catering Technology & Applied Nutrition, SIHM, FCI affiliated to National Council for Hotel Management & Catering Technology (NCHMCT)	Bachelor of Science in Hospitality & Hotel Administration, Diploma in Bakery & Confectionery Short term courses in Bakery & Confectionery	India
24	Indian Culinary Institute	Bachelor in Culinary Arts	India

Source: Authors Compilation

Table 8.7 Promotion of Asian Bakery in international Baking Events

Asia Bakery Show 2019, New Delhi	19th–21st Aug 2019	It has been designed to provide an audience where suppliers, bakers small and larger, and buyers of baked goods can meet on common ground and conduct business
BAKERY CHINA -International Exhibition for Bakery and Confectionery Trade, Shanghai (China)	April 27–30, 2021	BAKERY CHINA is Asia Pacific's top trading, sourcing, networking, and branding platform serving the complete bakery industry chain
MOBAC SHOW 2021, Osaka (Japan)	March 09–12, 2021	An event dedicated to New Generation Technologies for Bakery and Confectionery related Industry
FHA – Food & Beverage ASIA 2021, Singapore (Singapore)	March 02–05, 2021	International Exhibition of Food and Drinks, Hotel, Restaurant, Bakery & Foodservice Equipment, Supplies & Services Expo & Conf
Chocolate & Bakery Expo (CBEx), Bombay Exhibition Center, Mumbai, India.	11–13 June 2020	Chocolate & Bakery Expo focuses on new trends and innovation in bakery products
Bakery Business Trade Show, South Edition, Hyderabad (India)	10–12 Jul 2019	To harness the market potential of South India Bakery Business as "Torchbearer of the Indian Bakery industry"
Bakery Tech Expo, Bengaluru, India	1–3 May 2020	Bakery Tech Expo is ensured to achieve the slogan "Make In India" and also the best platform for the Micro Small Medium Enterprise units across the country

Source: Authors Compilation

content in vegetable oils, vegetable fats, and hydrogenated vegetable oil. Baking Industry is exploring new opportunities and entering the market with higher penetration. The target is urban as well as rural customers. With the help of the latest trends and technologies, higher profits can be achieved and higher satisfaction can be achieved. At the customer level, consumers get better products and get a wide variety of options to choose from. The innovations and technological advancements will work hand in hand with the bakers to provide benefits to the sellers as well as the buyers. With the use of standardized recipes, organic ingredients, sustainable practices, the Asian bakeries are targeting the global market segments. Asian bakeries are adopting various means and practices to go green and attain sustainability.

References

CVVM. (2014). Občané o způsobu zacházení s potravinami – duben 2014 (Citizens about the way of food handling – April 2014). Press release. The institute of Sociology, the Czech Academy of Sciences. http://cvvm.soc.cas.cz/ostatni-ruzne/postoj-obcanu-kplytvani-potravinami-duben-2014

Duchin, F. (2005). Sustainable consumption of food: a framework for analyzing scenarios about changes in diets. *Journal of Industrial Ecology, 9*(1–2), 99–114. https://doi.org/10.1162/1088198054084707
Garrone, et al. (2014). Opening the black box of food waste reduction. *Food Policy, 46*, 129–139.
Gusia, D. (2012). Lebensmittelabfälle in Musterhaushalten im Landkreis Ludwigsburg. Ursachen Einflussfaktoren – Vermeidungsstrategien. Diplomarbeit für den Studiengang. Umweltschutztechnik, Universität Stuttgart. Cited in Priefer et al. (2013).
Gustavsson, et al. (2011). *Global food losses and food waste – Extent, causes, and prevention. A study conducted for the International Congress SAVE FOOD! at Interpack 2011 Düsseldorf.* FAO. http://www.fao.org/docrep/014/mb060e/mb060e00.pdf
Katajajuuri, et al. (2014). Food Waste in the Finish Food Chain. *Journal of Cleaner Production, 73*(2014), 322e329.
OECD. (2011). A Green Growth Strategy for Food and Agriculture, Preliminary Report. *OECD, 2011*, 20–21. http://www.oecd.org/greengrowth/sustainableagriculture/48224529.pdf
Mont, O. K. (2002). Clarifying the concept of a product-service system. *Journal of Cleaner Production, 10*, 237–245. http://www.sciencedirect.com/science/article/pii/S0959652601000397
Mylan, J. (2014). Understanding the diffusion of Sustainable Product-Service Systems: Insights from the sociology of consumption and practice theory. *Journal of Cleaner Production*. https://doi.org/10.1016/j.jclepro.2014.01.065
Prieffer, et al. (2013). Options for Cutting Food Waste - Technology options for feeding 10 billion people, IP/A/STOA/FWC/2008–096/Lot7/C1/SC2-SC4, p. 12
Ratinger et al. (2016). Sustainable consumption of bakery products; A challenge for Czech consumers and producers. Agricultural Economics – Czech, 62(10), 447–458. https://doi.org/10.17221/244/2015-AGRICECON
Sanco, D. G. (2014). Stop food waste. Directorate Health and Consumers, http://ec.europa.eu/food/food/sustainability/index_en.htm
Ventour, L. (2008). Food waste report - The food we waste. In *Waste & Resources Action Programme (WRAP)*. Banbury.
Warde, A. (2005). Consumption and theories of practice. *Journal of Consumer Culture, 5*, 131–153. https://doi.org/10.1177/1469540505053090

Chapter 9
African Experience in Ensuring Sustainability in Baking

Umar Garba, Kabiru Ya'u Abdullahi, and Hadiza Kabir Bako

9.1 Introduction

Africa is the second largest continent in regards to area and population in the world. It covers 6% of the world's area with 30.8 million square meters and more than 24% of the world's land with more than 1.3 billion people and counting. The continent has 15% of the world's population. No doubt, the population growth puts a strain on food security. Feeding this population and supplying the food need in maximum, rest on the government's shoulder. In order to sustain this responsibility better, the government is expected to find a balance between the population growth and food production (supply). Wheat is widely consumed in Africa on daily basis and thousands of tons of wheat are used by bakeries and food establishments to make bread, noddle, biscuit, and other pastries in the continent. It was estimated that African countries spend more than USD 6 billion on the wheat import. Nigeria, South Africa, and Angola are said to be the biggest importer of the continent. Nigeria, for example, is the third biggest wheat customer of the United States (Andrae & Beckman, 1985). The demand for baked goods like bread and other baked food increases every day in Africa whose growing population increases daily as such increases the continent's demand for wheat and its dependence on foreign countries every year.

U. Garba (✉) · H. K. Bako
Department of Food Science and Technology, Faculty of Agriculture, Bayero University, Kano, Nigeria
e-mail: ugarbafst@buk.edu.ng

K. Ya'u Abdullahi (✉)
Department of Agricultural Economics and Extension, Aliko Dangote University of Science and Technology, Wudil, Kano, Nigeria

J. M. Ferreira da Rocha et al. (eds.), *Baking Business Sustainability Through Life Cycle Management*, https://doi.org/10.1007/978-3-031-25027-9_9

Historically, Wheat was first cultivated for bread production approximately over 9000 years ago in Mesopotamia (Iraq). The first bread was probably round flat loaves of crushed wheat which were unleavened. In any event, the baking of leavened or fermented dough became an accepted practice over time. Historical records indicate that bread baking was originally a purely domestic activity carried out by the women of the household. Bread preparation for the courts of the Pharaoh's, however, required specialized bakers. The Jewish people also developed bread baking as a trade and thoroughly understood the leavening process. When Rome colonised the known world at that time, bread bakers were brought from Macedonia (Iran). A *Collegium Pistorium*, a Bakers College or Guild, was established in Rome in the century before the birth of Jesus Christ. The bakers in Rome during this period also enjoyed a high status and were given many privileges. The art and skill of bread baking spread throughout Europe with many countries developing strong Bakery Guilds. Settlers in African countries from the Arab world, Holland, England, France, Germany and other European countries brought their bread baking traditions and preferences with them. Heritage, communication and immigration therefore resulted in Africa having a cosmopolitan taste for bread (Abass et al., 2018). Although bread was not the only baked food in Africa but it is the most dominant and steadily become a popular product as it adds variety to the diets, is tasty and nutritious as well as convenient and ready to eat.

9.2 History of Baking in Africa

The term baking refers to an operation of heating dough in an oven. But since there are many operations taking place before the baking, the broader definition encompasses all the science and technology that must be conducted prior to the oven-heating operation itself Baking goes back as far as 2600 B.C. Egyptians were first known to use baking techniques, yeast to make dough rise and ovens for baking (Doolittle et al., 2012). The origin of these ovens has dated back from about 5600 BC, discovered in archeological digs from Hacilar, Turkey to Jericho, Palestine. Baked goods such as bread have long been an important part of human diet. (Hansen & Jacobsen, 2013). Bread baking started in the Ancient Greek around 600 BC, which leads to the invention of enclosed ovens. (Adeniji & Adeniji, 2015). A bakery is an establishment that produces and sells flour-based food baked in an oven such as bread, cookies, cakes, pastries, and pies. Some retail bakeries are also cafés, serving coffee and tea to customers who wish to consume the baked goods on the premises. An example of traditional bread baking in clay oven is shown in Fig. 9.1. Baking industry is made of four different bakeries namely wholesale industrial/plant bakeries; in-store retail bakeries; independent stand-alone bakeries; and franchise bakeries (Fig. 9.2).

Fig. 9.1 Bread baking in traditional clay oven. (Picture by Britannica.com)

Fig. 9.2 Bread baking in traditional clay oven in Kano State – Nigeria - APPEALS – Project

9.3 Traditional Baked Products in Africa

There are quite a number of baked products that can be produced at small and medium scale industries such as cookies, biscuits, leavened and unleavened breads, cakes, pastries, pies, flans, pizzas, samosas and scones. They are made of different shapes and sizes, flavor, texture, and taste (XE "Bakeries" World History of Food Products and Bakery XE "Bakeries" Products, n.d.).

Bread is a dietary staple in north and northeastern Africa and is an important food for some Saharan and East African groups. African breads are baked, steamed, and sometimes fried to produce pancakes, flat breads, loaves or cakes using ovens, griddles, and moulds. Technical choice is associated with the baking properties of bread ingredients and should not be considered evolutionary stages.

Wheat and barley are Near Eastern cereals that were introduced into northeastern Africa in ancient times. Only Near Eastern cereals contain gluten, an elastic protein formed when their flours are mixed with liquid to produce dough. If yeast is added and ferments, it produces carbon dioxide gas which becomes trapped in the dough's elastic structure causing it to rise. Sourdough leavening is common in Africa and is produced with residue left in dough mixing containers or by adding dough saved from a previous batch (Lyons & D'Andrea, 2008). Some examples of African baked products are discussed as follows:

1. Nigeria, which hosts many different ethnic groups, is a very rich country in terms of food and beverage culture. The Nigerian cuisine is a world-renowned kitchen credit to migrants from Nigeria to the Middle East, America and Europe and vice visa. Despite small differences, Nigerian bread resembles the British loaf but differs in taste. It has soft, sticky and stretchy texture. Although the Northern part of Nigerian have a Saudi Arabian inspired bread called *Gurasa* which is a pot baked bread that was introduced to Kano by settlers from the Kingdom of Saudi Arabia, who settled at the ancient Dala hills. The production and sales of *'Gurasa'* are best described as a multi-million naira baking industry, in Kano State, providing employment opportunities for thousands of residents (Nafisat and Mustapha (n.d.)). Another Arab inspired bake good is the '*Gireba*' cookies (Naniya, 2000) that has been made for decades for special occasion like weddings or as an elite snack before they were hawked in knock and crannies of town and villages.
2. In Egypt, Bread is the king of the country's cuisine. It is ranked the first country in the consumption of bread, and one person consumes an average of 400 grams of bread per day in the country. Bread has been cooked in many different forms for thousands of years. Now, it has been cooked in a way to have a pocket so that foods can be put into it. Bread accounts for more than two-thirds of food consumption in North and West Africa and Egypt. Egyptians are fond of desserts and consume desserts like baklava and kadayif daily (Krondl, 2011). Among these desserts, one of the most important is "Om Ali" (Mother of Ali) made from nuts, milk, and phyllo dough.
3. Tunisians make their own bread varieties, like other Maghreb countries, they prefer the flatbread with whole grain and common flour that can be roasted in a pan (Othmani et al., 2015). Occasionally, this bread is made from yeast dough is also preferred as fluffy.
4. Botswana is a country of contradictions. From the marshlands of the Okavango and Chobe, to the Kalahari Desert, wheat is not always readily available. *Diphaphata* however, is a bread made from imported wheat that has somehow found itself as a staple of Botswana's cuisine. *Diphaphata,* though prepared in a manner similar to the English muffin, but differ in taste from the English muffin. The *Diphaphata* also has a denser texture and is much faster to prepare. The bread is charred on a griddle to give it a slightly Smokey flavor (Mogobe, 2020).
5. Bread baking was first established in the Cape Province with the arrival of the Dutch Settlers in the Seventeenth Century. *Roosterkoek* is a South African bread

that also referred to as *'Braari'* when eaten accompanied with barbeque. This bread is prepared by cooking balls of bread dough on a grill over an open flame. This results in a heavy bread with a doughy center and a smooth crust (Balcomb, 1998; Barnard et al., 2002). They are best eaten hot straight off the grill with a smattering of garlic butter and cheese.

6. The Mozambican *Pao* is the African version of Portuguese bread rolls and are made by local village bakers. *Pao* is baked by preparing a dough made of a mixture of water, yeast, sugar, flour and wheat which is placed in an oven and dusting with a generous helping of raw flour. The best way to eat this colonially influenced bread is with a generous filling of peri-peri beef or chicken.
7. During the colonization of Kenya, the British constructed a railway line spanning most of the country. To complete the line, the British brought in Indian laborers to assist in the effort. These Indian laborers then settled into the region and they brought with them their culture and food. *Chapati* is a hallmark of Indian cuisine that found new roots in Kenya. Pictures of some African baked products including *Chapati* are shown in the Fig. 9.3 below:

a

Chapati -East Africa (Tanzania, Kenya, Uganda, Ruwanda, Brundi and South Sudan). Adopted the India Chapati but made using white flour instead of whole wheat Atta

b

Lahoh – [Luhuh (Somali), Laxoox, Lahooh, Canjeero, Anjeero] Popular in Somalia,

c

Injera – [Injera (Amharic); (Bidenaa (Oromo)] Bread popular in Ethiopia and Eritrea.

d

Baghrir - North Africa. Popular bread in Magrib region (Algeriia, Morocco, Tunusia, Libya, Mauritania), Egypt, Sudan

Fig. 9.3 Picture's source: http://www.clovegarden.com/ingred/bk_africae.htm

9.4 Consumption of Baked Products in Africa

There is a growing demand for consumption of bread and other wheat-based products in Africa. This increases the wheat demand in the continents. For example, African countries spent over USD 6 Billion on wheat importation. Nigeria is the biggest wheat customer of the United State and together with South Africa and Angola formed the biggest importers of the continent.

Egypt is the first country to consumed bread with an Egyptian having an average consumption of 400 grams per day. Biscuit market in In Egypt has been developing since 2013 where the retail sales sector increases with 8% in volume and 11% in value, reaching up to 1 billion 31 million Egyptian pounds. In the North and West Africa, bread accounts for more than two-third of the food consumption. Tunisian eat bread with jam and butter and drink coffee with milk during their breakfast. This has been part of their eating habit that has remained with them from the French. Like many other Magrib countries, Tunisians are well successful in preparing different varieties of bread but prefer flatbread made from whole grains and common flour baked on pan (Magazinebbm, 2018).

Bread consumption in Africa is expected to be majorly in the form of packaged products in the coming years. Corn, sorghum and starchy plant roots are mainly preferred in Sub-Saharan Africa, particularly in rural regions, bread consumption is expected to grow slightly during the forecast period. Although the demand for healthier snacking options is increasing, sweet baked goods remain popular and the preferred item in African markets. Majority of bakery product launches in African countries belong to sweet segments strongly led by biscuits and cookie (Mordor intelligence, 2021).

9.5 Status of Baking Industries in Africa

9.5.1 Constrains

Some of the big barriers that keep companies from gaining or maintaining success is the lack of capital, brand recognition and high cost of materials and labor. Those companies that are able to function and survive despite these barriers can enjoy a lot of success as there is relatively few competitors. As stated, there are six key factors that can help to determine whether a baking company will be successful of not. (1) The ability to pass on cost increases, (2) Supply contracts in place for key inputs, (3) Proximity to key markets, (4) Use of the most efficient work practices, (5) Product differentiation, and (6) Establishment of brand names. If new entrants in the baking industry or if existing companies within the industry are able to utilize these factors as well as avoid some of the costly barriers in the industry, they will be able to bring joy to many homes for many years to come and the industry itself will enjoy many years of growth and success (Doolittle et al., 2012). Another important constrain in the African baking industries is the continent's insufficient production of the wheat required by the industry. For examples due to the small production capacity for wheat, African countries such as Angola, Cameroon, Côte d'Ivoire, Ghana, Guinea, Madagascar, Mozambique, Nigeria, Tanzania, Uganda and Zaire imported about two million metric tons of wheat each year for their baking industry. Although an attempt has been made to partially substitute wheat flour with other cereals, legumes or tuber crops, such composite flour usually requires at least 70% of wheat flour to obtained a good leavened product like bread. This is due to the lack of gluten in flour other than wheat which offer the unique functional viscoelastic properties (Adeniji & Adeniji, 2015).

9.5.2 Prospects

The manufacture of baked goods is a growing industry, and artisanal and in-store bakeries are on the increase. Higher-income consumers have variety and health requirements, while there are opportunities for emerging bakers to enter local informal markets. The market is primarily driven by the convenience, accessibility and nutrition profile associated with them. Moreover, as a result of urbanization and worldwide increase of working population, the growth of out-of-home consumption and increased demand on instant and nutritious products, this trend has been fueling the continuous increase of bakery products (Mordor intelligence, 2021). Africans Bread consumption is expected to be dominantly in the form of packaged products in the coming years. Corn, sorghum and starchy plant roots are mainly preferred in Sub-Saharan Africa, particularly in rural regions, bread consumption is expected to grow slightly during the forecast period (2020–2025).

Moreover, in Nigeria Kano state Gurasa (Local Bread) producers are among the second largest bakers after conventional bread makers the nature of their operation is purely baking using traditional oven made from sand. History have shown that, the gurasa baking occupation exist more than hundredth years in Kano State. The finding from the gurasa bakers visited found to be experienced in Gurasa Baking for years with an average of 13 years, minimum of 2 years and maximum of 45 years as shown in Fig. 9.4. The experience would help in ensuring business success and easy means of expanding the scope of customers.

Now with the Agro-processing Project and Livelihood Improvement Support, APPEALS – Project, a World Bank Assisted in the state. The baking is enjoying massive supports with the project fund, empowering about 1000 women with better equipment, inputs, utensil, kitchen wares among other things. The equipment includes gas oven which demonstration proved to be better than their old practice. While in an effort of the project to reduce the risk of fire outbreak that may cause damage and destruction and improve sanity and hygiene of the gurasa produce set of Personal Protective Equipment (PPE) provided for the beneficiaries among which are Kitchen Coat, Apron, Protective boot, head tie/ chef hat, disposable hand glove, hand glove for mixing and face masks (Figs. 9.5 and 9.6).

Fig. 9.4 Traditional clay oven for Gurasa (Local Bread) Bakers in Kano State, Nigeria

Fig. 9.5 Improved Baking Oven Developed by APPEALs Project – Kano State, Nigeria. (Picture by Britannica.com)

9.6 Sustaining African Baking Industry

Sustaining baking industry is beyond ingredient transparency but many other factors including commitment to the environment, reduction of food waste, dedication to employees, and care for consumers (Unrein, 2017). It's seen as less of a driver for bakers when compared with other key drivers such as health and fitness, convenience or good labels (Tucker, 2016). Consumer's demand for an alternative ingredients and packaging materials are also key players in sustaining a baking industry. The industry also needs to develop strategies to reduce the use of scarce raw

Fig. 9.6 Mobile Gurasa baking & Marketing Ban Developed by APPEALS Project – Kano State, Nigeria. (Picture by Britannica.com)

materials such as phosphate and be more innovative in designing packaging materials to meet the consumer's demand. Principles of sustainability such as reducing energy usage and water consumption and lowering food waste level are not only better for environment but for economic reasons of lowering overall production cost while increasing the bottom-line bakeries' green credentials (Tucker, 2016).

To sustain African baking industry, innovation to meet the dynamic nature of the health-conscious consumers must be given higher priority. Many consumers now demand for high fiber, low sodium, and gluten-free products. It is also important for bakers to be aware that for celiac disease and gluten intolerance, the only treatment is lifelong adherence to gluten-free diet. Therefore consumers with such diseases must avoid wheat, rye and barley and baker had to replace these major ingredient with other flours such as rice, oat flour and other flour that do not contain gluten (EYE, 2018).

Over the years, profitability, viability and sustainability of African baking industries are hindered by several factors such as unbearable cost of production and government regulations on importations. For example, 15% extra duty is charge for wheat importation making the cost of flour to rise. Additionally, the activities of unregistered and unlicensed producers who produces cheap products with poor quality and unacceptable ingredients such as potassium bromate attract higher patronage due to their lower price particularly for bread. For this reason, licensed and regulators baker who bear the high cost of production to meet the requirement of the government regulatory agencies find it difficult to compete with the unlicensed and unregulated producers in term of price (Nnah, 2018).

9.7 Conclusion

Baking is an oldest form of cooking used for producing everyday products such as bread, cakes, cookies, pastries, donuts and pies. These products are made using various ingredients like grain-based flour, water and leavening agents. They are consumed daily and are considered fast-moving consumer goods. The palatability, appearance and easily digestible nature of baked products made them highly preferred for both formal and informal occasions. Nowadays, most traditional baking methods have been replaced by modern machines. This shift has enabled manufacturers to introduce innovative bakery products with different ingredients, flavors, shapes and sizes. The market for baked products is primarily driven by the convenience, accessibility and nutrition profile associated with them. Moreover, due to urbanization and global increase of working population, the growth of out-of-home consumption and increased demand on instant and nutritious products, this trend has been fueling the continuous increase of bakery products. Higher-income consumers have variety and health requirements, while there are opportunities for emerging bakers to enter local informal markets. To sustain African baking industry, manufacturers has to dynamics and innovative to meet the consumers demand.

References

Abass, A. B., Awoyale, W., Alenkhe, B., Malu, N., Asiru, B. W., Manyong, V., & Sanginga, N. (2018). Can food technology innovation change the status of a food security crop? A review of cassava transformation into "bread" in Africa. *Food Reviews International, 34*(1), 87–102.

Adeniji, T., & Adeniji, T. (2015). Making : prospects for industrial application. *African Journal of Food, Agriculture, Nutrition and Development, 15*(4), 10182–10197.

Andrae, G., & Beckman, B. (1985). *The wheat trap: bread and underdevelopment in Nigeria*. Zed Books Ltd.

Balcomb, T. (1998). From liberation to democracy: theologies of bread and being in the new South Africa. *Missionalia: Southern African Journal of Missiology, 26*(1), 54–73.

Barnard, A. D., Labuschagne, M. T., & Van Niekerk, H. A. (2002). Heritability estimates of bread wheat quality traits in the Western Cape province of South Africa. *Euphytica, 127*(1), 115–122.

Chapter Four World History of Food Products and Bakery Products. (n.d.). Available at http://lib.unipune.ac.in:8080/xmlui/bitstream/handle/123456789/3216/11_chapter%204.pdf?sequence=11&isAllowed=y. Retrived on 3/4/2021

Doolittle, A., Jones, A., Pope, L., Vorontsov, O., & Wray, J. (2012). *Bakery industry industry analysis* (Vol. 66).

EYE. (2018). *Enhancing YOUTH (18–26) employability in baking sector 2017-1-tr01-ka205-039233* (pp. 1–55). Best Bread Production Handbook.

Hansen, Z. N. L., & Jacobsen, P. (2013). Challenges facing the food industry examples from the baked goods sector. In *International conference on the modern development of humanities and social science (MDHSS 2013) challenges, Internatio(Mdhss)* (pp. 430–433). https://doi.org/10.2991/mdhss-13.2013.113

Krondl, M. (2011). *Sweet invention: a history of dessert*. Chicago Review Press.

Lyons, D., & D'Andrea, A. C. (2008). Bread in Africa. In H. Selin (Ed.), *Encyclopaedia of the history of science, technology, and medicine in non-Western cultures*. Springer. https://doi.org/10.1007/978-1-4020-4425-0_9476

Magazinebbm. (2018). *Pasta, Biscuit and Bakery Products in Africa*. Available at https://www.magazinebbm.com/english/pasta-biscuit-and-bakery-products-in-africa/.html. Retrieved on March 14, 2021

Mogobe, S. S. (2020). *Exploring livelihood strategies employed by women street food vendors in Gaborone, Botswana*

Mordor Intelligence. (2021). *Middle East and Africa Bakery Products Market – Growth, Trends, Covid-19 Impact, and Forecasts (2021–2026)*. Available at https://www.mordorintelligence.com/industry-reports/middle-east-and-africa-bakery-products-market

Nafisat, N., & Mustapha, A. *Analysis of women participation in processing and marketing of Gurasa in, Kano State Nigeria*.

Naniya, T. M. (2000). Arab settlers in sub-Saharan Africa: a survey of their influence on some central Sudanese states. *Kano Studies, 1*(1), 1–12.

Nnah, M. (2018). *Salvaging the bread industry in Nigeria*. Thisday. Available at https://www.thisdaylive.com/index.php/2018/09/07/salvaging-the-bread-industry-in-nigeria/. Retrieved on March 23, 2021

Othmani, A., Mosbahi, M., Ayed, S., Slim-Amara, H., & Boubaker, M. (2015). Morphological characterization of some Tunisian bread wheat (TriticumaestivumL.) accessions. *Journal New Scientist, 15*.

Tucker, G. (2016). *Sustainability in Baking*. World Baker. Available at https://www.worldbakers.com/process/gary-tucker-sustainability-in-baking/. Retrieved on March 23, 2021

Unrein, J. (2017). *The Sustainable Bakery*. Available at https://www.bakemag.com/articles/5996-the-sustainable-bakery. Retrieved on March 14, 2021

Chapter 10
Indian Experience in Ensuring Sustainability in Baking Industry

Sharad Kumar Kulshreshtha, Ashok Kumar, and Deborah Rose Shylla Passah

10.1 Overview

Bakery products, are one of the food item which are consumed worldwide, due to its high nutritive value and affordability. Bakery products have gained popularity among people, due to the rising foreign influence, the rapid population growth, the emergence of a female work force and the changing eating habits of people. Bakery is a traditional activity and largely contributes to the food processing industry. Consumers are demanding newer and innovative options, in many bakery products, thereby leading the industry to experiment with fortification of bakery products in order to satiate the burgeoning appetite of the health-conscious Indian. Healthy bakery products have been launched in the bakery segment, and are being bought and tasted by many consumers. Entrepreneurs in the field of bakery have created chains of their bakery outlet in different locations within the country which has further triggered the growth in the sector. India has the potential of being the biggest producer of food next to China. Varieties of breads and biscuits, from the bakery section accounts for over 82% of the total bakery products produced in the country. The bakery segment in India can be classified into the three broad segments of bread, biscuits, cookies, and cakes. India's organised bakery sector produces about 1.3 millions tonne of bakery products (out of three million tonnes) while the balance is produced by unorganised, small-scale local manufacturers (Bhisel & Kaur, 2013).

S. K. Kulshreshtha (✉)
Department of Tourism & Hotel Management, North-Eastern Hill University, Shillong, India

A. Kumar
Department of Tourism & Hotel Management, North-Eastern Hill University, Shillong, India

Institute of Hotel Management Catering Technology & Applied Nutrition, Shillong, India

D. R. S. Passah
Institute of Hotel Management Catering Technology & Applied Nutrition, Shillong, India

J. M. Ferreira da Rocha et al. (eds.), *Baking Business Sustainability Through Life Cycle Management*, https://doi.org/10.1007/978-3-031-25027-9_10

Consumers still prefers locally made fresh bread and cookies over bread and biscuit which are being made in an automatic and semi-automatic bread and biscuit manufacturing units in India. With the franchise of multinational companies (MNC) introducing pizzas and burgers across the country, taste buds of the consumers have also change. Lately, consumers have had their taste to other derivatives of the bakery products like varieties of choux pastry, laminated pastry, Danish pastry and other International delicacies. Companies like Britannia, Biskfarm and Morish, are competing with one another to experiment with different innovative, organic ingredients and flavours to capture their consumers. Also, the Indian market is observing the establishment of bakery café chains in the form of Barista, Café Coffee Day and Monginis (Pradeep Raj et al., 2015). Glucose biscuits, Marie, cream biscuits, crackers, digestive biscuits, cookies and milk biscuits are among the variants which are popular and consumed in India. The shares of the branded and organised sector and the unbranded and unorganised sectors are 60% and 40% respectively, in the context of the Indian biscuit market. Bakery products from India, especially biscuits, are highly consumed and in great demand in developing countries. Major companies in this sector, like Parle, Britannia and ITC Foods, have captured the markets to a great extent, with Britannia holding the leadership position (Townsend, 2001). Britannia and Parle account for around 38% share each of the total volume of branded biscuits marketed in India, in terms of value. Britannia Industries, with an expected retail value share of 9% of the baked goods category, proved to be the most successful company in India. Organised and unorganised bread companies contribute around 45% and 55% of the total bread production, respectively (Bhisel & Kaur, 2013). In India, approximately 1800 small-scale bread manufacturers are in the organised sector, 25 medium-scale manufacturers and two large-scale industries. Three (03) or four (04) large-sized players (namely, Britannia, Parle, ITC and Cadbury's) comprises 75% of the market. The bread and cake market is much more diverse, with multiple regional and local players. Ninety percent of the market share in the bread segment includes Britannia and Modern Industries Ltd. Spencer's in South India, Kitty and Bonn in Punjab and Harvest Gold and Perfect in Delhi and the National Capital Region (NCR) are regional players. The bakery industry in India has seen an annual growth rate of more than 15 per cent during the past years which triggers the immense growth potential in the global and domestic markets. There is an estimated 75,000 bread bakers which is under the unorganised sector, mostly located in the residential areas of cities and towns. Thirty-five percent of the total production comes from the small scale sector, with about 1500–1800 units in operation (Maric & et al., 2009). The bread industry is a low-margin business activity, whereby cost control becomes crucial in sustaining profitability. The segment's innovation has powered the rise of newer café formats like bakery cafés, which reverse the concept of cafés by extending existing bakeries and chocolate retailers to offer complementary beverage items (Oguntona et al., 1999). A bakery café is only providing baked food, with the beverage being a complement to the menu. A typical bakery café menu includes a wide selection of breads, encompassing such exotic variants as wheat, rye, oats, five-grain, multigrain, cracked wheat, flute, baguettes(French Bread) and ciabatta, and other baked goodies like brioches, croissants, cookies, muffins, cakes, scones, strudels, brownies, pies and puffs (Serivastava,

2009). The menu in a bakery café may also feature soups, salads, and other dishes, made using wholesome, locally sourced ingredients, cooked and served fresh. These can include eggs preparations, varieties of sandwiches, rolls, and baked beans on toast (Smitha et al., 2010). The emerging concept of bakery cafés, mocktail bars which has grasp a good market share among the young consumers in recent years, is not new to India (Serivastava, 2009).

10.1.1 Indian Baking Industry

Bakery products in India are in common use. In India, the bakery industry developed largely to meet the needs of the British army. Thus, this industry developed at first at the seaports of Calcutta and Bombay to carter to the need of foreign travellers and army personnel. Biscuits were first made in India about 300 years ago. Up to the Second World War, there were only 7 factories producing bread and biscuits and also confectionary in India (Manley, 1991). During the post independence period, the growth of bakery industry was more scientific and with advance technology. However, still the traditional items such as, kulcha, nans, are produced in traditional cylindrical clay ovens known as *Tandoors*. Almost all the States in India has its own traditional bakery products, particularly in Jammu and Kashmir the traditional bakery industry continues to be popular. The traditional items of production are Girdha, Assorted Kulchas and Bakerkhani (Mohammed et al., 2008). Bakery Industry plays a vital role in the economic development of the country, utilization of its grains and cereal resources and in shaping the health of the citizens. The bakery Industry has the potential of generating additional employment and investment. The capital investment requirement per unit of additional employment is comparatively lower to most of the other capital-intensive industries in India (Selvaraj et al., 2002). Latest Trends in the industry includes E-retailing of bakery products; Expanding foothold; Improved packaging, and Innovation in ingredients, aggressive expansion plans, and technological advancements (Nare, 2002; Table 10.1).

Table 10.1 Types of Bakeries

S. No	Types of Bakery	Attributes
1	Online bakery	More economical, e-commerce platform is used for marketing, selling and delivery of the bakery product.
2	Traditional counter service bakery	Customers are catered when they approach the counter in any bakery store. Products are kept at display, and customers can choose from there.
3	Speciality bakery	Specialises in some bakery items. They bake on order and deliver the same to the customer. They may have special vehicles which are air conditioned.
4	Cafe' like bakery	They are emerging to cater the demand of road side light refreshment. Menu mostly includes baked products such as assorted sandwiches, puffs, croissants along with beverages such as coffee, shakes & soda based drinks.

Source: Authors

10.1.2 Emerging Trends in Indian Bakery Industry

- **E-Retailing of Bakery Products**: Indian bakeries are capitalizing on the emerging need for bakery products. Bakery products are also being sold through e-retailing which is a latest trend in the social media platform. Indian consumers are fond of e-marketing whereby they are willing to buy food and bakery products online. Several mobile apps and e-retailing sites are attracting the Indian buyers for bakery and food items.
- **Entry of Multinational Companies**: Multinational bakery chains are expanding their base in Indian cities. There are many companies like Monginis who has done market survey, feasibility study and after thorough research on Indian customers and their preferences, setup their units for production or sale at almost every city of the country. Bakery units which are under the organized sector along with the unorganized sector are gaining reputation and have come up with innovative ideas to stay in the market and stand the throat cut competition.
- **Technological Advancement**: With the emerging economy, Globalization, urbanization, and rising disposable incomes of the citizens, Indian bakeries has diversified the bakery products. The diversification of bakery products is due to the growing demand of superior products like gourmet breads and increasing focus on healthy products which has also led to the up gradation of technology in bakeries. With the technological advancement, machines like Combination Ovens, Heavy duty planetary mixers, chocolate machines and moulds, pizza oven, etc. and photo printing machines to print image directly on cream cakes for decoration has occupied space in the bakeries. Indian bakeries are using fuel efficient equipment such as the infrared burners which emits less pollution which are due to the technological advancements of these machineries. Technology is also helping the bakeries in attaining sustainable development and environment friendly practices.
- **Innovation in Ingredients**: Basic ingredients plays an important role in shaping and moulding bakery products. Indian bakery industry has been experimenting and using innovation in ingredients to enhance nutrients for health concern consumers. Introduction of Lame Quick and Spongolit, is a very recent and major innovation in the use of bakery ingredients. Lame quick are largely made from healthier mono and polysaturated fatty acids, and whipping properties which produce a pleasant mouth feel and light creamy texture in products like whipped desserts, cream fillings and cake decorations. Spongolit acts as an aerating emulsifier which allows cake to be baked faster, with consistent quality creating excellent stability, volume and crumb texture. Apart from these Government of India is encouraging the food processing sector to make use of indigenous ingredients to directly support the farmers and come up with the healthier products.
- **Improved Packaging Solutions**: The bakery industry has been experimenting with packaging solutions for their products. Packaging of bakery products plays a significant role in increasing shelf life, preventing mechanical damage, retaining nutritive value, displaying food safety related information and marketing. To

reduce the impact of packaging on the environment, use of sealants for low temperature seal initiation, outstanding hot-tack strength has been marked as a tremendously innovative effort for the preservation of baking products. Bakers have widely used vertical pouches or sachets for packaging and marketing the bakery products. Another trend that has emerged in recent years is the preference for homemade baked items (Fig. 10.1 and Table 10.2).

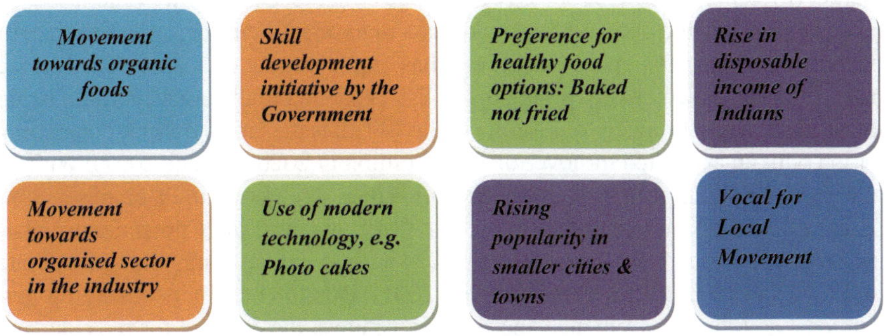

(Source: Authors)

Fig. 10.1 Trends in Indian Bakery Industry. (Source: Authors)

Table 10.2 List of Bakery Products in India

S. No.	Name of the product	Usage of the product
1	Bread loaf	Breakfast toast, sandwiches, bread pakoras, shahitukda, croutons
2	Bread rolls, soft rolls	Accompaniment for soup, burgers
3	Hot dog buns	Varieties of hotdogs as snacks
4	Croissant	Breakfast, sandwiches
5	Pizza Base	Varieties of pizzas (snacks)
6	Pita	Accompaniment for main course, sandwich
7	Varieties of cookies/biscuit	Snacks, crumbs for cheese cake, decorative pieces
8	Varieties of choux pastry	Snacks, desserts, cakes, savoury
9	Laminated pastry	Varieties of savoury food to serve as snacks and appetizers
10	Fatless sponge cake, Genoise sponge cake	Base for decorative cakes for different occasion, added to desserts
11	Varieties of pound cakes	Tea cake, high tea
12	Short crust pastry	Cookies, tart and pie shells
13	Doughtnuts	High tea snacks
14	Fruit cakes	Tea cake

Source- Authors

10.2 Review of Literature

The well-to-do biscuits and cookies industry is the reason behind the growth and spread of the bakery market in India. Among the Indian food processing sector, the bakery industry is the largest segments with an annual turnover of about $ 7.60 billion in the year 2020 (Raut, 2018). Due to their affordable prices and ready to eat category, the cookies and biscuits are in high demand both in the urban and the rural market (Mohammed et al., 2008). Further, due to the recent growth of the fast food chains and cafes, the demand for breads as a main ingredient for varieties of sandwiches, buns, burgers, cheese toast, etc. has exponentially increased. The market base for value added bakery products too has expanded in the country. India being the second largest producer of biscuits in the world, is the largest biscuit consuming nation with an estimated turnover of $ 4.65 billion in 2020 (Baisya, 2007). With an abundant supply of primary ingredients and skilled workforce, India enjoys a comparative advantage in manufacturing sector. Cake, pastries, and Confectionary market is about to reach $ 882.24 million by 2024 with an annual compound growth rate of 12.5% between last 5 years (Bhagvat, 2001). Due to consumers' changing perceptions, and their convenience and health attributes, the demand for the cake and confectionary products, are rising sharply. Hassan et al. (2013) suggested that proper production planning, standardized recipe is a necessary for an effective production system. The consumption culture and consumer preference towards bakery products has been studied by (David, 2013) and further analysed the consumers perception towards the bakery and bakery products. Sakhale et al. (2012) made mixed fruit toffees from fig and mango pulp at different proportions, and evaluated for physicochemical and sensory quality characteristics. The basic trends in bread, bakery, and pastry innovations related to health, pleasure, convenience and quality has been studied and documented by (Martínez-Monzó et al., 2012). The study highlights, how the recent culinary trends are impacting in product innovation with regard to bread and similar products. The basic food science aspects of Carbohydrates, Minerals, Proteins, Vitamins, and lipids can help bakers and similar practitioners of the bakery and confectionery to innovate newer products and the processing of baked foods (Pyler & Gorton, 2008). Nambiar and Parnami (2008) studies has helped to standardize and evaluate the addition of freshly blanched green leafy vegetables in bakery products based on organoleptic attributes. The creative ideas on newly set standards are based on quality standards such as weight, standard portions, deformations, cracks and faults (Barbiroli, 2002; Dubey, 2002).

10.3 Research Objectives

The main objectives of the study are:

(a) To study about the Indian Baking Industry, consumption culture of Bakery products in India.
(b) To highlight the sustainable practices involved and the emerging trends that help to flourish the bakery industry.

10.4 Research Methodology

This study is an explorative research based on information, facts & figures on the bakery processing & sustainable practices adopted in the bakery Industry. The paper has been structured on the basis of review of related literature on history of baking industry, and consumption culture of Bakery products in India. The sustainable practices involved, emerging trends in the industry, different approaches adopted at macro and micro level in order to attain sustainability have been highlighted. This study involves personal experience, close observation and some secondary sources of information such as research papers, food blogs, Bakery & Confectionary books and some useful websites.

10.5 Findings and Discussion

10.5.1 Production of Bakery Product During Colonials and Post Colonial Period

The Roman Empire has been credited for the emergence and development of art of baking. Baked goods have been around for thousands of years. Baking was introduced as a decent profession for Romans, due to the fame and yearning that the art of baking received, around 300 BC. Bakery and confectionary industry is one of the major food processing industries in India. It has played a significant role in the economic development of the country. The bakery items such as Pastries, cakes, savouries, breads, biscuits, buns, etc. are manufactured from both the organized and unorganized sectors. The bakery industries were worst affected after the end of World War II. Due to the war and mass destruction the quality of raw materials were not available and baking schools were closed during this time which resulted non availability of skilled bakers & confectioners. This resulted in the eradication of old baking techniques and new methods and techniques were developed to satisfy the need for breads, and other related bakery products. High demands of bakery products resulted in the inclusion of chemical agents and food additives to the dough, pre-mixes and invention of specialized and customized tools and equipments. The unorganized sector in the Indian Bakery Industry contributes more than 3/4th of the total bread production, and more than half of the total biscuit production. The consumption of bakery products in India is very low, about 1 to 2 kg per person annually, as compared to the developed countries where it is between 10 and 50 kg per annum. The different types of bakery products are processed as per the specific palatable characteristic, and the quality standards laid by Food and Drug Administration (FDA), USA (U.S. Department of Agriculture, Food and Nutrition Service, & National Food Service Management Institute (1996) and FSSAI, India (Table 10.3).

Table 10.3 List of Indian Baked Breads

Name of the Bread	Region of India	Speciality
Roti (Chapati)	Indian subcontinent	*Flat unleavened breads mainly made up of whole wheat flour, baked on flat tawa.*
Assorted parathas such as onion, aloo, paneer, gobhi,	Indian subcontinent	*Paratha is made with two words-parat and atta which means layered bread. A perfect paratha is the one which separates into different layers, and can be made in triangular or round shape. Parathas are slightly thick and can be stuffed with anything from meat, potatoes, paneer, cheese, sugar, Jaggery and even dals and vegetables.*
Naan	North India	*Naan in Persian means bread. This soft, spongy leavened bread came to India from Central Asia and became an inseparable part of Mughlai food. It is cooked in tandoor and while kneading the dough, usually milk is used instead of water, to make it soft.*
Kulcha	North India	*Punjabi variant of naan, it is a leavened flatbread made with refined flour and cooked in tandoor. It is made with or without stuffings.*
Appam	South India	*It is made with rice and coconut milk batter, it has several variations and can be sweet as well as savoury.*
Puran Poli	Western India	*A popular Maharashtrian bread, it is a kind of paratha stuffed with chana daal, jaggery and coconut mixture and then cooked with some ghee.*
Coconut Poli	South India	*A popular south Indian delicay, is a kind of paratha stuffed with coconut mixture and then cooked with some ghee.*
Sheermal	North India	*It is inseparable part of traditional Awadhi and Nizami cuisine. It is a leavened bread baked in tandoor (earthen clay oven). It is flavoured with saffron and has a mildly sweet taste.*
Bakarkhani	North India	*It is a leavened bread and has a biscuit like texture. It is mildly spiced, is sweet and is flavoured with saffron. It is a part of Eid celebrations and tastes great with meats and gravies.*
Pathiri	South India	*This white, thin pancake made with rice flour, is popular with Malabari Muslims and is an essential part of Iftaar during the fasting of Ramzaan*
Parotta	South India	*It is a layered flatbread, midway between laccha paratha and tikona paratha and is popular in Kerala.*
Bhakri	Western India	*It is a hard and crisp, unleavened variety of bread made with flour and is popular in Maharashtra, Goa and Gujarat.*
Tandoori roti	North India	*It is similar to naan, and is made of whole wheat flour and is cooked in clay oven at a high temperature. It is usually smeared with butter before serving.*
Baati	North-West India	*It is a preparation from Rajasthan, and it is round and small and is served with daal and choorma.*

(continued)

Table 10.3 (continued)

Name of the Bread	Region of India	Speciality
Litti	Eastern India(Bihar)	*A popular dish from Bihar, it looks like a ball which is baked in clay oven and is stuffed with sattu (roasted gram flour). It is served with chokha made with roasted brinjal, potatoes, and spices.*
Thalipeeth	Western India	*Very popular in Maharashtra, Gujarat and some other parts of Western India, it is made with rice, wheat, jowar, bajra (millets), urad and chana. It is spicy and served as a savoury.*
Thepla	Western India(Gujarat)	*It is made with whole wheat flour, gram flour, fenugreek leaves and spices, this is soft and has a longer shelf life than most breads.*
Taftan	North India (Uttar Pradesh)	*Aftan came to India from Iran and is a leavened bread made with eggs, flour, saffron, spices and kneaded in milk.*
Roomali roti	North India	*Very thin bread flattened with the help of hand and baked on roomali kadhai. Derived from the word Roomal (handkerchief).*

Source: Authors

10.5.2 Traditional and Modern Bakeries in India

Baking is the art and science of processing food using long-standing dry heat action generally on hot stones in a close chamber or in an oven. Bakery is primarily used for the preparation of breads, buns, pastries, cakes, rolls, Cookies, tarts, quiches, and pies. The dry heat action of baking transforms the carbohydrates present in the food, giving it an attractive appearance, shape, size, taste and causes its outer surfaces to change the colour to brown. This change in colour to browning is either caused by the Maillard's reaction or due to caramelisation of sugars present in the dough. Varieties of bakery products are produced around the world using different types of ovens, ingredients and recipes. Processed Baking, and Artisan Baking are the two major trends being followed in the global bakery industry. Processed Baking involves a lot of chemicals, processes, shelf life, whereas, Artisan Baking does not involve chemicals; it is organic in nature and requires a lot of skills.

The Indian bakery industry is well known for some of the largest food categories such as breads, biscuits, cakes etc. Among the biscuits and cookies sector, 4 large players such as Britannia, Parle, ITC, and Cadburys encompass about 75 per cent of the Indian market. The breads, cakes, and Pastries market is much more disjointed with manifold regional players. Few international players such as Unibic, and United Biscuits have gained distinction in the last few years in their specific product and market segments (Figs. 10.2, 10.3 and Table 10.4).

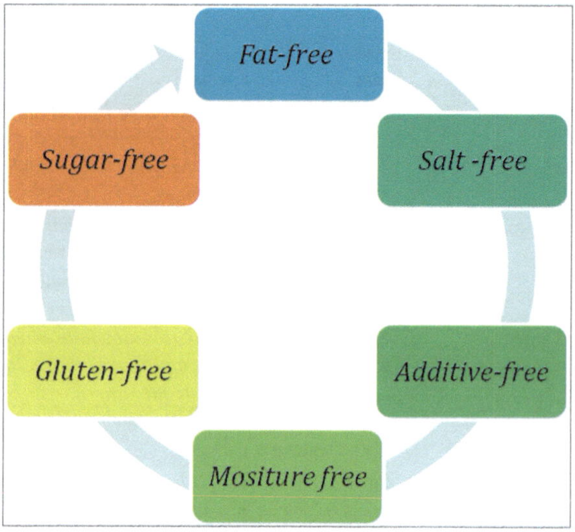

Fig. 10.2 Food and health and nutrition conscious bakery products

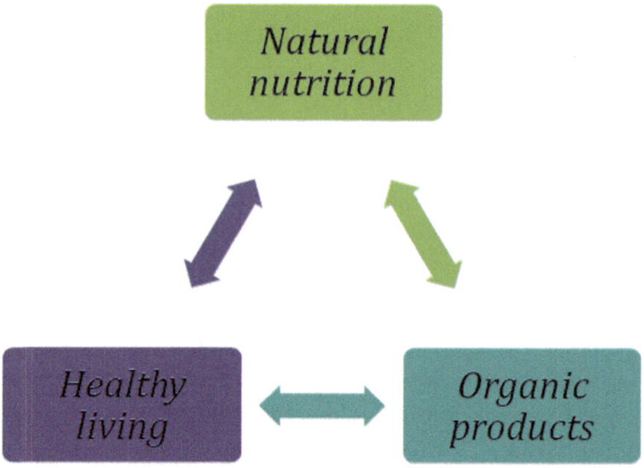

Fig. 10.3 Healthy life style

10.5.3 Demand and Consumption Culture of Bakery Products in India

In terms of revenue and employment generations among the processed food sector, the bakery industry has achieved third position globally. Every Government is putting their best effort to promote bakery industry and related food processing sector to curb the menace of unemployment and hunger. The per capita consumption of bakery products in India is very low as compared to the developed countries. Café like Bakery cafes are emerging to cater the demand of road side light refreshment. There are small bakery cafes, also which offers sitting arrangement for their

Table 10.4 Famous and Old Bakeries in India

Name of the bakery	Founded	Specialities
Mambally's Royal Biscuit Factory, Thalassery, Kerala	It was founded in 1880 by Mambally Bapu.	Varieties of buns, bread, biscuits, and cakes
Smith Field Bakery, Chennai, Tamil Nadu	It was founded in 1885 by Ponnuswamy N in the heart of erstwhile Madras	Chicken and vegetable puff pastries, fresh assorted breads, cookies and biscuits.
Albert bakery, Bengaluru	It was established 1902 and founded by Mohammed Suleman	Assorted cocktail samosas, stuffed minced meat samosas, chicken seekh rolls, chicken Malai cutlet, varieties of puffs, khova (reduced dry milk) naan.
Nahoum's, Kolkata, West Bengal	It was founded in 1902 by Nahoum Israel Mordecai, a Baghdadi Jew	Baklava, Jewish cheese samosas and patties, Almond pastry, marzipan, Franzipan, date sticks, etc.
Ahdoo's, Srinagar, Kashmir	It was started by Abdul Ahmad in the year 1918	Varieties of cakes, vegetable and chicken patties, assorted kebabs, and the famous Kashmiri barbeque, *seek-e-tujj*.
Wenger's, New Delhi	Designed in 1926 by British architect sir Robert tor Russell. It was owned by Swiss couple Jeanne Sterchi Wenger and H C Wenger.	Soft sugar and chocolate coated doughnuts, sandwiches, waffles, apple pies, and varieties of smoothies and milkshakes
Yazdani bakery, Mumbai, Maharashtra	It was established in 1951 by Zend Meherwan Zend.	Shepherd's pie, apple pies, rum cake, Baba au rum, and ginger biscuits.
Glenary's, Darjeeling, West Bengal	Started by an Italian, Jater Pilva, in the year 1935.	Assorted open and closed sandwiches, cinnamon buns, meat pies, cookies, bread and cakes, assorted non-alcoholic stimulating beverages.
Jila bakery, Loutolim, Goa	Started in 1972, by a couple–Jose Inacio and Ludovina Antao (JILA)–in their home.	Plum cake, assorted macaroons, caramel-butter cake, Geneva pastry, apple strudel, and eclairs.
Kayani bakery, Pune, Maharastra	It was started by two brothers Khodayar and Hormazdiar Kayani, in 1955.	Assorted cookies and biscuits.

Source: Compiled by Authors

customers. Their menu mostly includes baked products such as assorted sandwiches, puffs, croissants along with beverages such as coffee, shakes & soda based drinks. Biscuits & Cookies are the most popular bakery items followed by Breads in the popularity list. Majority of Indians have chosen assorted breads as their common breakfast item. Whole wheat flour (WWF), Multigrain breads, as more healthier options are also available in the market. Cakes & Pastries are widely used during different occasion and celebration in the country. They are available in many flavours and can be customised as per the choice of the customers. Indians have the tradition of consuming varieties of soft and hard toast such as cake rusks, puffs, and savoury toast etc. as High tea snacks or as a breakfast items.

Table 10.5 Famous Biscuits & Cookies Brands in India

1. Parle	2. Britannia
3. Priya gold	4. Anmol
5. Cremica	6. Sunfeast
7. Biskfarm	8. Dukes
9. Unibic	10. McVities
11. Patanjali biscuits	12. Horlicks

Source: Authors

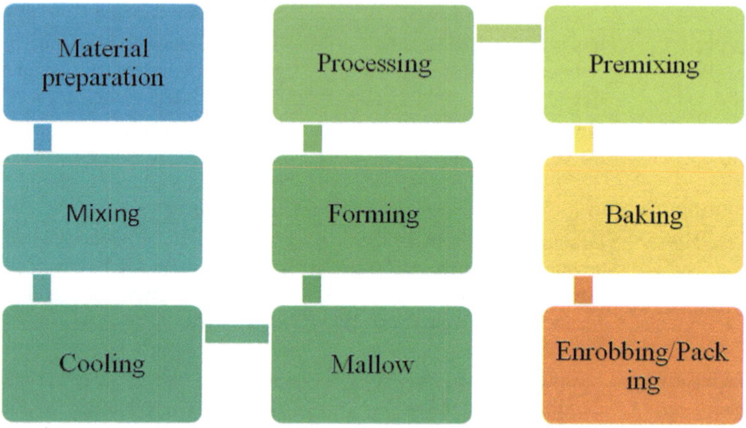

Fig. 10.4 Food processing in Bakery. (Source: Authors Compilation)

In India, traditionally people are very much fond of baking foods items. Bread is a staple food included in the meals, prepared from a dough of flour and water, and baked (*Breads known as Roti in India), Pav, Cream Rolls, Buns, Toasts, Biscuits, Cakes and Cream Puff, Muffins, Pastries, Fruits cakes,* Cup cakes, Patties, Brownie, Tarts, etc. (Table 10.5 and Fig. 10.4).

10.5.4 Trainings and Development Programmes Associated with Baking Industry

The demand and supply for the indigenous bakery items is increasing day by day and for this reason the growth rate of baked items has also surged in both urban and rural sector. The reason behind this tremendous growth are availability of better ingredients such as local grown grain and cereals, chocolate, toppings, fillings, flavours etc., and education in terms of training and skill development. New generation is taking bakery as a profession and several entrepreneurial ventures have been

Table 10.6 Institutions offering Bakery Education & Training Programs in India

S. No	Name of the Institution	Courses offered
1	Central Institute of Hotel Management Catering Technology & Applied Nutrition (CIHM), (Sponsored by Ministry of Tourism, Government of India)	BSc in Hospitality & Hotel Administration Trade diploma in Food Production & Patisserie Craftsmanship Certificate course in Food Production & Patisserie Bakery & Confectionery training programme under Hunar Se Rojgar Tak scheme (HSRT) Skill testing and certification program for Bakers Destination Based skill development training program
2	State Institute of Hotel Management Catering Technology & Applied Nutrition (SIHM), (Sponsored by State Government of respective state) and Food Craft Institutes (FCIs)	BSc in Hospitality & Hotel Administration Trade diploma in Food Production & Patisserie Craftsmanship Certificate course in Food Production & Patisserie Bakery & Confectionery training programme under Hunar Se Rojgar Tak scheme (HSRT) Skill testing and certification program for Bakers Destination Based skill development training program
3	Indian Culinary Institute (ICI) (Sponsored by Ministry of Tourism, Government of India)	BSc in Culinary Arts MSc in Culinary Arts

Source: Authors

started from the house itself, after doing short/long courses. The Ministry of Tourism, Government of India and Ministry of Labour, Government of India has started several skill based training programmes in the field of Bakery education. There are several training programmes organised by Government and private institutions which are sponsored by the State and Central Government. Along with the training input, the trainees are provided with free meals, uniforms, tool kit, course materials and stipend as well (Table 10.6).

10.5.5 Micro, Small and Medium Enterprises (MSME's) in Bakery

Due to augmented consumption of biscuits and cookies, and related bakery products, the Indian bakery market attained a net value of USD 7.60 billion in 2020. With an enhancement in the bakery training facilities, and female working

Table 10.7 Reasons to Invest in Indian Bakery Industry

Market Growth opportunities	The bakery sector in India is growing at a rate of 7–8%, per annum, which shows the enormous growth opportunity in the industry. The mammoth effort by the Ministry of Food Processing and Skill development has resulted in growth and expansion of bakery sector in the country.
Consumers food habits, and taste preferences	Indians prefer fried snacks such as assorted samosa, pakoras, kachori etc. however, this trend is changing rapidly, as people are moving towards bakery food options like sandwiches, puffs, pizzas and varieties of breads and biscuits.
Type of investment required	A bakery start-up does not require a very huge capital investment, or space requirement. It may be started on a small scale from home or maybe by renting a place for operation as small business unit(SBU).
Simpler business model	A complex business plan is not required for initiating a bakery operation. A team of qualified and trained personnel can do the job using simpler business model. Uses of organic ingredients, efficient tools and equipments, and modern technology, will be very useful to start a bakery.

Source: Authors

population, as well as the availability of healthier ingredients and end products in the bakery segment, the industry is expected to grow at a CAGR of 8.5% in the upcoming forecasted period of 2021–2026. The major regional markets for bakery are North India, West and Central India, South India, and East India. According to its market segmentation and distribution channel, the Indian bakery market can be segmented into artisanal bakeries, and online bakeries, supermarkets and hypermarkets and, independent retailers. There are few major players in the Indian Bakery industry which includes Britannia Industries Ltd., Parle Products Pvt. Ltd., Surya Food & Agro Ltd., and ITC Limited. (https://www.expertmarketresearch.com/reports/indian-bakery-market) (Table 10.7).

10.5.6 Entrepreneurial Development and New Start-Ups in Bakery

Modern is an iconic brand that pioneered the concept of bread, and literally created the bread category in India. Launched in 1965 as Modern Bakeries (India) Limited, the company was rechristened Modern Foods India Ltd. (MFIL) in 1982. For generations of Indians Modern Bread's iconic Blue & Orange Waxed Paper Bread pack was virtually synonymous with bread. Modern Foods was the first company to be privatised by the Government of India in 2000, and it was sold to Hindustan Unilever Ltd. (HUL).

In April 2016, Modern changed hands again. This time to Everstone Capital. Today, the company is called Modern Foods Enterprises Pvt. Ltd. and is 100% owned by Everstone Capital. Modern is now in an all new avatar. The re-launch of Modern Foods has been crafted with intensive efforts over the last 12 months, with significant investments in manufacturing, and R&D at Modern's new innovation hub in Chennai, as well as thorough efforts on consumer insights, strategy and design.

10.5.7 Sustainability Approach and Environmental Practices in Bakery Industry

Sustainability means different things to different people. For some, it's all about minimizing emissions or reducing the carbon footprint. For others, it's strengthening local economies and communities. Bakeries, and snack producers and other stakeholders are taking strong steps toward minimizing their eco-footprints while still maximizing growth. At present so many bakeries and related units are emphasizing more on organic or natural ingredients. Sustainability is no longer a "bonus" for any food manufacturer. It's a mandate, and a mandate that's coming directly from consumers. According to Nielsen's "Global Corporate Sustainability Report," up-and-coming consumers—namely, Millennials and Generation Z—are the most willing to pay extra for goods they perceive to be sustainable and environmental friendly. Large bakery chains are investing in waste management programs designed to convert non-recycled plastics into energy, which helps to meet the energy need of the plant as well as in fueling nearby areas. For the large bakery houses sustainability is all about a local focus. They procure their ingredients from local suppliers and maximize the usage of local ingredients as well to attain the food security. Large bakery houses are shifting their focus on sustainable and eco friendly packaging, labeling and brand building. Waste disposal management, cleanliness, Hygiene & Sanitation of the operational areas, Health Status of the employees and Visitors are the other important practices adopted by the Indian Bakery industry.

10.5.8 Laws Governing the Indian Food Industry

- *The Prevention of Food Adulteration Act, 1954*
- *The Fruit Products Order, 1955*
- *The Meat Food Products Order, 1973*
- *The Vegetable Oil Products (Control) Order, 1947*
- *The Edible Oils Packaging (Regulation) Order, 1998*
- *The Solvent Extracted Oil, De oiled Meal, and Edible Flour (Control) Order, 1967*
- *The Milk and Milk Products Order, 1992*
- *Essential Commodities Act, 1955 (in relation to food)*
- *The Food Safety and Standards Act, 2006 (FSSAI Licence)*

10.6 Challenges Faced by the Bakery Industry

The growth & development of any sort of industry invites its own challenges. The most important among all the challenges are diverse the production capacity to fulfill the demand and satisfy the tang of the young generation. In order to congregate the standards of food safety and human health, the bakery stores have to endow in enriching their facilities in more hygienic manner and also hiring skilled and knowledgeable manpower. Innovation and creativity for newer products is another major challenge for cut throat competition in the market. The need of raising awareness about the digital technologies and convenience of social media platforms among the unorganized bakery sector is another challenge to reach a wider market.

Major Challenges are as follows:

1. Government regulations – Like Food & Beverage outlets, bakeries have to obtain permits & licenses from Food Safety and Standard Authority of India (FSSAI), Municipal Board, etc (FICCI, 2007).
2. Brand Loyalty – Emergence of numerous bakeries in the market offering more or less similar products have raised confusion among the consumers. Thus making it difficult for the customers to maintain a loyal customer base, and brand loyalty.
3. Market Inflation Rate – The rate of market inflation, rising prices of major ingredients such as Refined flours, and other raw materials such as refined oils, margarine, butter, Food additives, fat and eggs has resulted in the rise of prices or reduction in profit margin.
4. Innovation and Creativity – Bakery industry demands lots of innovation and creativity in order to meet the demand of newer and better products. Bakeries unable to come up with innovative products and retain the quality standards may peril and shut down.
5. Concern about rising health issues – Saturated and Unsaturated Fats, Refined flour, bakery additives are claimed to be unhealthy for prolong consumption and may result in flabbiness and weight gain. However, the introduction of healthier form of raising agents and other variants like whole wheat flour(WWF) & uses of multigrain flour has resolved the issue up to larger extent.
6. Research & Development – Like every sector, the bakeries have to invest in research and development of newer and healthier products to attract and retain customers. R&D may help the bakeries to retain a good market share and compete in cut throat market.
7. Availability of Skilled Manpower – Skilled manpower has always been a major constraint in Catering Industry. Bakery industry being a highly skilled based sector has to face the shortage of skilled manpower resulting in incompatible products and dissatisfaction among the consumers.
8. Perishability in nature – The bakery items either raw or in finished form are perishable in nature. In case of improper storage conditions and non consumption of the bakery products, it will result in spoilage and loss of revenue.

Case Study: Karachi Bakers in Hyderabad

Karachi Bakery has emerged as a *"True Icon of Hyderabad Baking"* with a reputation that is unmatched to the core. Always a step ahead of competition, we focus on innovation, developing new products while improving our existing products to evolving consumer needs.

History: The bakery was founded by *Sri Khanchand Ramnani*, a Sindhi Hindu refugee who migrated from Karachi to Hyderabad in 1947 during the partition of India.

Logo title: *Karachi Bakery: Quality & Goodness since 1953.*

Located: Karachi Bakery is located in Hyderabad, Hyderabad, Telangana.

Bakery products: Biscuits, and cakes, gluten and sugar free cakes & biscuits, vegan biscuits, Pastries, cup cakes, chocolates, macron, artisan bread, rusks, special baked delights etc.

Quality: Quality is an integral part of Karachi Bakery's working philosophy. The quality norms are following with strict global standard hygiene and health norms throughout the preparation, baking, storing, packaging, and distribution of these bakery products. In line with USPH, HACCP and ISO standards, periodical audit are conducted and quality tests are constantly monitored.

Outlets: They have 23 stores in Hyderabad alone. Now Karachi bakery is in five cities Hyderabad, Bengaluru, Chennai, Mumbai and Delhi outlets at railway stations, airports, shopping malls, and retails shops, sweets and confectionary shops etc.

Marketing and Branding: The Karachi Bakery has proper focus on new edge marketing strategies including, bulk supply to businesses or corporate events or meets, gift pack or place a customized order, online shopping, E-mail and Telephonic orders, social media marketing etc. Branding and with brand logo and slogan with committed quality taste, attractive packaging, customise quantity and availability.

Source: https://www.karachibakery.com/home

10.6.1 Bakery in North Eastern Region of India

North East India comprises of 8 states, Arunachal Pradesh, Assam, Manipur, Meghalaya, Mizoram, Nagaland, Sikkim, and Tripura. The indigenous tribes of these north eastern states include bakery items in their daily diet and it is a part of their consumption culture. Due to demand and consumption NE states have lots of bakeries operating in the region. Most of the bakeries in North East produce almost all types of bakery products. A larger number of the bakeries uses both traditional and modern techniques for making their bakery products. As far as purchasing raw materials and ingredients is concerned majority of the bakeries procure them within state, and a slight majority even purchase locally and very few purchase from outside state. The bakery units choose to sell their finished products in their locality. While a few number sell their products within the district. The bakery units mostly market their products in their shops at the counter and few units distribute their products with help of distributor. The bakeries are mostly run by the family members as a family business and may consist of less than 5 employees which include skilled and unskilled labours. A comparatively lesser percentage of the bakeries advertise their products, and they prefer using newspapers and magazines as their mode of advertisement.

10.6.2 Future Strategies for Bakery Industry in India

Based on the present scenario it can be submitted that the available bakery units should be upgraded to cater the needs of their customers. Special schemes in terms of financial and Training & Development should be made available by the government to the traditional bakery units so that they can survive in the existing cut throat competition. More number of Training Institutions should be set up relating to bakery education where people mainly youngsters can be trained about modern and traditional techniques of operating a bakery. The bakery units should use different social media and e commerce platforms in promoting their business and increase their sales. More of Food testing laboratories should be set up and FSSAI officials should be deployed to manage the licensing and inspections of the bakery units. This strategy will help in checking their quality control measures and benchmarking (Tables 10.8 and 10.9).

10.7 Conclusion

Indian bakeries were set up as a result of the British influence during pre independence era. The Bakery products in India are in common use. In India, the bakery industry developed largely to meet the needs of the British army. Thus, this industry developed at first at the seaports of Calcutta and Bombay to carter to the need of foreign travellers and army personnel. Concepts like Wenger's in New Delhi and other metro cities provided the pace for the growth of bakeries and bakery cafés in the Independent India. During the post independence period, the growth of bakery industry was more scientific and with advance technology. Almost all the States in India has its own traditional bakery products, particularly in North East India; the traditional bakery industry continues to be popular. Bakery Industry plays a vital role in the economic development of the country, utilization of its grains and cereal

Table 10.8 Ten I's Future Roadmaps for Bakery Industry in India

1. Improvement	The improvement of bakery products from production to consumption, production level, bakery business, customer satisfaction.
2. Investment	There is urgent need to focus proper investment opportunities through public private investment
3. Income	The will be more focus increase the income of bakery producer, labours and entrepreneurs
4. Innovations	Continues bakery products development with innovative ideas and techniques.
5. Incentive	There will more need to incentivise this bakery sector to revive and
6. Ingredients	Availability of all essential ingredients in the country
7. Indigenous	Sustain and encourage local indigenous bakeries and theirs bakery products.
8. International	Global market reach and competitiveness of local bakery products.
9. Institutions	Establish new institutions of skill development in bakery products training research and capacity building of local youths and women.
10. Industry	Bakery Industry growth and development by increase production and profits margin of this industry.

Source: Authors

Table 10.9 Other Future Strategies of Bakery Industry

Taste	Maintain real and authentic Indian bakery taste, flavours.
Quality	More focus on quality of production and packaging
Preservation	Adopt preservation good and health conscious preservation methods
Health and hygiene	Better take care health and hygiene by providing pure, chemical free ingredients, sugar-free products, and natural
Packaging	New design of packaging for dry and moist bakery products i.e., cellophane, oriented polypropylene film, (OPP) combinations (pearlised or metallised), wrapped or fold wrapped
Distribution	More focus of bakery distribution in rural and urban market including international markets
Promotion and advertising	New methods and techniques showcase bakery products, digital and social media marketing some innovative promotional advertising.
Branding,	Branding through maintain quality, price, availability, packaging and health and hygiene standards.
Marketing	Effective marketing of bakery products by distribution channels.

Source: Authors

resources and in shaping the health of the citizens. The bakery Industry has the potential of generating additional employment and investment. The capital investment requirement per unit of additional employment is comparatively lower to most of the other capital-intensive industries in India. Latest Trends in the bakery and confectionery industry includes high safety and license regulations; E-retailing of bakery products; use of digital marketing for market expansion; standard packaging technology, Innovation in use of ingredients, aggressive expansion plans, and overall technological advancements.

References

Baisya, R. K. (2007). Bakery industry in India. *Processed Food Industry, 10*, 13.

Barbiroli, G. (2002). *Classification and standardization of bakery products and flour confectionery in relation to quality and technological Progress*. National Food Service Management Institute. Measuring success with standardized.

Bhagvat, N. B. (2001). *Study of Bakery Industry in Kolhapur district with special reference to production, financial and marketing problems*. Unpublished phd Thesis submitted to Shivaji University, Kolhapur.

Bhisel, S., & Kaur, A. (2013). Department of Food Science & Technology, Punjab Agricultural University, Ludhiana-141004, *International Journal of Advanced Scientific and Technical Research, 1*(3). ISSN 2249-9954

David, M. (2013). A study on the consumption pattern of bakery products in southern region of Tamil Nadu. *International Journal of Research in Commerce, IT & Management, 3*(02), 101.

Dubey, S. C. (2002). *Basic baking*. Society of Indian Bakery.

FICCI Study on Implementation of Food Safety and Standard Act. (2007). *An industry perspective report* (p. 3).

Hassan, N., Afifah Hanim, M., Pazil, N. S., & Razman, I. A. N. F. (2013). A goal programming model for bakery production. *Advances in Environmental Biology, 7*(1), 187–190.

Manley, D. (1991). *Technology of biscuits, crackers and cookies*. Woodhead Publishing Ltd.

Maric, A., & et. al. (2009). Contribution to the improvement of products quality in bakery industry. *International Journal for Quality research UDK, 378.014.3(497.11)*.

Martínez-Monzó, J., García-Segovia, P., & Albors-Garrigos, A. (2012). *Trends and innovations in bread, bakery, and pastry*. Journal Of Culinary Science & Technology.

Mohammed, A., Pawar, G. T., & Khan, I. D. (2008). Performance of bakery and confectionary food processing units in Maharashtra, India. *International Journal of Agricultural Science, 4*(1), 132–137.

Vanisha S. Nambiar And Shilpa Parnami (2008), Standardization and organoleptic evaluation of drumstick (Moringa Oleifera) leaves incorporated into traditional Indian recipes standardization recipes, Trees for Life Journal 3:2

Nare, P. D. (2002). *A study of bakery industry in Belgaum district (with special reference to financial and marketing problems)*. Unpublished Ph.D Thesis submitted to Shivaji University Kolhapur.

Oguntona, C. R. B., Odunmbaku, J. A., & Ottun, B. O. (1999). Proximate composition of ten standardized Nigerian dishes. *Nutrition & Food Science, 99*(6), 295–302.

Pradeep Raj, T., Ramkumar, R., & Subramani, A. K. (2015). Customer satisfaction towards good day biscuits, Avadi, Chennai. *International Journal of Multidisciplinary Management Studies*. ISSN 2249- 8834.

Pyler, E. J., & Gorton, L. A. (2008). *Baking science & technology* (Vol. 1).

Raut, T. S. (2018). A study of bakery businesses in the state of Goa. *International Journal of Research & Analytical Reviews, 5*(2), 908–912.

Sakhale, B. K., Chalwad, R.U. And Pawar, V. D. (2012) Standardization of process for preparation of fig-mango mixed toffee International Food Research Journal 19(3)

Selvaraj, A., Balasubramanyam, N., & Rao, H. (2002). Packing and storage studies on biscuit containing finger millet(ragi) flour. *J Food Sci Technol, 39*(1), 66–68.

Serivastava, A. (2009). *India per capita biscuit consumption*. July 04,. India Retailing Bureau.

James P. Smitha, Daphne Phillips Daifasa, Wassim El-Khourya, John Koukoutsisa & Anis El-Khourya (2010), Shelf life and safety concerns of bakery products—A review19-55

Townsend, G. M. (2001). Cookies, cakes and other flour confectionery. In D. A. V. In Dendy & B. J. Dobraszczyk (Eds.), *Cereals and cereal products: Chemistry and technology* (pp. 233–252). ASPEN Publishers, Inc..

U.S. Department of Agriculture, Food and Nutrition Service, & National Food Service Management Institute. (1996). Choice plus: a food and ingredient reference guide. :

Web References

http://timesofindia.indiatimes.com/business/india-business/Medium-scale-dairy-farming-an-emergingsector-trend-Rabobank-report/articleshow/51744088.cms. Retrieved on 25/07/2021

http://www.foodprocessingbazaar.com/articles/99-bakery-industry-present-and-future-prospects.html3. Retrieved on 15/07/2021

http://www.niir.org/books/book/bakery-industry-in-india-bread-biscuits-other-products-present-futureprospects-market-size-statistics-trends-swot-analysis-forecasts-upto-2017/isbn9789381039366/zb,,18b71,a,3,0,3e8/index.html2. Retrieved on 15/07/2021

http://www.researchandmarkets.com/reports/2041431/indian_bakery_industry_2011_20154. Retrieved on 11/07/2021

http://www.soyconnection.com/newsletters/soy-connection/food-manufacturing/articles/BakingIndustry-Trends-Consumers-Demand-Healthier-Baked-Goods. Retrieved on 25/07/2021

https://en.wikipedia.org/wiki/List_of_baked_goods

https://fmtmagazine.in/bakery-industry-in-india/

https://www.bizencyclopedia.com/article/an-overview-of-the-bakery-industry-in-india

https://www.expertmarketresearch.com/reports/indian-bakery-marketretrieved on 05/08/2021

https://www.thebetterindia.com/227895/india-best-bakeries-iconic-kayani-pune-kolkata-nahoum-glenary-wengers-delhi-ang136/. Retrieved on 19/08/2021

Part III
Economic Feasibility and Efficiency in the Bread Industry

Chapter 11
Measuring Baking Business Performance

Anatoliy G. Goncharuk

11.1 Introduction

Speaking about business performance, as a rule, it considers absolute and/or relative performance indicators that characterize various aspects of the business – individual processes, divisions and the business as a whole. For instance, Goncharuk (2014b) proposed a system of measuring business performance that includes three metrics of performance: key result indicators, productivity indicators and key performance indicators. This system covers all levels of management and permeates the entire enterprise from workplaces and departments up to senior management and owners, giving to external environment only general information about the performance, which is usually included in the annual and quarterly reports of the enterprise. This predominantly absolute approach enables to evaluate the performance of any business from the inside. However, sometimes it is important to look at a business from the outside and evaluate how well it is performing compared to other businesses in the industry. In this case, it talks about the relative performance of the business, which is important for the success in the competition and sustainability of the business (Chang et al., 2019). For the baking business, like any other food business, both absolute and relative performances are important, as it usually operates in a highly competitive environment.

In this chapter, we will look at a possible methodology for measuring performance for the baking business and consider cases of its application to bakeries from opposite ends of the European continent.

A. G. Goncharuk (✉)
Hauge School of Management, NLA University College, Kristiansand, Norway

J. M. Ferreira da Rocha et al. (eds.), *Baking Business Sustainability Through Life Cycle Management*, https://doi.org/10.1007/978-3-031-25027-9_11

11.2 Literature Review

Since performance is a prerequisite for survival and success, studies and methods for measuring business performance are very common in the research literature. However, till now there is not common approach to the understanding and measuring business performance.

So, Taouab and Issor (2019) wrote that common models for firm performance are the *Balanced Scorecards* (BSC), *Performance Prism* (Neely et al., 2001), *Malcolm Baldrige Model* (Wilson, 1997), and the *Performance Pyramid* (Wedman & Graham, 1998).

The BSC was developed by Kaplan and Norton (1992) as a tool to describe, develop and implement a firm's vision and strategy in the form of fixed goals and a clear set of financial and non-financial performance indicators. The introduction of BSC means that goals, indicators and strategic actions are linked to specific perspectives. The BCS transforms an organization's mission and strategy into a set of performance indicators that provide a model for measuring performance.

The model at Fig. 11.1 demonstrates the results of the organization's activities from four points of view: from a financial point of view, from the point of view of the client, from the point of view of innovation and learning, as well as from the point of view of the internal process.

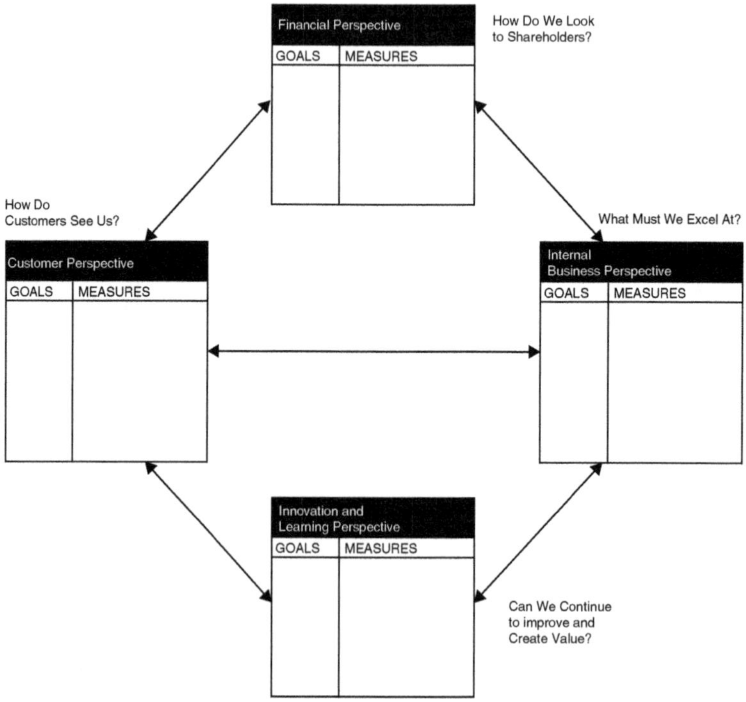

Fig. 11.1 The balanced scorecard links performance measures (Kaplan & Norton, 1992)

Nowadays BSC actively uses for performance measurement and management. So, Taye (2018) considered the BSC as the main tool for performance measurement. Apriansyah et al. (2019) used the BSC and key performance indicators (KPIs) for measuring small seed business performance. Bigliardi and Bottani (2010), using BSC, first identified the relevant financial and non-financial indicators suitable to be used for companies belonging to the food industry.

Furthermore, Munisamy et al. (2015) integrated the BSC approach with the *Data Envelopment Analysis* (DEA). The DEA was created by Charnes et al. (1978) to compute a relative efficiency of the businesses. Since the concept of DEA was proposed this method has been widely developed in the works of their followers, which is expressed in many different models of DEA, proposed by them. The original model developed by the authors of the method – DEA-model with constant returns to scale was called the CCR-model in honor of the founders. Later Banker et al. (1984) proposed a model with variable returns to scale, called the BCC-model. The CCR and BCC models are the basic DEA-models.

Nowadays there are many modifications of tasks and their solutions by DEA, whose selection is primarily determined by the nature of production technology of the analyzed companies. Existing models of the DEA can be classified according to:

(a) orientation of optimization, it distinguishes input-oriented model, i.e. models focused to minimize the inputs, output-oriented models – models aimed to maximize output, and the models without orientation;
(b) a scale effect and a type of efficiency frontier, it distinguishes CCR-models or CRS – constant returns to scale models, the BCC-models or VRS – variable returns to scale models, the NIRS-models – with non-increasing returns to scale models, NDRS-models – with non-decreasing returns to scale models;
(c) a form of the production function, it distinguishes linear, additive and multiplicative models;
(d) a measurement system, it distinguishes the models with a radial, non-radial and the hyperbolic efficiency measurement.

Various DEA models can calculate reserves for inputs reduction (Slack-Based model), consist a full ranking of the observed units (Super-Efficiency model), etc. Over the past decade, the DEA has been actively applied to various industries, e.g. for electricity distribution (Goncharuk et al., 2020a), natural gas distribution (Goncharuk & Lo Storto, 2017), winemaking (Goncharuk, 2017; Goncharuk, 2018; Goncharuk, 2019), higher education (Figurek et al., 2019), healthcare (Lo Storto & Goncharuk, 2017), dairy industry (Goncharuk, 2014a), etc. However, we have found only one study of the efficiency of baking industry using the DEA: applying the DEA, Malmquist productivity index (Goncharuk, 2013) and the non-parametric Kruskal–Wallis test (Goncharuk et al., 2020b), Chang et al. (2019) found increased productivity change of 22 self-owned stores of a famous bakery company in Taiwan from 2011 to 2016. Thus, DEA can be used both to measure the relative performance of different bakeries, and for divisions of the same bread producer.

The Malmquist productivity *index* (MPI) suggested by Caves et al. (1982) does not require any price data and is constructed by measuring the radial distance of the vectors of inputs and outputs in periods s and t with the appropriate technologies. As the distances can be focused on output and inputs, and MPI indexes divided into output-oriented M_o and input-oriented M_i:

$$M_o\left(x^s,y^s,x^t,y^t\right)=\left[\frac{D_o^s\left(x^t,y^t\right)}{D_o^s\left(x^s,y^s\right)}\cdot\frac{D_o^t\left(x^t,y^t\right)}{D_o^t\left(x^s,y^s\right)}\right]^{1/2} \tag{11.1}$$

$$M_i\left(x^s,y^s,x^t,y^t\right)=\left[\frac{D_i^s\left(x^t,y^t\right)}{D_i^s\left(x^s,y^s\right)}\cdot\frac{D_i^t\left(x^t,y^t\right)}{D_i^t\left(x^s,y^s\right)}\right]^{1/2} \tag{11.2}$$

where $D_o^s\left(x^t,y^t\right)$, $D_i^s\left(x^t,y^t\right)$ are, respectively, Shephard's (2015) output and input distance functions that reflect the distance from observation in period t to the technology in period s (Balk, 1993).

In fact, the MPI is the geometric mean of two indices of productivity, based on the technology of period t and s. So, if its value is greater than 1, there was a positive productivity growth between periods t and s. If it is 1, the productivity changes were absent, and if less than 1, the productivity was reduced. The Malmquist TFP index can be used to determine differences in performance between the two companies or one company between the two periods.

Search for ways to extract additional information for efficiency analysis led to the decomposition of MPI. Nishimizu and Page (1982) decomposed the growth of productivity into two components – efficiency change and technological change, and later Färe et al. (1994) have formalized this decomposition for the MPI as follows:

$$M_o=EC\cdot TC=\frac{D_o^t\left(x^t,y^t\right)}{D_o^s\left(x^s,y^s\right)}\cdot\left[\frac{D_o^s\left(x^t,y^t\right)}{D_o^t\left(x^t,y^t\right)}\cdot\frac{D_o^s\left(x^s,y^s\right)}{D_o^t\left(x^s,y^s\right)}\right]^{1/2} \tag{11.3}$$

where EC is the change of efficiency that equal to the ratio of technical efficiency in period t to the technical efficiency in period s (expression in square brackets); TC is the value of technological change that is equal to the geometric mean of change in technology between the two periods, evaluated in x^t и x^s(expression in square brackets to the power 1/2).

Change of technical efficiency (EC) takes place when the use of available labor, capital and other inputs that can produce more (less) amount of the same output. Graphically, it is expressed in the approximation or distancing of enterprise from the production frontier (Fig. 11.2).

Technological change (TC) is the development of new products and new technologies to improve the production methods that results in an upward shift of production frontier. This includes new production processes or new production methods

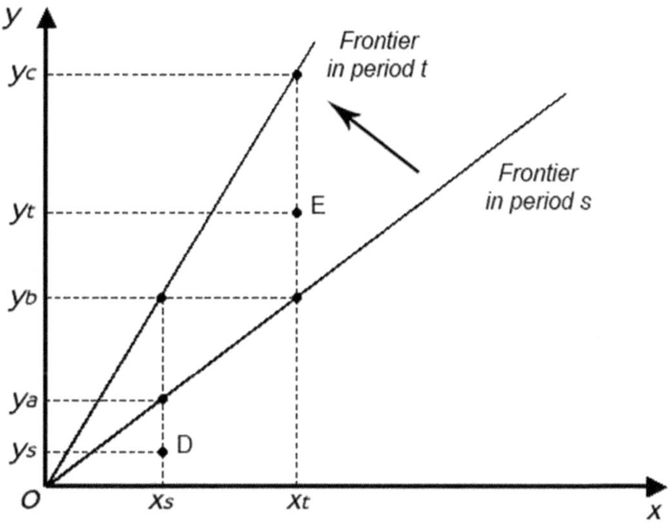

Fig. 11.2 The Malmquist TFP index and shift of production frontier (Figurek & Goncharuk, 2019)

that enable to accelerate the growth of output compared with the growth of input, and creation of new types of products (goods) (Squires & Reid, 2004).

Recently, Shahraki et al. (2021) applied the *Stochastic Frontier Analysis* (SFA) (see in Kumbhakar & Lovell, 2003) for assessing technical efficiency of the bakeries of one Iranian city and found out that bakers' age and experience had a negative relationship with their technical inefficiency, but their educational level had no significant effect. Actually, one of the main goals of the SFA is to estimate the impact of inefficiency, which is determined by the value of technical efficiency for enterprises of the sample by the following formula:

$$TE_i = \frac{\exp\left(\beta_0 + \sum_n \beta_n \ln x_{ni} + v_i - u_i\right)}{\exp\left(\beta_0 + \sum_n \beta_n \ln x_{ni} + v_i\right)} = \exp\left(-u_i\right). \tag{11.4}$$

This measurer of technical efficiency is limited by the interval [0…1] and evaluates the output of the *i*-th enterprise regarding the output, which can be produced by fully efficient enterprise using the same vector of inputs. However, Kumar and Singh (2020) also applied the SFA to estimate the productivity growth of Indian bakeries and defined that the bakery industry needs to define its innovation strategies, as these strategies lead to different outcomes that can be achieved only through the management of resources dedicated to the generation and implementation of innovations.

Considered above and the other methods for measuring business performance have different advantages and disadvantages described by Figurek and Goncharuk

(2019). Hence, depending on the study objectives and the available data, one method or another can be selected to measure the baking business performance.

11.3 Methodology

Measuring baking business performance depends on the objectives and the available data. So, we tried to select appropriate methods to demonstrate the opportunities of business performance measurement for different bakeries.

Based on experience of the authors in studying business performance, for this study the following 10-stages methodology for bakeries can be suggested (Fig. 11.3).

11.3.1 Stages for Internal Performance Measuring

Stage 1 – collecting the data on bakery internal business processes (internal business perspective), customers (customer perspective), cash flows (financial perspective), innovation and staff skills (innovation and learning perspective) to implement the BSC method.

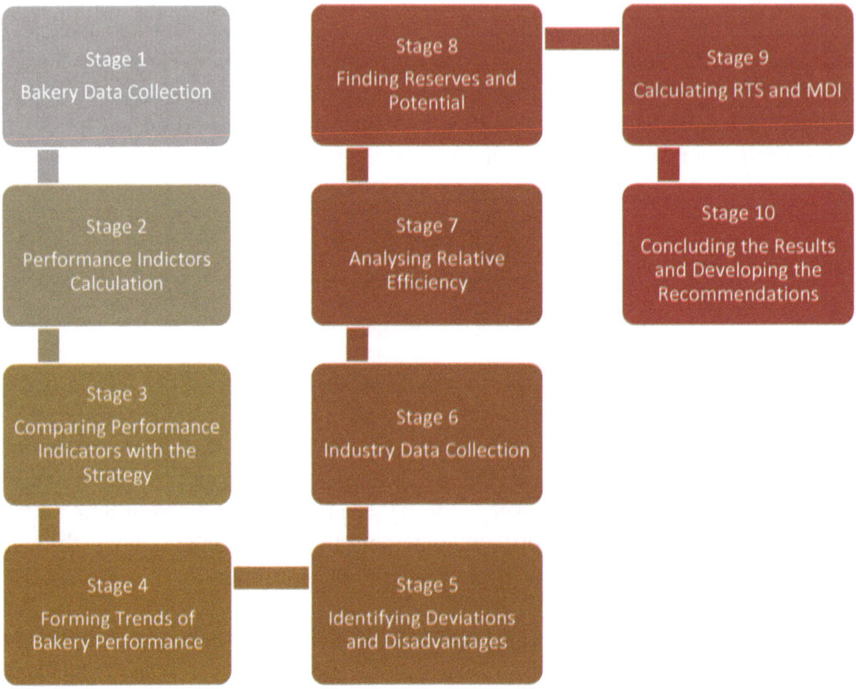

Fig. 11.3 Stages for studying baking business performance

Stage 2 – calculate the performance indicators for every of the four BSC perspectives.
Stage 3 – comparing performing indicators with the strategy and main goals of the bakery to realize to see if the bakery is heading in the right direction.
Stage 4 – comparing performing indicators with previous years (periods) to figure out the trends in bakery performance.
Stage 5 – identifying the deviations from the strategy and preparation of a list of the main disadvantages hindering the bakery performance improvement.

11.3.2 Stages for External (Relative) Performance Measuring

Stage 6 – collecting the data (inputs and outputs) from the other bakeries.
Stage 7 – analysing the relative efficiency of the baking businesses.
Stage 8 – finding the inputs reduction reserves and efficiency growth potential for the bakery.
Stage 9 – calculating returns to scale and the MPI for the bakeries.

11.3.3 Summarizing Stage

Stage 10 – concluding the measuring results and developing the recommendations for improving the baking business performance.

The Stages from 1 to 5 can be performed on the one bakery. Here it is proposed to apply the BSC method with key performance indicators, as well as the calculation of generalized performance indicators by years (periods), labour productivity, profitability, material-output ration, value-added, etc.

The Stages from 6 to 9 need the data from the other bakeries. The minimum required number of bakeries under consideration depends on the requirements of a particular efficiency analysis method that will be applied. The author's experience suggests that the DEA method is more comfortable and effective here. However, if a researcher has large data sets and tasks to separate the influence of certain efficiency factors, then the SFA method would be more preferable.

The Stage 10 is common and enables to develop appropriate recommendations for the baking business management for improving bakery performance.

11.4 The Case Study

To demonstrate how the above methodology works, we studied the case of bakeries in Portugal and Ukraine. These countries are on opposite ends of Europe and comparing them can show how different baking business performance can be.

Stage 1. Getting the data from bakery management including top-managers and financial manager, it was collected the set of indicators for three last years (2018–2020) that further used for calculating the performance indicators.

Stage 2. Based on the data collected from Portuguese bakery – *Pão de Gimonde*, we calculated the performance indicators for every of the four BSC perspectives (Fig. 11.4).

Stage 3. Comparing performing indicators with the strategy and main goals of the bakery showed that the bakery has a number of lacks. The main strategic goals of observed Portuguese bakery are making the innovative healthy products and world expansion with these products. However, foreign trade needs language and international trade skills from employees, which currently mainly speak Portuguese only. Besides, the bakery has a lot of equipment and bought more one during the pandemic that enable to produce a lot of innovative healthy food. However, it is not loaded, the production division works only several hours per day when sales go down. Hence, during the pandemic 2020 the bakery is not yet close to achieving its strategic goals. Nevertheless, new equipment can provide higher potential for productivity growth and production expansion.

Stages 4 and 5. Comparing main performing indicators with previous 2 years helped us to figure out the trends in bakery performance (Fig. 11.5).

The bakery hired employees and bought new equipment during the pandemic, when net sales decreased. This led to declined labour productivity and process efficiency. So, new staff and fixed assets did not bring better results that caused lower profitability and even losses. However, the bakery held consumers during the pandemic, but sales per consumer declined. Hence sales efficiency decreased. Moreover, the

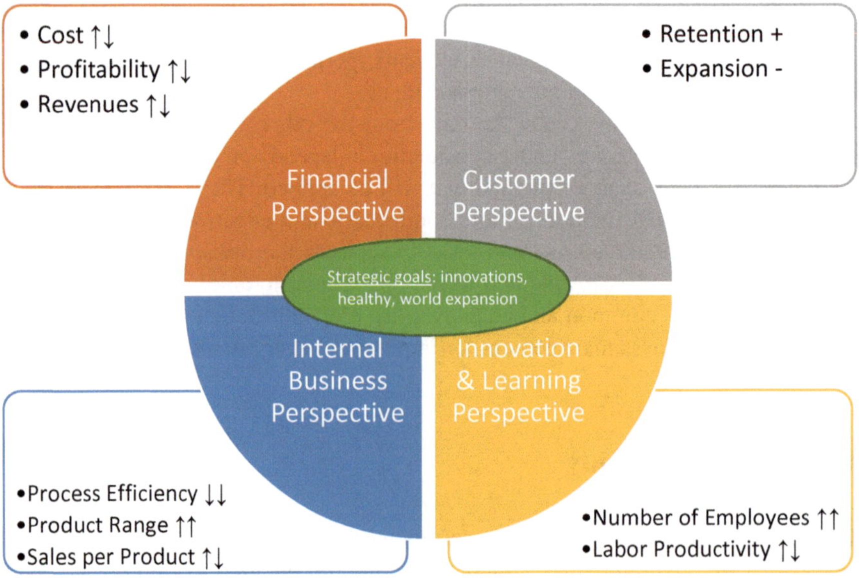

Fig. 11.4 Performance Indicators for Portuguese Bakery

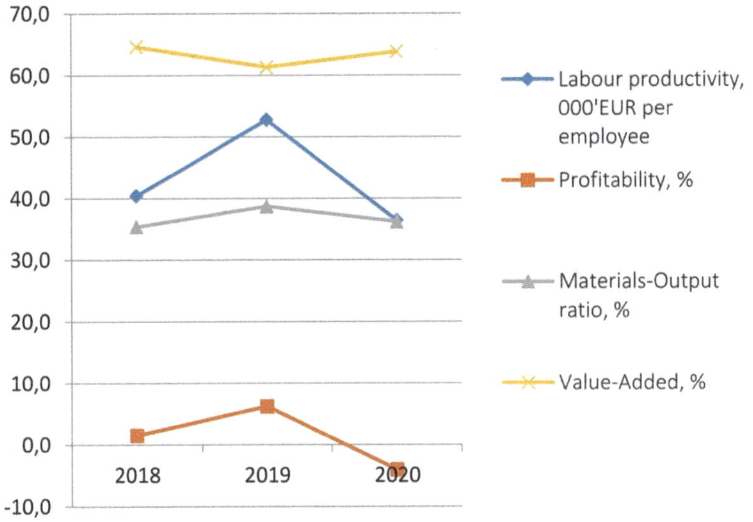

Fig. 11.5 Bakery's Business Performance Changes in 2018–2020

bakery's management declined materials-output ratio in crisis 2020 that helped it to increase value-added in per cent of net sales. Lower prices on fuel and right steps by management made it possible to decline material cost of the products. This helped the bakery to save working places and salaries on the same level. Nevertheless, jointly with the decline in sales, this led to a drop in labour productivity and losses.

Thus, the considered case demonstrates how, in a pandemic and crisis, not to be led by the economic interests of the owner, but to preserve the jobs of employees and social standards. This should provide good headroom for future productivity and business performance gains after the pandemic ends.

Stage 6. For the further analysis of relative efficiency, the data on inputs and outputs were collected from the annual reports of 25 Ukrainian bakeries. The reliability of all of them is confirmed by audit reports. A balanced panel sample for 2019 and 2020 was formed. Descriptive statistics of the sample are presented in Table 11.1.

As you can see, in 2020, there was a decrease in all sample averages. The sample includes 9 large bakery companies with over 250 employees and 16 SMEs. In general, the sample covers over 10% of the entire bakery business in Ukraine, which employed 8654 people in 2020 (in 2019–2220) and whose net sales exceed 207 million euros (in 2019 – almost 250 million euros).

The number of employees, material cost and depreciation were selected as inputs, because they reflect the basic resources for the production of bread: a labour, raw materials and fixed capital. Net sales were taken as an output, since they consolidate the total bakery production result.

Stage 7. Applying the CCR DEA-model, the efficiency score for every bakery of the sample has been calculated (Table 11.2).

Table 11.1 Descriptive statistics for bakeries' sample

Variables	Mean	Median	Stand. dev.
25 Ukrainian bakeries (2019)			
Number of Employees, people	449	190	133
Material cost, EUR	5400518	1317567	1548157
Depreciation, EUR	248321	37718	98230
Net Sales, EUR	9980680	1785862	3215003
25 Ukrainian bakeries (2020)			
Number of Employees, people	346	166	112
Material cost, EUR	4098662	1154546	1213965
Depreciation, EUR	222355	39626	81709
Net Sales, EUR	8297339	1438587	2567779

Table 11.2 Efficiency scores for Ukrainian bakeries in 2019 and 2020

	Efficiency score		
Bakery name	2019	2020	Change 2020/2019
Stryi bakery	0,536	0,428	−0,108
Kyivkhlib	0,852	0,770	−0,082
Toretsk bakery	0,618	0,619	+0,001
Mykolaiv bakery no. 1	0,959	1000	+0,041
Kyiv BKK	**1000**	**1000**	**0**
Concern Khlibprom	0,654	0,616	−0,038
Drohobych bakery	0,748	0,758	+0,01
Novograd-Volynsk bakery	0,713	0,797	+0,084
Ovruch bakery	0,508	0,396	−0,112
Korosten bakery	0,653	0,236	−0,417
Dnipropetrovsk Bakery no.9	0,581	0,556	−0,025
Dolyna bakery	1000	0,425	−0,575
Shostka bakery	1000	0,738	−0,262
Konotop bakery	0,548	0,547	−0,001
Izyaslav bakery	0,479	0,490	+0,011
Nikopol bakery	0,463	0,408	−0,055
Novomoskovsk bakery	0,735	1000	+0,265
Kryvyi Rih Bakery no. 1	0,719	0,481	−0,238
Kryvorizhkhlib	0,385	1000	+0,615
Pavlogradkhlib	0,417	0,388	−0,029
Bread (Slavyansk)	0,608	0,599	−0,009
CPF Roma	**1000**	**1000**	**0**
Enzyme company	0,996	1000	+0,004
TerA	0,468	0,404	−0,064
Cherkasykhlib LTD	**1000**	**1000**	**0**
Average	*0,674*	*0,619*	*−0,055*

As can be seen, only 3 bakeries stayed relatively efficient during 2019 and 2020: Kyiv BBC, CPF Roma, and Cherkasykhlib LTD. All of them represent large bakery business. However, if in 2019 only 5 bakeries of sample were efficient, then in 2020 already 7 of them had the highest efficiency scores. Nevertheless, the sample's average efficiency score has declined.

Separating the average efficiency by the size of the business, it can be noted that small bakery business having the highest average efficiency level in 2019 (0.853), in 2020 significantly lost in the efficiency level (0.727). At the same time, large business in 2020 significantly improved its efficiency on average: from 0.748 to 0.806. The medium bakery business in 2020 kept the average efficiency at the same low level as in 2019–0.564.

Thus, we can conclude that the COVID-19 pandemic in 2020 negatively affected only the relative efficiency of the small bakery business, while the big bakeries benefited from it. The relative efficiency of medium-sized bakeries, on the average, has not changed.

Considering the additional factor of business profitability, it can be noted that it slightly increased on average in the sample from 3.0% to 3.1%. However, while small businesses mostly suffered losses in 2020, large bakeries increased their profitability.

Analysing the above-mentioned Portuguese bakery jointly with the Ukrainian sample, its very efficiency scores were obtained: in 2019–1.000 and in 2020–0.922. Given its small size, such a decrease is consistent with the general trend of deterioration in the relative efficiency of baking SMEs during the COVID-19 pandemic.

Stage 8. Applying the DEA Slack-Based model, the inputs reduction reserves and efficiency growth potential for the bakeries have been calculated (Table 11.3).

The calculations showed that in 2020 the bakeries of the sample used significantly the reserves that they had in 2019, namely: the reserve for reducing material costs was used almost completely; the reserves on employees and depreciation were used up by about half. At the same time, the potential for growth in net sales has increased up to 27.9%, which allows maintaining a high potential for efficiency growth (34.5%) in the observed bakeries.

Comparing the average performance indicators for the Ukrainian sample with the same indicators for the Portuguese bakery (see Fig. 11.5), one can notice a twofold superiority of the latter in terms of profitability and threefold superiority in labour productivity in 2019. However, already in 2020, this international gap in labour productivity becomes less noticeable (only 52%), and the profitability of the

Table 11.3 The input reduction reserves and efficiency growth potential for the bakeries in 2019 and 2020

	Reserves of inputs reduction, %			Potential growth, %	
Year	Material cost	Depreciation	Employees	Output	Efficiency
2019	4.5	43.9	9.9	26.4	40.2
2020	0.2	20.7	4.9	27.9	34.5

Portuguese bakery has become completely negative (−6.1%), which indicates the need to take measures to overcome the efficiency crisis caused by the COVID-19 pandemic.

In addition, having superiority in labour productivity and profitability, the Portuguese bakery lags far behind Ukrainian bakeries in terms of capital productivity, investing significant funds in expensive equipment without adequate returns in the form of sales growth. However, it should be noted that the material-output ratio of the Portuguese bakery is significantly superior in comparison with the average Ukrainian level. This reflects the relatively higher value-added of Portuguese bread products, which reached 176% of the material input by the end of 2020. Only two Ukrainian bakeries of the sample in 2019 and three ones in 2020 had value-added above this level.

Stage 9. Applying CCR and BCC DEA-models, the returns to scale and MPIs for every observed bakery have been calculated (Table 11.4).

The calculation results show a weak positive trend in the total factor productivity for the observed sample. Despite the decrease in efficiency of the total sample (0.918), due to positive technological changes, the average MDI increased slightly. Likewise, the Portuguese bakery, having an increase in technology change (+ 4.1%), showed a decrease in MPI by 4.0% due to a drop in relative efficiency of 7.8%. However, having increasing returns to scale, it has a chance to increase efficiency and productivity by increasing sales in the post-pandemic period up to 7.8% as scale efficiency shows. Moreover, since the scale efficiency (0.922) of this bakery (Pão de Gimonde) coincides with the decline in its efficiency score (0.922), it can be argued that in 2019 it had the optimal size of operations and all its current inefficiency is associated only with a decrease in sales in 2020.

Thus, the calculated indicators enable to figure out what changes have occurred with the total factor productivity (MPI) in the sample as a whole and for individual bakeries, identify the reasons for these changes (due to efficiency change (EC) and/or technology change (TC)), as well as whether the size of this business is optimal (SE) and in which direction it needs to move for higher efficiency (RTS).

Stage 10. Summing up the results of the measuring baking business performance, it can make several recommendations for increasing the efficiency of both a separate bakery business (Pão de Gimonde) and for the bakery industry as a whole.

Having new equipment for innovative healthy baking production, observed bakery has a high potential for productivity growth and production expansion. However, for this, expanded foreign trade needs with hiring prospective staff with English/French/German/Spanish languages, communication and marketing skills that help this bakery to find and enter new markets for its products. This path will drive sales growth and bring the Portuguese bakery closer to the efficiency frontier, thanks to growing RTS and potential for scale efficiency gains.

In addition, the bakery should constantly look for new recipes and opportunities to renew products through healthier and more innovative products. The COVID-19

Table 11.4 MPI, scale efficiency and Returns to scale for the bakeries from 2019 to 2020

Bakery name	Malmquist productivity index, MPI	Efficiency change, EC	Technology change, TC	Scale efficiency, SE	Returns to scale, RTS
Stryi bakery	0,837	0,799	1048	0,816	↗
Kyivkhlib	0,966	0,905	1068	0,770	↘
Toretsk bakery	1027	1003	1024	0,975	↗
Mykolaiv bakery no. 1	1414	1042	1357	1000	→
Kyiv BKK	0,974	1000	0,974	1000	→
Concern Khlibprom	0,998	0,942	1060	0,616	↘
Drohobych bakery	1157	1014	1141	0,910	↘
Novograd-Volynsk bakery	1026	1118	0,918	0,994	↗
Ovruch bakery	0,887	0,781	1137	0,926	↘
Korosten bakery	0,390	0,361	1080	0,767	↗
Dnipropetrovsk Bakery no.9	0,910	0,957	0,951	0,996	↗
Dolyna bakery	0,515	0,425	1212	0,503	↗
Shostka bakery	0,642	0,738	0,870	0,885	↗
Konotop bakery	0,929	0,998	0,930	0,980	↗
Izyaslav bakery	1077	1024	1051	0,991	↗
Nikopol bakery	0,920	0,882	1043	0,995	↗
Novomoskovsk bakery	1885	1361	1385	1000	→
Kryvyi Rih Bakery no. 1	0,992	0,670	1481	0,481	↗
Kryvorizhkhlib	4816	2601	1852	1000	→
Pavlogradkhlib	0,970	0,930	1043	0,959	↗
Bread (Slavyansk)	1080	0,985	1096	0,954	↘
CPF Roma	1083	1000	1083	1000	→
Enzyme company	1169	1008	1160	1000	→
TerA	0,880	0,864	1018	0,987	↗
Cherkasykhlib LTD	0,869	1000	0,869	1000	→
Pão de Gimonde	0,960	0,922	1041	0,922	↗
Average	*1005*	*0,918*	*1095*	*0,885*	↗

pandemic not only forced consumers to count their money, but also seriously forces them to think about health, primarily through healthy eating. However, it should be noted that bread is not given much space in health-improving diets. Hence, the bakery needs to develop and implement innovations that can make bread products healthy and affordable for key consumer groups. Here help can be provided by the SOURDOMICS research project and technologists who create healthy food products, as well as marketers who are able to identify the necessary segments and convince their representatives to purchase products of the bakery.

Considering the industry as a whole, one can note some kind of discrimination against small bakery business during a pandemic, which does not receive appropriate support from the state and is forced to reduce production volumes, lacking such financial, communication and resource opportunities as a large bakery business has. Small bakeries only need to switch to a survival strategy, reducing purchases, staff and production. Successful in non-pandemic times, they desperately need the support of the authorities, as their business creates jobs and closer to customers, feeling their needs and wishes better than others.

11.5 Conclusions

Measuring baking business performance enables to identify weaknesses and strengths and prepare appropriate management decisions that contribute to increased efficiency and business development. After presenting the existing methods from appropriate literature, a 10-stages methodology was suggested for measuring and analysing the performance of bakery business. The work of the proposed methodology was demonstrated on the example of existing 1 Portuguese and 25 Ukrainian bakeries according to data for recent years, including 2020, the year of the onset and spread of the COVID-19 pandemic.

An in-depth study of the efficiency of the Portuguese bakery Pão de Gimonde made it possible to figure out the reasons for the decline in its efficiency in 2020. Comparison of the efficiency indicators of this bakery with a sample of Ukrainian bakeries shows that there is a common problem that led to a decrease in efficiency in the small bakery business. Moreover, it was identified that small bakeries, having the highest average efficiency level in 2019, in 2020 significantly lost in the efficiency level. However, large business in 2020 significantly improved its efficiency on average, and the medium bakery business in 2020 kept the average efficiency at the same low level as in 2019.

Even the use of reserves to reduce material, capital and labour costs did not save the sample bakeries from a decrease in efficiency in 2020. However, due to technological changes, their total factor productivity even increased slightly (by 0.5%), which adds optimism when assessing the prospects of the bakery industry in two countries. An in-depth analysis of efficiency revealed that the Portuguese bakery has a high potential for productivity growth and production expansion, thanks to new equipment for innovative production of healthy baked goods. However, mismanagement in human resources management and marketing prevented this potential from being realized in 2020 and resulted in losses for this bakery. However, the Portuguese bakery, as well as most of the observed Ukrainian bakeries, showed increasing returns to scale, which gives good prospects for the developing and improving baking business performance in these countries.

Developed recommendations for improving bakery business performance include constantly looking for new recipes and opportunities to renew products through healthier and innovative products, as well as covering new segments and markets to sale bakery products. Moreover, unequal conditions for small and large bakeries during the pandemic led to a significant reduction and even the closure of a number of small, but very important for society and the economy, businesses, which create jobs and are friendly to consumers. Therefore, appropriate recommendations were made to the government to support small bakeries and the management of such bakeries according to a strategy during the pandemic.

References

Apriansyah, M., Sukiyono, K., & Chozin, M. (2019). Performance measurement of small breeding business in North Bengkulu regency: Application of balanced scorecard (BSC) method. *Journal of Agri Socio-Economics and Business, 1*(2), 59–72.

Balk, B. M. (1993). Malmquist productivity indexes and fisher ideal indexes: Comment. *The Economic Journal, 103*(418), 680–682.

Banker, R. D., Charnes, A., & Cooper, W. W. (1984). Some models for estimating technical and scale inefficiencies in data envelopment analysis. *Management Science, 30*(9), 1078–1092.

Bigliardi, B., & Bottani, E. (2010). Performance measurement in the food supply chain: A balanced scorecard approach. *Facilities, 28*(5/6), 249.

Caves, D. W., Christensen, L. R., & Diewert, W. E. (1982). Multilateral comparisons of output, input, and productivity using superlative index numbers. *The Economic Journal, 92*(365), 73–86.

Chang, C. W., Wu, K. S., Chang, B. G., & Lou, K. R. (2019). Measuring technical efficiency and returns to scale in Taiwan's Baking Industry—A case study of the 85 °C company. Sustainability, 11(5), 1268.

Charnes, A., Cooper, W. W., & Rhodes, E. (1978). Measuring the efficiency of decision making units. *European Journal of Operational Research, 2*(6), 429–444.

Färe, R., Grosskopf, S., Norris, M., & Zhang, Z. (1994). Productivity growth, technical progress, and efficiency change in industrialized countries. *The American Economic Review, 84*(1), 66–83.

Figurek, A., & Goncharuk, A. G. (2019). *Agri-food management : Textbook.* Phoenix.

Figurek, A., Goncharuk, A., Shynkarenko, L., & Kovalenko, O. (2019). Measuring the efficiency of higher education: Case of Bosnia and Herzegovina. *Problems and Perspectives in Management, 17*(2), 177–192.

Goncharuk, A. G. (2013). What causes increase in gas prices: The case of Ukraine. *International Journal of Energy Sector Management, 7*(4), 448–458.

Goncharuk, A. G. (2014a). Competitive benchmarking technique for "the followers": A case of Ukrainian dairies. *Benchmarking: An International Journal, 21*(2), 218–225.

Goncharuk, A. G. (2014b). Measuring enterprise performance to achieve managerial goals. *Journal of Applied Management and Investments, 3*(1), 8–14.

Goncharuk, A. G. (2017). Exploring the factors of efficiency in German and Ukrainian wineries. *Journal of Wine Research, 28*(4), 294–312.

Goncharuk, A. G. (2018). Wine business performance benchmarking: A comparison of German and Ukrainian wineries. *Benchmarking: An International Journal, 25*(6), 1864–1882.

Goncharuk, A. G. (2019). Winemaking performance: Whether the crisis is over. *British Food Journal, 121*(5), 1064–1077.

Goncharuk, A. G., & Lo Storto, C. (2017). Challenges and policy implications of gas reform in Italy and Ukraine: Evidence from a benchmarking analysis. *Energy Policy, 101*, 456–466.

Goncharuk, A. G., Horobets, T., Yatsyshyn, V., & Lahutina, I. (2020a). Do high tariffs provide high efficiency: A case of Ukrainian electricity distribution companies. *Polityka Energetyczna, 23*(3), 125–134.

Goncharuk, A. G., Lewandowski, R., & Cirella, G. T. (2020b). Motivators for medical staff with a high gap in healthcare efficiency: Comparative research from Poland and Ukraine. *The International Journal of Health Planning and Management, 35*(6), 1314–1334.

Kaplan, R. S., & Norton, D. P. (1992). The balanced scorecard—Measures that drive performance. *Harvard Business Review, 70*(1), 71–79.

Kumar, S., & Singh, C. (2020). Productivity growth in India's bakery manufacturing industry. *Journal of Agribusiness in Developing and Emerging Economies.* https://doi.org/10.1108/JADEE-12-2019-0204

Kumbhakar, S. C., & Lovell, C. K. (2003). *Stochastic frontier analysis*. Cambridge University Press.

Lo Storto, C., & Goncharuk, A. G. (2017). Efficiency vs effectiveness: A benchmarking study on European healthcare systems. *Economics & Sociology, 10*(3), 102–115.

Munisamy, S., Fon, C. Z., & Wong, E. S. (2015). Innovation and technical efficiency in Malaysian family manufacturing industries. *Journal of Economic & Financial Studies, 3*(04), 50–67.

Neely, A., Adams, C., & Crowe, P. (2001). The performance prism in practice. *Measuring Business Excellence, 5*, 6–13.

Nishimizu, M., & Page, J. M. (1982). Total factor productivity growth, technological progress and technical efficiency change: Dimensions of productivity change in Yugoslavia, 1965-78. *The Economic Journal, 92*(368), 920–936.

Shahraki, J., Sardar Shahraki, A., Ali Ahmadi, N., & Radmehr, Z. (2021). The economic assessment of the production and technical efficiency of bakeries with a focus on social factor using stochastic frontier analysis. *Iranian Economic Review*. https://doi.org/10.22059/ier.2021.82971

Shephard, R. W. (2015). *Theory of cost and production functions*. Princeton University Press.

Squires, D., & Reid, C. (2004). Using Malmquist indices to measure changes in TFP of purse-seine vessels while accounting for changes in capacity utilisation, the resource stock and the environment. SCTB17 forum Fisheries Agency. In *SCTB17 forum Fisheries Agency, working paper* (pp. 1–15).

Taouab, O., & Issor, Z. (2019). Firm performance: Definition and measurement models. *European Scientific Journal, 15*(1), 93–106.

Taye, G. T. (2018). *Balanced scorecard as the main tool for strategic mapping and performance measurement: Literature review* (Working paper no. DML/10/00103/2/2018). University of Africa.

Wedman, J., & Graham, S. W. (1998). Introducing the concept of performance support using the performance pyramid. *The Journal of Continuing Higher Education, 46*(3), 8–20.

Wilson, D. D. (1997). *An empirical study to test the causal linkages implied in the Malcolm Baldrige National Quality Award*. The Ohio State University.

Chapter 12
Forming the Efficient Business Model for Bakery

Anatoliy G. Goncharuk, Aleksandra Figurek, and Marinos Markou

12.1 Introduction

Being a highly competitive industry, bakery is developing despite any crises and pandemics. According to IMARC (2022), having a value of US$ 478.4 billion in 2021, the global bakery products market can reach US$ 612.4 billion by 2027 with 4.2% of annual growth rate. However, to survive in a highly competitive and increasingly sophisticated consumer environment, bakeries need efficient business models to create value and grow sustainably.

The concept of a business model is widely used. However, it often has different meanings. Examples of the most interesting definition of business model in research publications are presented in Table 12.1.

Such a diverse understanding of the business model that acts as a design, *architecture,* level, a mechanism, plan, logic, the way, recipe, etc. makes this concept multi-valued and, as Baden-Fuller and Morgan (2010) noted, multiple role concept. However, what most authors agree on is that such a model should create value for its stakeholders, with a reasonable cost and high profit, i.e., it should be ***efficient***.

The typology of business models includes several dependent types. So, Schweizer (2005) distinguished the following types of business model (Fig. 12.1):

Integrated model that uses existing value chain and is characterized by low market power of innovators vs owners' complementary assets and high total revenue potential, e.g. ExxonMobil, Procter&Gamble, Nestle models;

A. G. Goncharuk (✉)
Hauge School of Management, NLA University College, Kristiansand, Norway

A. Figurek
Gnosis Mediterranean Institute for Management Science School of Business, University of Nicosia, Nicosia, Cyprus

M. Markou
Department of Agric. Economics, Agricultural Research Institute, Nicosia, Cyprus

J. M. Ferreira da Rocha et al. (eds.), *Baking Business Sustainability Through Life Cycle Management*, https://doi.org/10.1007/978-3-031-25027-9_12

Table 12.1 Definitions of business model

No.	Authors	Definition
1	Höflinger (2014)	It is the design of organizational structures for converting technological potentials into economically valuable outputs by exploiting business opportunities
2	Zott and Amit (2010)	It is the level of aggregation (of activities) affecting the ability to create and capture value
3	Gambardella and McGahan (2010)	*It is a mechanism for turning ideas into revenue at reasonable cost*
4	Casadesus-Masanell and Ricart (2010)	*It is the logic of the firm, the way it operates and how it creates value for its stakeholder*
5	Teece (2010)	*It is how a firm delivers value to customers and converts payment into profits*
6	Timmers (1998)	*It is an architecture for the products, service and information flows, including a description of the various business actors and their roles*
7	Itami and Nishino (2010)	*It is a profit model, a business delivery system and a learning system*
8	Demil and Lecocq (2010)	*It is the way activities and resources are used to ensure sustainability and growth*
9	Venkatraman and Henderson (1998)	*It is a coordinated plan to design strategy along three vectors (customer interaction, asset configuration, and knowledge leverage)*
10	Chesbrough (2007)	It is the "heuristic logic" that converts ideas and technical potential into economic value
11	Baden-Fuller and Morgan (2010)	It is not a recipe *or* scientific model *or* scale and role model, but can play any – or all – of these different roles for different firms and for different purposes: And will often play multiple roles at the same time

Orchestrator model that uses innovating value chain and is characterized by low market power of innovators vs owners' complementary assets and high total revenue potential, e.g. Nike, Adidas, Benetton models;

Layer player model that uses existing value chain and is characterized by high market power of innovators vs owners' complementary assets and low total revenue potential, e.g. Intel, Microsoft, Valmet models;

Market maker model that uses innovating value chain and is characterized by high market power of innovators vs owners' complementary assets and low total revenue potential, e.g. Amazon, Autobytel models.

However, recently Baden-Fuller (2021) divided all the business models on the Dyadic and Triadic types that further may be subdivided into sub-types like Dyadic-product, Dyadic-(servitized)-solution, Triadic-market-matchmaker, and Triadic-multisided-platform. Every of these types has a certain set of features regarding the way that value is created and the potential revenue-profit pools that can be exploited. At the same time, any business model must be ***efficient***, otherwise the company that uses it will fail sooner or later.

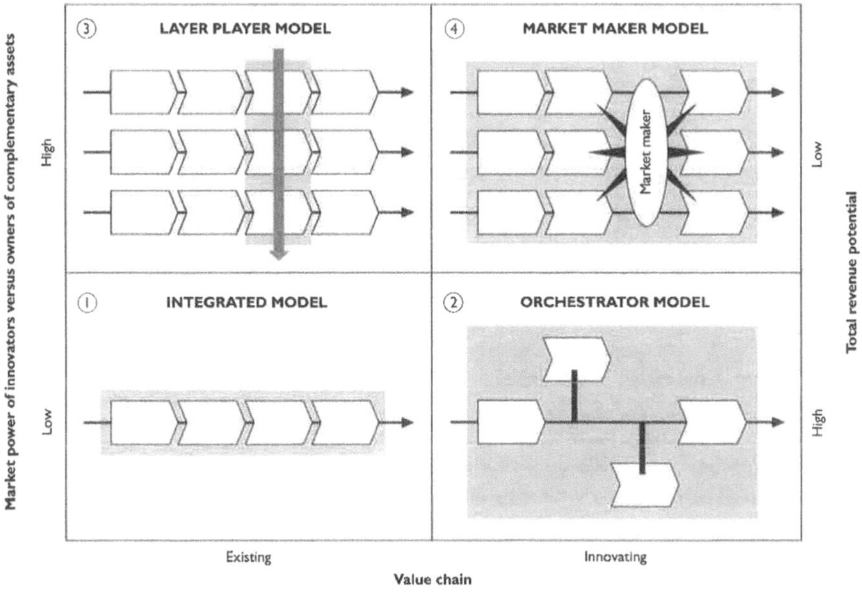

Fig. 12.1 Typology of business models (Schweizer, 2005)

Therefore, this study is aimed at creating theoretical foundations and developing practical recommendations for the formation of efficient business models in one of the most important and developing industries - baking.

12.2 Efficiency of Business Models

In the research literature and practice, there are several approaches to the definition and understanding of the concept of efficiency. Among them we can distinguish two key approaches to realising efficiency that can be applied to business models.

The first one is *economic approach*, when efficiency is considered as the ability of business to produce goods and services with minimum possible wasting materials, time, or energy. This ability makes it possible to maximize profits for a business operating in a competitive environment, which is bakery. Indeed, working in a competitive market does not allow for a significant influence of producers on price and income. However, limiting the cost of time and resources, by minimizing their waste, can lead the business model to higher profit and success. In turn, the economic approach has technical, scale and allocative components that allow you to control and identify reserves to reduce resources and time, as well as optimize a scale of business and a range of goods and services produced. This approach is widely used in assessing the relative efficiency of a business in comparison with competitors and companies operating in the same market. In particular, Kumar and Singh (2020) used it to evaluate the efficiency of the Indian bakery.

The second one is *managerial approach*, when efficiency is considered as the ability of business to achieve its goals. In this approach, the business model can be considered efficient if it leads to the achievement of the main business objectives. However, one of the main goals of business is to make a profit, which links the managerial approach with the economic approach. At the same time, while having common end goals, both approaches have different means of achieving them. In particular, the managerial approach deals with strategy, organizational behavior, staff motivation and other managerial functions, processes, and tools to ensure business efficiency. Considering managerial approach, it can be noted that Wilson et al. (2018) conducted systematic literature review and suggested a conceptual governance model for exploring the efficiency of business modeling to occupy the missing link between business strategy, processes, and software tools.

In the process of forming an efficient business model, reasonable questions arise, like how efficient is the business model and how to determine it? This requires efficiency criteria and appropriate measurement methods. In general, we can divide all methods of measuring the efficiency into individual and relative ones.

Individual methods are associated with internal evaluations of business performance and can be structural, dynamic, and static. These include various financial ratios based on the results of the business, for example, liquidity, profitability, etc. If they are evaluated for one period for the business, then these are static indicators. If they are compared for two or more periods, then they are dynamic. The ratios of the work of individual departments or business processes are structural indicators. The most appropriate examples of individual indicators for measuring efficiency of business model can be found in Table 12.2.

The collection and calculation of financial data is mainly carried out by the accounting and/or analytical department and is reflected in the form of aggregated indicators in the financial statements. Non-financial indicators, as a rule, are included in the KPI system, and are also elements of the business strategy and are controlled as part of business performance management, for example, using a Balanced Scorecard.

Methods of measuring the relative efficiency of business models are mostly statistical ones. Generally, they can be divided into frontier and non-frontier methods. The essence of the frontier methods of measuring efficiency is in the fact that the

Table 12.2 Individual indicators for measuring efficiency of business model

No	Groups of indicators	Examples of indicators	Examples of applying
1	Financial ratios	Net profit margin, accounts receivable turnover, accounts payable turnover, current liquidity ratio, inventory turnover	Goncharuk and Karavan (2013), Tsiouni et al. (2022), Zimon and Tarighi (2021)
2	Non-financial indicators	Employees' productivity rate, customer satisfaction rate, employee engagement index, product range	Bayne and Wee (2019), Bini et al. (2018), Mota et al. (2020)

business model efficiency is estimated in relation to the efficiency frontier or production possibility curve, which is determined by the most efficient businesses represented in the sample. In contrast to the frontier methods, non-frontier methods are based on a comparison with some sample average level, defined by calculating the index or by using the least squares method. In research and business practice both kinds of methods are widely used.

Besides, due to other classification of the methods of measuring efficiency can be divided into parametric and nonparametric ones, which include both frontier and non-frontier methods.

Parametric methods are based on econometric analysis and require the determination of the functional form of the production function of the business (or a function of costs, or profits). To create this function the regression analysis is used. The following are the most common parametric methods:

1. Ordinary least squares.
2. Corrected ordinary least squares.
3. Stochastic frontier analysis.

Ordinary Least Squares (OLS) is a common statistical parametric method of non-frontier analysis, which enables to estimate the average production function, or an average cost function for the group (sample) of homogeneous businesses. As well as all parametric methods, OLS requires the determination of the functional form of the simulated function.

General view of the model for the cost function is as follows:

$$C_i = F(Y_i, w, z, \beta)\exp(v_i) \tag{12.1}$$

where C_i is the cost of the company I; Y_i is the vector of output (produced by the company i); w is a vector of input prices for production factors (materials, labor, etc.); β is the vector of estimated parameters; z is external environmental factors (climatic, political, etc.) that are not controlled by the company; v_i is an error term (OLS residual).

As a measure of cost efficiency of the company OLS uses the difference between its actual costs and evaluated average costs (the value of average cost function by substituting in it the production volumes, prices of production factors and environmental factors for the company). This method enables to estimate the statistical significance and influence of the factors included in the model on the function of cost (production). So, it can be used to make decisions about the reallocation of production factors, changes in the functioning of the environment, etc., in accordance with the objectives of the company.

OLS method is one of the oldest methods of modern statistics, which was first published by Legendre (1805). During its two-century history it has received wide development and distribution in various fields of science and public life. The simplest and most frequently used form of the OLS is based on the linear regression associated with the problem of finding the line (curve), which is as close as possible to all the elements of a data set, i.e., is an average function that minimizes the sum of deviations between the theoretical and empirical values of the dependent variable.

Modern technique of OLS to measure the business model efficiency includes the following steps:

1. the choice of the indicator of cost (or production) and the external factors;
2. evaluation of cost function (or production function) for a sample of business models (companies);
3. calculation of an efficiency ratio for each business model (company) within a sample (Berg, 2006).

Corrected ordinary least squares (COLS) is frontier method that is derived from the OLS. In contrast to the OLS method COLS is not based on the determination of the average function, but on the construction of the efficiency frontier (Fig. 12.2).

This method was first proposed by Winsten (1957) as a procedure for constructing a deterministic model of the production function, which includes 2 phases:

1. by using the OLS we obtain the consistent and unbiased scores of the slopes of the lines and the consistent but biased value of the intercept;
2. we shift biased OLS intercept for such value that allows us to limit the data above, i.e. establish their upper frontier.

Subsequently the COLS method has been adapted and is justified to estimate the relative efficiency of companies (Aigner et al., 1977). It is assumed that at least one company in the sample is on the efficiency frontier: for the cost function it is a company with the largest negative value of the error term; for the production function it is a company with the most positive value of the error term. OLS line is shifted by this value (see Fig. 12.2) so that COLS line passes through the point corresponding to the "efficient" company and are the efficiency frontier for all other companies.

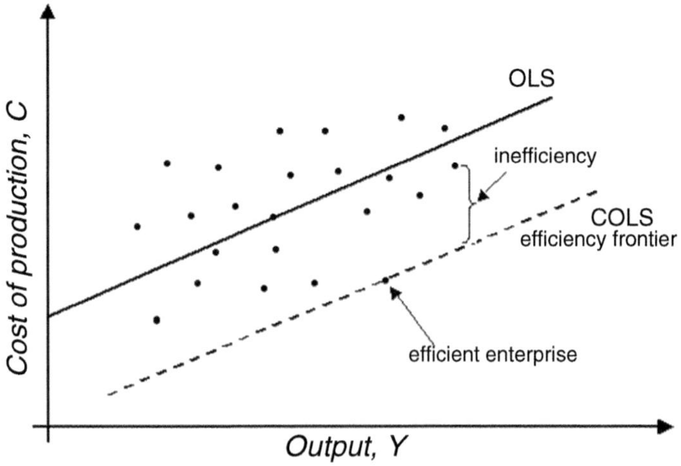

Fig. 12.2 OLS and COLS lines for evaluation of the cost efficiency

Then, for all other companies in the sample, the deviations from the efficiency frontier explain their ineffectiveness. Efficient company is assigned an efficiency rate equal to 1, and the coefficient of technical efficiency of any other company respectively for a cost function and production function is defined by (12.2–12.3):

$$TE_i = \exp\left\{\hat{v}_i - \min_i\left\{\hat{v}_i\right\}\right\} \quad (12.2)$$

$$TE_i = \exp\left\{\hat{u}_i - \max_i\left\{\hat{u}_i\right\}\right\} \quad (12.3)$$

where $\hat{v}$, $\hat{u}$ are the error terms of OLS respectively for a cost function and production function; $\hat{v}_i - \min_i\left\{\hat{v}_i\right\}$, $\hat{u}_i - \max_i\left\{\hat{u}_i\right\}$ are error terms of COLS respectively for a cost function and production function.

Thus, COLS is biased average function of OLS, the shift of which depends on the gap between the highest and average level of efficiency in the considered sample of companies. COLS suggests that: at least one company from the sample is efficient; there is no measurement error; the marginal effects of cost carriers are similar for the highest efficient companies and the average efficient, i.e. when constructing the COLS line we shift OLS line vertically without changing the slope (gradient). Hence the COLS technique is completely dependent on the OLS method, and all disadvantages of the OLS computational procedure of the COLS automatically takes over.

Moreover, a parallel shift of the function line from the OLS to the COLS denotes that the structure of production technology of the best (efficient) company is similar to the structure of production technology in the sample average, i.e. the average company. This drawback limits the accuracy of the COLS procedure, since it is likely that the structure of production technology of the best business model can differ from the production technologies of other less efficient models of the sample. This fact casts doubt on the accuracy of measuring efficiency obtained by the COLS method.

The method of constructing the production frontier and calculating the companies' efficiency scores by means of these stochastic models is named *Stochastic frontier analysis (SFA)*. Coelli et al. (2005) defined SFA as "… method for frontier evaluation that assumes a given functional form for the relationship between inputs and an output".

Considering the log-linear form of Cobb-Douglas function, the model of stochastic production frontier can be written as follows:

$$\ln y_i = \beta_0 + \sum_n \beta_n \ln x_{ni} + v_i - u_i \quad (12.4)$$

where y_i represents the output of the i-th company; β_0,β_n are unknown parameters; x_{ni} is n-th input of the i-th company; v_i is statistical noise; u_i is non-negative random variable associated with technical inefficiency of the i-th company (Kumbhakar & Lovell, 2003).

Model (13.4) is called stochastic frontier production function, because value of output is limited to the top of the stochastic variable $\exp\left(\beta_0 \sum_n \beta_n \ln x_{ni} + v_i\right)$.

Random variable v_i can be positive or negative, which provides a deviation of stochastic production frontier from the deterministic part of the model $\exp\left(\beta_0 \sum_n \beta_n \ln x_{ni}\right)$.

The most important properties of the stochastic frontier model can be represented graphically by the example of companies that produce one output y_i from one input x_i (Fig. 12.3).

Among all the considered business models the models A and B are selected. For them the inputs and outputs are defined respectively as x_A,x_B and y_A,y_B. If these models would be efficient, i.e., the impact of inefficiency is absent (u_A= 0 и u_B= 0), then their position on the diagram would change accordingly from A and B to A' and B'. The location of A' over the deterministic part of production frontier indicates the presence of positive impact of statistical noise (v_A> 0), while location of B' below this frontier indicates the negative effect of noise (v_B< 0).

One of the main goals of SFA is to estimate the impact of inefficiency, which is determined by the value of technical efficiency for business model of the sample by the following formula:

$$TE_i = \frac{\exp\left(\beta_0 + \sum_n \beta_n \ln x_{ni} + v_i - u_i\right)}{\exp\left(\beta_0 + \sum_n \beta_n \ln x_{ni} + v_i\right)} = \exp(-u_i). \quad (12.5)$$

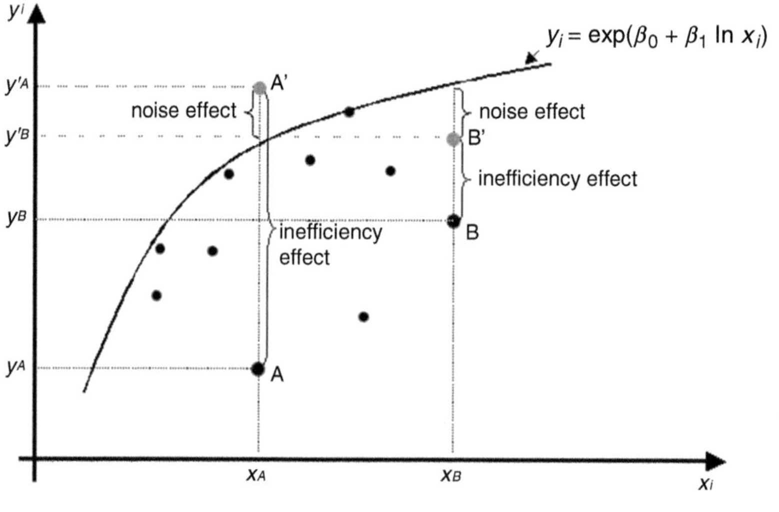

Fig. 12.3 Measuring Efficiency by Stochastic frontier analysis

This measurer of technical efficiency is limited by the interval [0…1] and evaluates the output of the *i*-th company regarding the output, which can be produced by fully efficient business model using the same vector of inputs.

Non-parametric methods for efficiency measurement use a mathematical programming and does not require the determination of the form of production function (costs, etc.), which is one of their main advantages over parametric methods. These include the following methods:

1. Total Factor Productivity Index;
2. Data Envelopment Analysis (DEA);
3. Free Disposal Hull (FDH).

In the science and practice the most common method to estimate the changes in various economic values is the *index method*, which consists in determining the various indexes. Efficiency analysis is no exception. There are widely used indices to characterize the change in productivity and efficiency in time and space.

Theory of indices has a century-old history. An important contribution to the creation of this theory was made by Laspeyres (1871) and Paasche (1874), whose formulas are still widely used by national statistical offices around the world. Since this theory has received a comprehensive development, and in terms of efficiency evaluation the basis are the works by Fisher (1922) and Törnqvist (1936). A detailed historical description of the index method is given by Diewert and Nakamura (2001).

Set of indices used to measure the business efficiency are merged under the name of *Total factor productivity indexes (TFPI)*, which generally estimate a change in total output relative to the change of total input for the periods (or companies) s and t, and have the following general form:

$$TFP_{st} = \frac{\text{Output Index}_{st}}{\text{Input Index}_{st}} = \frac{OI_{st}}{II_{st}}. \tag{12.6}$$

The TFP indexes have several disadvantages associated with the need for additional data, the complexity of a direct calculation, constraints in the number of factors and observations, etc. However, some of them, e.g., Malmquist TFP index, effectively uses to study changes in efficiency and productivity in various industries, for instance in food industry (Goncharuk, 2007; Goncharuk, 2011; Goncharuk, 2019), energetics (Goncharuk, 2013), etc.

At the same time, most of the disadvantages of TFP indexes can be solved using the *Data envelopment analysis (DEA)*. DEA uses linear programming methods to construct non-parametric piecewise surface (frontier) for a sample of companies, and the calculation of the efficiency with respect to this surface (Coelli et al., 2005). DEA methodology is based on the approach of piecewise-linear convex envelope to calculate the frontier proposed by Farrell (1957). For more than two decades this approach has remained in the shadow and only some authors attempted to solve the problem with mathematical programming methods, e.g., Shephard (1970). And only after the publication of an article by Charnes et al. (1978), where was first used the term “Data Envelopment Analysis” and the model of linear programming to solve the problem of frontier constructing and efficiency evaluation, this method has received recognition and development.

DEA study a decision-making unit (DMU) j with the set of inputs x_i and outputs y_i, and analyze its activity in the environment, i.e., in comparison with other DMUs (companies) that have a similar product (technology, industry, type of activity). Efficiency here is defined as the quotient of a weighted sum of all output parameters on the weighted sum of all input factors:

$$TE = \frac{\sum_{N}^{i=1} u_i y_{ij}}{\sum_{M}^{i=1} v_i x_{ij}}, \tag{12.7}$$

where u_i and v_i are, respectively, weights for the i-th output and input; N and M are, respectively, the number of observed inputs and outputs.

For each j-th DMU in the set J we define vectors of the weights u' and $v'$and the value of technical efficiency TE by comparing the observations using the efficiency frontier, which is constructed through a linear programming problem. Multiplicative form of this problem is as follows:

$$\begin{aligned}
&\max_{u',v'} \left(u'y_i\right) \\
&subject\ to: v'x_i = 1, \\
&u'y_j - v'x_j \le 0, \\
&u', v' \ge 0, \\
&j = 1, 2, 3, \ldots, J.
\end{aligned} \tag{12.8}$$

For ease of computations (to reduce the number of constraints) and the presentation it is often applied the dual problem, which has the following form:

$$\begin{aligned}
&\min_{\theta,\lambda} \theta, \\
&subject\ to: -y_i + Y\lambda \ge 0, \\
&\theta x_i - X\lambda \ge 0, \\
&\lambda \ge 0,
\end{aligned} \tag{12.9}$$

where θ is a scalar that determines the efficiency of the i-th company; λ is a vector of constants with dimension J × 1, representing the weights.

Formulation of the problem in the dual form (13.9) assumes fewer restrictions than in the multiplicative form (13.8), i.e. N + M < J + 1. The value of θ is a measure of the efficiency of the i-th company: it cannot exceed 1; if the value of θ is equal to 1, then the company is located on the frontier of productive capacity and, hence, is technically efficient by Farrell; if θ is less than 1, then the company is inefficient.

Note that this linear programming problem must be solved for each company of the sample, i.e. J times. Thus, the value of θ is determined for each company. Set of estimated coefficients λ enables to establish a hypothetically efficient company, i.e. the optimal ratio of output and input for each company. The obtained set of solutions indicates how output can be increased or input can be reduced to make company efficient.

Free disposal hull (FDH) is nonparametric method for the efficiency evaluation that first was proposed by Deprins et al. (2006). FDH has a special approach and the properties that distinguish it from the DEA models:

1. FDH rejects the hypothesis of concavity (convexity) of the frontier of production capacity (efficiency), while for DEA a concavity (convexity) is the main hypothesis;
2. FDH-analysis relies on the relationship of domination between the input-output sets to measure the efficiency of using a special algorithm (Ray, 2004);
3. FDH efficiency frontier does not include the points on the lines connecting the most efficient companies (Harker & Zenios, 2000);
4. FDH as a rule calculates the higher estimates of average efficiency than DEA due to the fact that the FDH frontier either equals or is inside (below) the DEA frontier (Tulkens, 2006);
5. FDH does not have limitations related to production technology (Deprins et al., 2006).

FDH method is based on the representation of production technology, which is defined by the production plans that assume a strict allocation of input and output, but without the assumption of convexity (Leleu, 2006). The main purpose of developing this method was to provide assurance that the efficiency scores are the result of the actual parameters, but not hypothetical (Cooper et al., 2006). For example, such points as Q 'at the Fig. 12.4 does not take into account, since they are not actually observable indicators. They were computed and are hypothetical.

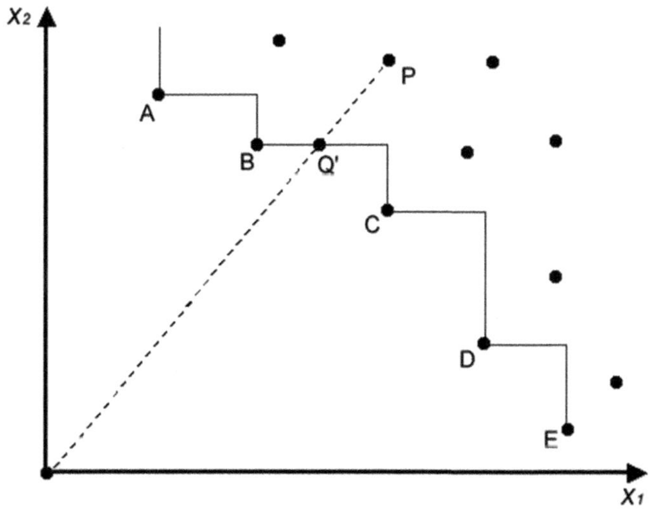

Fig. 12.4 Efficiency measuring by FDH method

Figure 12.4 demonstrates a simplest example of an efficiency frontier creation by FDH in the case of production of one output of two inputs. This frontier is a hull that defines a minimal set of companies, which includes all production capabilities that can be obtained from observations. Formally, it looks as follows:

$$P_{FDH} = \{(x,y) \mid x \geq x_j, y \leq y_j, x, y \geq 0, j = 1,\ldots,n\}, \tag{12.10}$$

where $x_j(\geq 0)$, $y_j(\geq 0)$ are actually observed parameters of companies $j = 1,...,n$ (Cooper et al., 2006). In other words, point at the Fig. 12.4 is a part of the set of production capabilities, if all its input-coordinates are at least as great as their correspondents on the vector of considered values x_j for each of companies $j = 1,...,n$ and if its output-coordinates are no more than their correspondents on the vector of considered values$y_j$for the same company j. This makes the presence of a step function, which is displayed in Fig. 12.4. None point that lies below the function (frontier) meets the properties of a set of production possibilities P_{FDH}. Moreover, this frontier includes a minimal set with these properties. For example, the connection of points B and C in the Fig. 12.4 would lead to the construction of the frontier with a larger set of production possibilities.

Thus, FDH determines the actual points (companies) that are included in the frontier as efficient. Efficiency of the other points (companies) can be calculated by computing the relative vertical distance from the frontier.

Having many advantages, FDH is one of the most preferable methods for measuring efficiency of business models and finding appropriate benchmark for improvement if your business model is not efficient. It is actively used for efficiency measurement and performance benchmarking for food industry companies (Goncharuk, 2014) and SMEs (Horobets, 2020). That is why we can advise to use FDH in forming efficient business model for bakeries.

12.3 Discussion and Conclusions

As Kim et al. (2018) noticed, it is important for small business owners to choose an efficient business model because of constraints such as technical problems. The principle of business model helps in building and assess the value that a company offers and delivers to its consumer (Nosratabadi et al., 2019). The business model describes a company's position in the production chain (Mosleh and Nosratabadi, 2015).

Kähkönen (2012) presents ways for innovating the value proposition in order to developing a business model in the food sector. Vojtovic et al. (2016) introduce a new method for creating a sustainable business model for food and beverage processors. Inspired by the business model canvas, they presented a ten-pillar business model to establish a sustainable business strategy for food and beverage processors. They recommend that the business concept should defined in the first stage, where key aspects and values that the firm gives to clients, long-term benefits to society and the environment, and the corporate vision and long-term goals are clearly stated. The next phases relate to discovering clients, creating connections, branding, habit formation, defining a distribution channel, articulating resource planning, designing critical activities (i.e., operational, support, and development), and also establishing a sophisticated support system. The remaining two pillars of this approach are devoted to estimation the cost structure and determining the income model.

According to Pölling et al. (2017), urban farming and agriculture necessitate diverse business models from rural locations. They identify three urban farming business models: differentiation, diversification, and low-cost specialisation, with the 'differentiation' business model connected with niche production and differentiation. Barth et al. (2017) suggested the business model which is built on four pillars: (1) value proposition, (2) value creation and delivery, (3) value capture, and (4) value intention

Martinovski (2016) argues that consumer behaviour is the most important factor to consider when developing a business model. As a result, he presents the idea of developing a business model based on customer behaviour while purchasing food goods. This finding reveals that such a approach is an instrument for decision-makers to use in designing a sustainable business model in which, on the one hand, businesses can use this customer-centric approach to get feedback from customers in order to develop corresponding value propositions for their potential customers, and on the other hand, society and customer benefits are considered, and healthy proper food productions are supplied to them based on their feedback.

Di Gregorio (2017) presents a novel business model for merchants in the food market, where items are delivered to customers. He introduces a strategy in which location-specific resources are exploited to produce and capture value using the notion of a placed-based business model. According to his findings, the food industry's-based business model will contribute to social context resilience, sustainability, and prosperity by restoring enthusiasm for traditional food cultures and increasing supply and demand for local food items.

According to Nosratabadi et al. (2020), customer needs and product quality are two major aspects influencing the business models of all enterprises functioning independent of their location in the chain.

When developing an effective business model, valid issues emerge, such as how efficient the business model is and how to determine it. This necessitates efficiency standards as well as proper measurement methodologies. In general, all techniques of assessing efficiency may be classified as either individual or relative.

Individual techniques are related with internal corporate performance evaluations and can be structural, dynamic, or static. These include numerous financial statistics depending on company performance, such as liquidity, profitability, and so on. These are static indicators if they are analysed for a single business period. They are dynamic when compared over two or more time periods.

The majority of methods for determining the relative efficiency of business models are statistical in nature. In general, they are classified as frontier or non-frontier approaches. The basis of frontier techniques of evaluating efficiency is that the business model efficiency is calculated in respect to the efficiency frontier or production possibility curve, which is set by the sample's most efficient enterprises.

Non-frontier approaches, in contrast to frontier methods, are based on a comparison with some sample average level, which is specified by computing the index or using the least squares method.

Furthermore, the techniques of assessing efficiency may be classified as parametric or nonparametric, which includes both frontier and non-frontier methods.

Parametric approaches focus on econometric analysis and need determining the functional form of the business's production function (or a function of costs, or profits). The regression analysis is utilised to develop this function.

Non-parametric approaches for measuring efficiency employ mathematical programming and do not need determining the shape of the production function (costs, etc.), which is one of their primary benefits over parametric methods. These approaches include Total Factor Productivity Index - TFP Index; Data Envelopment Analysis (DEA); and Free Disposal Hull (FDH).

The index approach, which consists of determining numerous indexes, is the most prevalent method in science and practice for estimating changes in various economic variables. Indicators are commonly used to quantify changes in productivity and efficiency over time and place.

FDH is one of the most preferred approaches for analysing the efficiency of business models and determining acceptable benchmarks for change if company model is inefficient. The FDH technique is based on the representation of production technology, which is characterised by production plans that require rigorous input and output allocation but do not assume convexity.

References

Aigner, D., Lovell, C. K., & Schmidt, P. (1977). Formulation and estimation of stochastic frontier production function models. *Journal of Econometrics, 6*(1), 21–37.

Baden-Fuller, C. (2021). *Building better business models for the digital economy report march 2021*. City, University of London.

Baden-Fuller, C., & Morgan, M. S. (2010). Business models as models. *Long Range Planning, 43*(2–3), 156–171.

Barth, H., Ulvenblad, P.-O., & Ulvenblad, P. (2017). Towards a conceptual framework of sustainable business model innovation in the Agri-food sector: a systematic literature review. *Sustainability, 9*, 1620.

Bayne, L., & Wee, M. (2019). Non-financial KPIs in annual report narratives: Australian practice. *Accounting Research Journal, 32*(1), 7–19.

Berg, S. (2006). *Water benchmarking support system: survey of benchmarking Methologies*. University of Florida.

Bini, L., Simoni, L., Dainelli, F., & Giunta, F. (2018). Business model and non-financial key performance indicator disclosure. *Journal of Business Models, 6*(2), 5–9.

Casadesus-Masanell, R., & Ricart, J. E. (2010). From strategy to business models and onto tactics. *Long Range Planning, 43*(2–3), 195–215.

Charnes, A., Cooper, W. W., & Rhodes, E. (1978). Measuring the efficiency of decision making units. *European Journal of Operational Research, 2*(6), 429–444.

Chesbrough, H. (2007). Business model innovation: It's not just about technology anymore. *Strategy and Leadership, 35*(6), 12–17.

Coelli, T. J., Rao, D. S., O'Donnell, C. J., & Battese, G. E. (2005). *An introduction to efficiency and productivity analysis*. Springer.

Cooper, W. W., Seiford, L. M., & Tone, K. (2006). *Introduction to data envelopment analysis and its uses*. Springer.

Demil, B., & Lecocq, X. (2010). Business model evolution: in search of dynamic consistency. *Long Range Planning, 43*(2–3), 227–246.

Deprins, D., Simar, L., & Tulkens, H. (2006). Measuring labor-efficiency in post offices. In *Public goods, environmental externalities and fiscal competition* (pp. 285–309). Springer.

Di Gregorio, D. (2017). Place-based business models for resilient local economies: cases from Italian slow food, agritourism and the albergo diffuso. *Journal of Enterprising Communities, 11*, 113–128.

Diewert, W. E., & Nakamura, A. O. (2001). *Essays in index number theory. Volume 1.Contributions to economic analysis*. Butterworth Heinemann.

Farrell, M. J. (1957). The measurement of productive efficiency. *Journal of the Royal Statistical Society: Series A (General), 120*(3), 253–281.

Fisher, I. (1922). *The making of index numbers: a study of their varieties, tests, and reliability*. Houghton Mifflin Company.

Gambardella, A., & McGahan, A. M. (2010). Business-model innovation: general purpose technologies and their implications for industry structure. *Long Range Planning, 43*(2–3), 262–271.

Goncharuk, A. (2007). Impact of political changes on industrial efficiency: a case of Ukraine. *Journal of Economic Studies, 34*(4), 324–340.

Goncharuk, A. G. (2011). Benchmarking for investment decisions: a case of food production. *Benchmarking: An International Journal, 18*(5), 694–704.

Goncharuk, A. G. (2013). What causes increase in gas prices: the case of Ukraine. *International Journal of Energy Sector Management, 7*(4), 448–458.

Goncharuk, A. G. (2014). Competitive benchmarking technique for" the followers": a case of Ukrainian dairies. *Benchmarking: An International Journal, 21*(2), 218–225.

Goncharuk, A. G. (2019). Winemaking performance: whether the crisis is over. *British Food Journal, 121*(5), 1064–1077.

Goncharuk, A. G., & Karavan, S. (2013). The investment attractiveness evaluation: methods and measurement features. *Polish Journal of Management Studies, 7*(1), 160–166.

Harker, P. T., & Zenios, S. A. (Eds.). (2000). *Performance of financial institutions: efficiency, innovation, regulation*. Cambridge University Press.

Höflinger, N. F. (2014). The business model concept and its antecedents and consequences–towards a common understanding. In *Academy of management proceedings* (Vol. 201416705). Academy of Management.

Horobets, T. A. (2020). Measuring performance of the SMEs: a case of Ukraine. *Journal of Applied Management and Investments, 9*(4), 162–168.

IMARC. (2022). *Bakery products market: Global industry trends, share, size, growth, opportunity and forecast 2022–2027*. Available at: https://www.imarcgroup.com/bakery-products-market/toc

Itami, H., & Nishino, K. (2010). Killing two birds with one stone: profit for now and learning for the future. *Long Range Planning, 43*(2–3), 364–369.

Kähkönen, A.-K. (2012). Value net—a new business model for the food industry? *British Food Journal, 114*, 681–701.

Kim, H., Lee, D., & Ryu, M. H. (2018). An optimal strategic business model for small businesses using online platforms. *Sustainability, 10*(3), 579.

Kumar, S., & Singh, C. (2020). Productivity growth in India's bakery manufacturing industry. *Journal of Agribusiness in Developing and Emerging Economies, 12*(1), 94–103.

Kumbhakar, S. C., & Lovell, C. K. (2003). *Stochastic frontier analysis*. Cambridge University Press.

Laspeyres, E. (1871). Die Berechnung einer mittleren Waarenpreissteigerung. *Jahrbücher für Nationalökonomie und Statistik, 16*, 296–315.

Legendre, A. M. (1805). *New methods for the determination of orbits of comets*. Courcier.

Leleu, H. (2006). A linear programming framework for free disposal hull technologies and cost functions: primal and dual models. *European journal of operational research, 168*(2), 340–344.

Martinovski, S. (2016). Nutrition business models of consumer behaviour when purchasing self-explanatory food products. *Journal of Hygienic Engineering and Design, 16*, 53–58.

Mosleh, A., & Nosratabadi, S. (2015). Impact of information technology on Tehran's tourism agencies' business 9 Model's components. *International Journal of Business and Management, 10*, 107.

Mota, J., Moreira, A., Costa, R., Serrão, S., Pais-Magalhães, V., & Costa, C. (2020). Performance indicators to support firm-level decision-making in the wine industry: a systematic literature review. *International Journal of Wine Business Research, 33*(2), 217–237.

Nosratabadi, S., Mosavi, A., Shamshirband, S., Zavadskas, E. K., Rakotonirainy, A., & Chau, K. W. (2019). Sustainable business models: A review. *Sustainability, 11*(6), 1663.

Nosratabadi, S., Mosavi, A., & Lakner, Z. (2020). Food supply chain and business model innovation: a review. *Food, MDPI, 9*, 132.

Paasche, H. (1874). Ueber die Preisentwicklung der letzten Jahre nach den Hamburger Börsennotirungen. *Jahrbücher für Nationalökonomie und Statistik, 23*, 168–178.

Pölling, B., Prados, M.-J., Torquati, B. M., Giacche, G., Recasens, X., Paffarini, C., Alfranca, O., & Lorleberg, W. (2017). Business models in urban farming: a comparative analysis of case studies from Spain. *Italy and Germany Moravian Geographical Reports, 25*, 166–180.

Ray, S. C. (2004). Data envelopment analysis: theory and techniques for economics and operations research.

Schweizer, L. (2005). Concept and evolution of business models. *Journal of General Management, 31*(2), 37–56.

Shephard, R. W. (1970). *Theory of cost and production functions*. Princeton University Press.

Teece, D. J. (2010). Business models, business strategy and innovation. *Long Range Planning, 43*(2–3), 172–194.

Timmers, P. (1998). Business models for electronic markets. *Electronic Markets, 8*(2), 3–8.

Törnqvist, L. (1936). The Bank of Finland's consumption price index. *Bank of Finland Monthly Bulletin, 10*, 1–8.

Tsiouni, M., Konstantinidis, C., Pavloudi, A., & Giovanis, N. (2022). Financial ratio and efficiency analysis as a competitive advantage of wine manufacturing firms. The case of Greece. *Theoretical Economics Letters, 12*(1), 229–239.

Tulkens, H. (2006). On FDH efficiency analysis: some methodological issues and applications to retail banking, courts and urban transit. In *Public goods, environmental externalities and fiscal competition* (pp. 311–342). Springer.

Venkatraman, N., & Henderson, J. C. (1998). Real strategies for virtual organizing. *Sloan Management Review, 40*(1), 33–48.

Vojtovic, S., Navickas, V., & Gruzauskas, V. (2016). Sustainable business development process: the case of the food and beverage industry. *Zeszyty Naukowe Politechniki Pozna ́nskiej, 68*, 225–239.

Wilson, M., Wnuk, K., Silvander, J., & Gorschek, T. (2018). A literature review on the effectiveness and efficiency of business modeling. *E-Informatica Software Engineering Journal, 12*(1), 265–302.

Winsten, C. B. (1957). Discussion on Mr. Farrell's paper. *Journal of the Royal Statistical Society, 120*(3), 282–284.

Zimon, G., & Tarighi, H. (2021). Effects of the COVID-19 global crisis on the working capital management policy: evidence from Poland. *Journal of Risk and Financial Management, 14*(4), 169.

Zott, C., & Amit, R. (2010). Business model design: an activity system perspective. *Long Range Planning, 43*(2–3), 216–226.

Chapter 13
Assessment of the Sustainable Competitiveness of Agricultural Enterprises on the Grain Market: Case of Ukraine

Anatolii Kucher and Lesia Kucher

13.1 Introduction

Assessment of the sustainable competitiveness of agricultural enterprises on the grain market plays an important role in the managing baking business sustainability. This is especially true for those countries that play a leading role in the world grain market and guarantee global food security. Ukraine is one such country, which recently exported about 75% of its grain production annually.

According to Skrypnyk, Klymenko, et al. (2021a), the Ukrainian agrarian business (in particular, crop production) is developing its own way, not following the development path of Germany or the USA. After a rather long recession, the agricultural sector has embarked on a path of stable growth with an annual increase in wheat yields of around 0.1 t/ha. Furthermore, the extremely low proportion of forested land in Ukraine by European standards significantly increases the impact of weather and climate risks. The article by Skrypnyk, Zhemoyda, et al. (2021b) confirm that if the boundaries of soil-and-climatic zones change, the conditions of growing crops and their yield will consequently change as well. Based on current global forecasts, the impact of weather on Ukraine's agriculture will increase, and the most negative effects can be expected in the Steppe zone, where the likelihood of weather and climate risks increases. It is obvious that the shortage of water resources will play an important role in shaping the sustainable development of

A. Kucher (✉)
Lviv Polytechnic National University, Lviv, Ukraine

National Scientific Center "Institute for Soil Science and Agrochemistry Research named after O. N. Sokolovsky", Lviv, Ukraine
e-mail: kucher@karazin.ua

L. Kucher
Lviv Polytechnic National University, Lviv, Ukraine
e-mail: kucher@btu.kharkov.ua

J. M. Ferreira da Rocha et al. (eds.), *Baking Business Sustainability Through Life Cycle Management*, https://doi.org/10.1007/978-3-031-25027-9_13

grain production. According to Fedulova et al. (2021), after reaching a certain water scarcity threshold, the country begins to demand for grain imports, which increases as water resources decrease. Therefore, a further sustainable intensification of grain production is probably the way to the formation of sustainable competitiveness of enterprises and sustainable development of agriculture.

Analysis of recent publications shows that different aspects of sustainable competitiveness are covered in the works of such scientists as: Atta Mills et al. (2021)), Biazzin et al. (2021), Castro Oliveira et al. (2022), Choudhary (2022), Danileviciene and Lace (2021), Delgosha et al. (2021), Dziembała (2021), Farinha et al. (2021), Hategan et al. (2021), Karman and Savanevičienė (2021), Kucher et al. (2021), Liu and Wang (2022), Liu et al. (2022), López-Fernández et al. (2022), Rajnoha and Lesnikova (2022), Shahbaz et al. (2022), Yahya et al. (2022). The scientific developments of these scientists are aimed primarily at assessing the sustainable competitiveness of enterprises, regions and countries various industries (tourism, e-commerce, high-tech industries, etc.), identifying and modeling factors for its improvement.

According to the search and analysis of the documents in Scopus (search object – "sustainable competitiveness"; search scale – title), 112 documents were found (as of June 1, 2022). Based on the review of articles on sustainable competitiveness indexed by Scopus during 2005 and 2022 (Fig. 13.1), we identified (i) a growth trend in the number of publications in the world (leaders – China, Portugal, Spain, United Kingdom and Lithuania), and (ii) research gaps in Ukrainian literature (only one document has been published by Ukrainian scientists (Kucher et al. (2021)), that have not yet been addressed.

Analysis of the distribution of documents by type shows that the largest share (66.1%) is made up of articles. The second position was occupied by book chapter (15.2%), the third place – conference papers (13.4%). Analysis of the distribution of documents by subject area shows that the largest number of publications fell on business, management and accounting (20.0%), social sciences (17.8%) and economics, econometrics and finance (13.5%).

In terms of the number of publications on the sustainable competitiveness, the most influential organizations in the world are the Chinese Academy of Social Sciences, Universitat Politècnica de València, and Vilniaus Gedimino Technikos Universitetas. Analysis of the distribution of documents in the context of the main sources of publication indicates that the largest number of studies was published in the following journals: Sustainability (15 documents), Information Systems Frontiers, Innovation Technology and Knowledge Management, Journal of Business Economics and Management, Journal of Environmental Planning and Management, Structural Survey (2 documents each).

However, to date, the problem of comprehensive assessment of the sustainable competitiveness of agricultural enterprises on the grain market is insufficiently covered (none of the analyzed 112 documents does not relate to the grain market), which determines the relevance of this study. Therefore, the proposed work is a

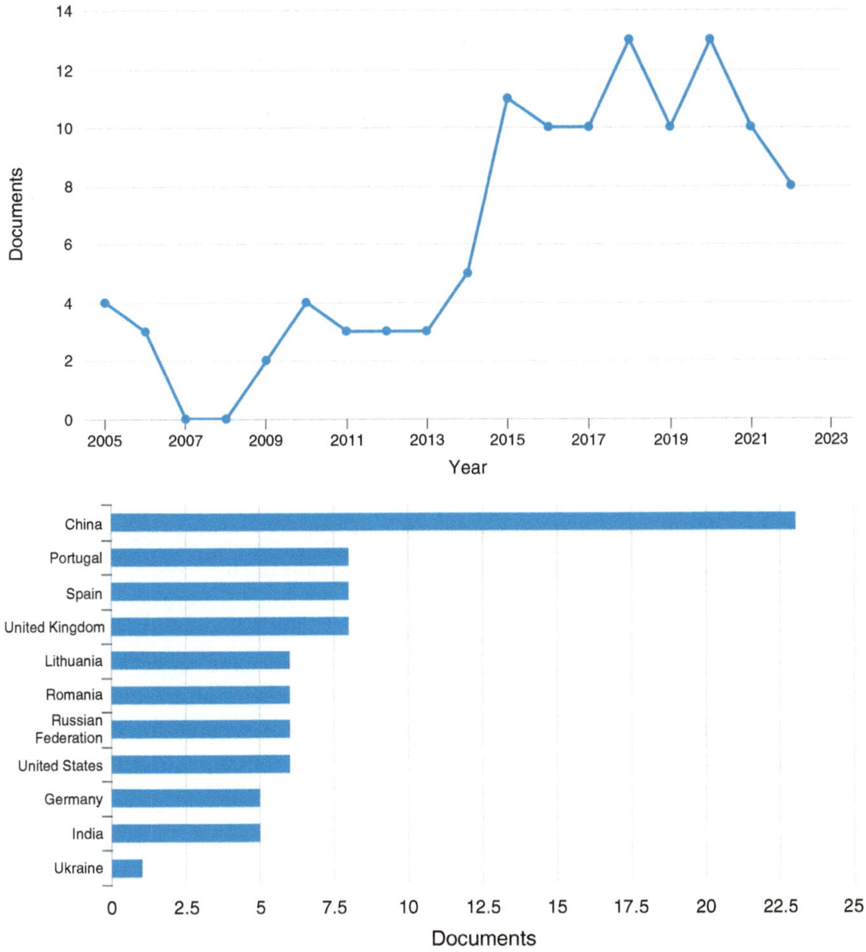

Fig. 13.1 Number of documents on "sustainable competitiveness" published in journals indexed by Scopus in dynamics and in the TOP–10 countries and Ukraine. (*Source:* built on the basis of Scopus database)

logical continuation of the authors' research on this topic (Kucher, 2019, 2020; Kucher et al., 2021).

The purpose of this section is to highlight the results of the study of zonal features of the formation and reserves of increasing the sustainable competitiveness of agricultural enterprises on the Ukrainian grain market. From a financial point of view, these reserves can be combined into two groups: (i) reserves, the practical implementation of which does not require the attraction of additional financial resources, but requires a more rational use of them; (ii) reserves, the practical implementation of which requires the attraction of additional financial resources.

13.2 Methodology

In this study, we used such methods: econometric modeling (to develop an econometric model of the dependence of the Sustainable Competitiveness Index (SCI) of agricultural enterprises on the main sub-indices); economic-statistical and monographic (for the assessment and analysis competitiveness of enterprises); abstract-and-logical (for theoretical generalization and analysis of the research results); grapho-analytic (for the visual representation of the obtained results). The economic database of the 5597 agricultural enterprises located in Ukraine, which represent all the soil-climatic zones, was used as the empirical basis.

The time period of the study covers 2016, since this is the last year when the system of agricultural statistics functioned in Ukraine, which allows collecting data on the main economic indicators of the activities of agricultural enterprises in full. Subsequently, as a result of reforming the program of statistical observations, the possibilities for collecting and statistical processing of mass data decreased significantly.

At the first stage, the research assessed sustainable competitiveness of all agricultural enterprises of Ukraine in terms of three soil-climatic zones, i.e. Steppe, Forest-steppe, and Polissya. The study determined the zonal features (zone particularities) of competitiveness formation, and thus, outlined the priorities of its improvement and therefore financial support. The second stage suggested development of econometric models of dependence of the SCI of agricultural enterprises of Ukraine from the main subindexes. Their approbation has secured forecasting of a rise of the competitiveness of agricultural enterprises by employing the internal reserves in the medium-term prospect.

13.3 Results

Results of the research demonstrate that most agricultural enterprises are concentrated in the Steppe soil-climatic zone (48.3%), the least share (10.8%) – in the zone of Polissya, others (40.9%) – in the Forest-steppe zone, which is characterized by the highest level of sustainable competitiveness among the analyzed zones (Table 13.1). Analyzing the zone peculiarities, one should note that small differences in the average indexes of competitiveness confirm the opportunity to achieve its high level in any soil-climatic zone. Moreover, basing on the comparative analysis of all indexes and subindexes of competitiveness, it is possible to conclude that agricultural enterprises of the Forest-steppe zone are competitive, although at the average level, and business entities of Polissya and Steppe are not competitive. However, the zone of Polissya exceeds the indexes of the Steppe zone by all points. It is interesting that by some subindexes of competitiveness, e.g. by the annual average number of persons employed in agriculture per 1000 ha of agricultural land (employment level), share of labor costs with tax-deductions in the structure of

Table 13.1 Zonal features of formation of sustainable competitiveness of agricultural enterprises of Ukraine in 2016, coefficient

Competitiveness indexes and subindexes	Soil-and-climatic zones		
	Steppe	Forest-steppe	Polissya (Forest)
Number of enterprises, units	2706	2289	602
Integral sustainable competitiveness index (SCI)	**0.813**	**1.031**	**0.878**
Competitiveness indexes: Technological (ITC)	***0.745***	***1.036***	***0.920***
Subindexes by the: Conditional yield per 1 ha	0.719	1.017	0.882
Operating (production) expenses per 1 ha	0.771	1.056	0.957
Financial (IFC)	***0.910***	***1.053***	***0.818***
Cash revenue per 1 ha	0.799	1.056	0.892
Gross value added per 1 ha	0.813	1.044	0.732
Payback (profitability)	1.109	1.030	0.945
Covering of production expenses with cash revenue	1.154	1.031	0.918
Profit per 1 ha	0.919	1.014	0.721
Marginal income per 1 ha	0.924	1.036	0.593
Expert monetary value of 1 ha of land	0.868	1.040	0.711
Conditional normative monetary value of 1 ha of land	0.691	1.172	1.028
marketing (IMC)	***0.789***	***1.073***	***0.755***
Relative share of the market	0.849	1.039	0.828
Competitiveness on the land rental market	0.728	1.108	0.683
Ecological (IEcolC)	***0.802***	***0.955***	***0.890***
Balance of humus per 1 ha	0.146	1.048	0.740
Greenhouse gas emissions from fuel combustion per 1 ha	1.262	1.094	1.209
Coefficient of compliance with the structure of sown areas	0.734	0.671	0.502
Coefficient of erosion risk (hazard)	1.068	1.007	1.108
Social (ISC)	***0.819***	***1.039***	***1.006***
Annual average number of persons employed in agriculture per 1000 ha of agricultural land	0.827	1.213	1.265
Labor costs with tax-deductions per 1 ha	0.666	1.057	0.945
Share of labor costs with tax-deductions in the structure of production costs	0.962	0.988	1.017
Average wage of 1 employee	0.821	0.898	0.796

Source: authors' calculations based on the data of the form No. 50-s.g

production costs, and the factor of erosion danger, enterprises of Polissya exceeded the indexes of both Steppe and Forest-steppe zones.

The work confirms that payback of production costs by means of the conditional yield was the highest in the zone of Polissya, a little lower – in the Steppe zone, and the lowest – in the Forest-steppe zone. It can be explained by the effect of the law of diminishing returns and differentiation of the level of moisture supply. The lowest figure of the competitiveness subindex by the humus balance (0.146) argues that enterprises in the Steppe zone employ land the least rationally, whereas in the

Forest-steppe zone, land use is balanced above the average level. However, the recommended structure of cropping area was not followed by the enterprises in all zones. Thus, to improve sustainable competitiveness of the enterprises in the Steppe zone, it is necessary to increase their technological competitiveness, whereas for the enterprises in the Forest-steppe zone, it concerns the ecological competitiveness, and for the zone of Polissya – the marketing one, as they are their weakest aspects.

Results of the analysis of graphic-analytical models (Fig. 13.2) demonstrate a similarity of the zone peculiarities of formation of the competitiveness of enterprises at some segments of the agricultural market referring to the general situation. A visual analysis of the diagrams proves a relative balance of the different kinds of competitiveness, expect for sunflower, where the vector of ecological competitiveness in the polygon is reduced in all soil-climatic zones, demonstrating the necessity of its improvement.

The zonal distribution of enterprises, engaged in winter wheat growing, appeared to be similar to the general situation, i.e. most of them were in the Steppe zone (52.5%), less (39.5%) – in the Forest-steppe zone, and the least number (8.0%) – in the zone of Polissya (Table 13.2).

According to the SCI value, enterprises of the Forest-steppe zone (1.100) were competitive, as they achieved its average level, and the enterprises of the zone of Polissya (0.997) and Steppe (0.917) were uncompetitive, although they approached the threshold level.

According to the technological (1.022) and ecological (1.166) criteria, enterprises of the zone of Polissya reached the competitive level at the market of winter wheat grain, whereas for the enterprises of the Steppe zone, such situation characterized the financial competitiveness. In the Forest-steppe zone, enterprises utilized land in the most rational way, and in the Steppe zone – at the lowest level of rationality. To get a high level of sustainable competitiveness of enterprises in the Forest-steppe zone, it is primarily necessary to increase financial competitiveness, in the Polissya zone – marketing one, and in the Steppe zone – technological competitiveness. By those indexes, they demonstrate the worst figures, comparing to the other kinds. The principal ways to reach those goals suggest rising of land use profitability, improvement of competitiveness at the market of land lease, and growth of the yield capacity by means of sustainable production intensification respectively.

Besides winter wheat, more than a half (51.8%) of enterprises produced maize in the Forest-steppe zone, 37.5% of enterprises were engaged in such production in the Steppe zone, and 10.7% – in the Polissya zone (Table 13.3).

It is significant that by the SCI value, enterprises of each zone have not achieved the competitive level, whereas preserved the previous regularity, particularly, better positions are reached in the Forest-steppe zone (0.973), worse ones (0.913) – in the Polissya zone, and the worst fig. (0.815) – in the Steppe zone. Thus, climatic changes, caused by warming and the increased moisture deficit, have resulted the Steppe zone is getting less attractive for organization of a competitive production of maize grain. That conclusion is completely confirmed by the subindex of competitiveness by the yield, which demonstrates that in the Steppe zone, it is by 36.1% lower than the average level. Enterprises of the Forest-steppe zone have managed to

Fig. 13.2 Graphic-analytical models of sustainable competitiveness of agricultural enterprises in the context of soil-and-climatic zones of Ukraine, 2016. (*Note. ITC* technological competitiveness index, *IFC* financial competitiveness index, *IMC* marketing competitiveness index, *IEcolC* ecological competitiveness index, *ISC* social competitiveness index. *Source:* built by the authors based on own calculations)

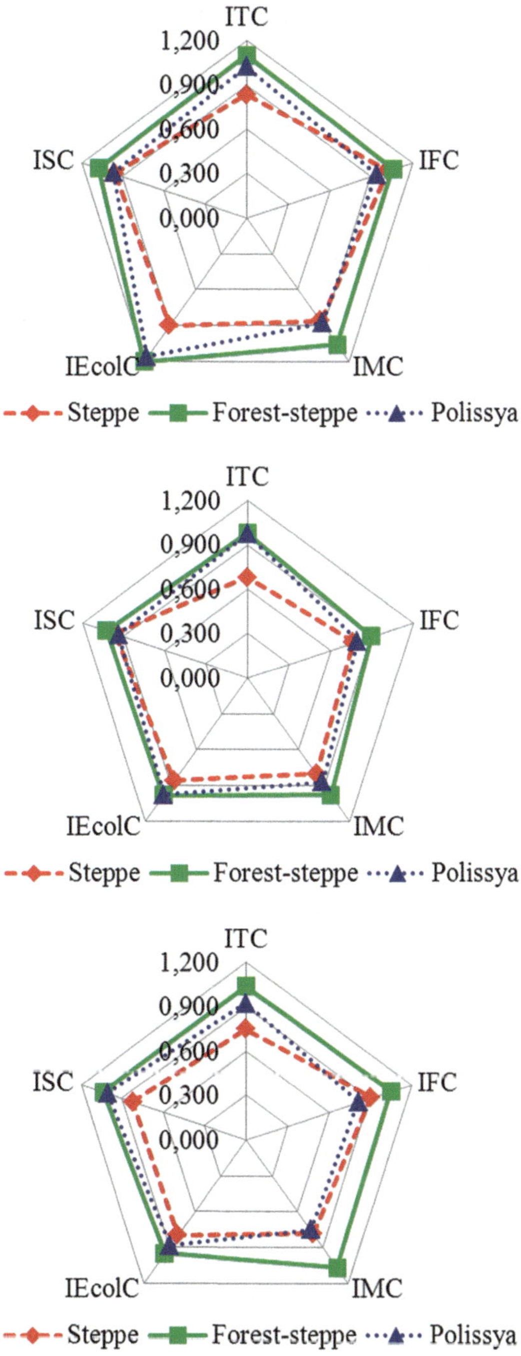

Table 13.2 Zonal features of formation of the sustainable competitiveness of agricultural enterprises of Ukraine on the winter wheat grain market in 2016, coefficient

Competitiveness indexes and subindexes	Soil-and-climatic zones		
	Steppe	Forest-steppe	Polissya (Forest)
Number of enterprises, units	2238	1683	343
Integral sustainable competitiveness index (SCI)	**0.917**	**1.100**	**0.997**
Competitiveness indexes: ***Technological (ITC)***	***0.831***	***1.100***	***1.022***
Subindexes by the: Yield per 1 ha	0.832	1.092	1.019
Operating (production) expenses per 1 ha	0.829	1.107	1.026
Financial (IFC)	***1.025***	***1.059***	***0.943***
Cash revenue per 1 ha	0.876	1.058	0.927
Gross value added per 1 ha	0.901	1.028	0.794
Payback (profitability)	1.079	1.013	1.000
Covering of production expenses with cash revenue	1.094	0.986	0.905
Profit per 1 ha	0.992	0.984	0.803
Marginal income per 1 ha	0.914	0.980	0.691
Expert monetary value of 1 ha of land	0.906	1.034	0.780
Conditional normative monetary value of 1 ha of land	1.309	1.414	1.432
Production cost per 1 centner	1.057	1.035	1.049
Total cost per 1 centner	1.119	1.062	1.052
marketing (IMC)	***0.866***	***1.062***	***0.883***
Relative share of the market	0.891	1.093	0.953
Competitiveness on the land rental market	0.737	1.133	0.736
Selling price per 1 centner	0.970	0.958	0.960
Ecological (IEcolC)	***0.903***	***1.203***	***1.166***
Balance of humus per 1 ha	0.554	1.253	1.093
Greenhouse gas emissions from fuel combustion per 1 ha	1.253	1.152	1.240
Social (ISC)	***0.958***	***1.076***	***0.969***
Annual average number of persons employed in the crop industry per 1000 ha of arable land	0.922	1.175	1.037
Labor costs with tax-deductions per 1 ha	0.897	1.118	0.934
Share of labor costs with tax-deductions in the structure of production costs	1.132	1.055	1.024
Average wage of 1 employee	0.882	0.957	0.883

Source: authors' calculations based on the data of the form No. 50-s.g

reach a competitive level only by the social criterion. According to the humus balance, better positions are taken by the enterprises of the Forest-steppe zone and Polissya, while, by the greenhouse gas emission, the leading positions are taken by the enterprises of the Steppe zone. At the market of land lease, a competitive level is achieved by the enterprises of the Forest-steppe zone, whereas the entities of other zones stay uncompetitive.

Table 13.3 Zonal features of formation of the sustainable competitiveness of agricultural enterprises of Ukraine on the maize grain market in 2016, coefficient

Competitiveness indexes and subindexes	Soil-and-climatic zones		
	Steppe	Forest-steppe	Polissya (Forest)
Number of enterprises, units	1213	1675	348
Integral sustainable competitiveness index (SCI)	**0.815**	**0.973**	**0.913**
Competitiveness indexes: ***Technological (ITC)***	***0.677***	***0.975***	***0.969***
Subindexes by the: Yield per 1 ha	0.639	0.956	0.949
Operating (production) expenses per 1 ha	0.714	0.994	0.990
Financial (IFC)	***0.767***	***0.899***	***0.800***
Cash revenue per 1 ha	0.708	0.896	0.812
Gross value added per 1 ha	0.649	0.883	0.745
Payback (profitability)	0.979	1.019	0.992
Covering of production expenses with cash revenue	1.024	0.942	0.872
Profit per 1 ha	0.623	0.840	0.709
Marginal income per 1 ha	0.679	0.754	0.514
Expert monetary value of 1 ha of land	0.655	0.884	0.734
Conditional normative monetary value of 1 ha of land	0.388	0.678	0.557
Production cost per 1 centner	0.927	1.012	1.014
Total cost per 1 centner	1.040	1.081	1.055
marketing (IMC)	***0.805***	***0.979***	***0.875***
Relative share of the market	0.704	0.957	0.860
Competitiveness on the land rental market	0.763	1.030	0.817
Selling price per 1 centner	0.948	0.950	0.847
Ecological (IEcolC)	***0.862***	***0.983***	***0.982***
Balance of humus per 1 ha	0.464	0.939	0.930
Greenhouse gas emissions from fuel combustion per 1 ha	1.260	1.028	1.034
Social (ISC)	***0.963***	***1.028***	***0.937***
Annual average number of persons employed in the crop industry per 1000 ha of arable land	0.916	1.105	0.988
Labor costs with tax-deductions per 1 ha	0.804	1.026	0.919
Share of labor costs with tax-deductions in the structure of production costs	1.230	1.071	0.947
Average wage of 1 employee	0.903	0.912	0.895

Source: authors' calculations based on the data of the form No. 50-s.g

It is necessary to point on identification of new peculiarities of competitiveness formation due to climatic changes. Thus, global warming has caused the "maize belt" of Ukraine is shifted towards the north. Hence, nowadays, it is possible to conduct economically beneficial growing of maize for grain under conditions of the Polissya zone. For instance, the Chernihiv cluster of the agroholding "Industrial Dairy Company" got a high yield of maize, i.e. 82 c/ha, on the area of 14.3 thousand ha (in the region, the average figure accounts for 72 c/ha, while generally in Ukraine – it is 62 c/ha).

Thus, to achieve sustainable competitiveness at the market of maize grain, the primary steps expect improvement of financial competitiveness for the enterprises of the Forest-steppe zone and Polissya, and technological competitiveness for the enterprises of the Steppe zone. It can be reached by increasing the marginal revenue, conditional normative monetary value of land, and yield capacity of maize due to cultivation of the crop under conditions of irrigation in the Steppe zone, respectively.

The next stage of the research expected development of the econometric models of dependence of the SCI of agricultural enterprises on the main subindexes in total and at the studied segments of the market. The hypothesis is based on the assumption about the existing relationship between the subindexes value and the SCI level of agricultural enterprises, whereas the largest positive impact on the SCI is made by the subindex of competitiveness by the yield. At the first stage, a correlation analysis was performed concerning the aggregate of the enterprises, and description of its variables was presented in the Table 13.4. Basing on the results of a pair correlation analysis, the authors of the research chose five key subindexes (one for each kind of competitiveness), which demonstrated the closest correlation relation with the resultant value, whereas little correlation with one another (less, than with the result).

It is determined that the SCI value (y) has a high direct correlation relation with the subindexes of competitiveness by the conditional yield (x_1, $r = 0.793$), gross added value per 1 ha (x_4, $r = 0.769$), relative share of the market (x_{11}, $r = 0.710$), whereas a significant relation with the subindex by the labor costs with tax-deductions per 1 ha of agricultural lands (x_{18}, $r = 0.603$). A choice of the subindex by the coefficient of conformity with the structure of cropping area (x_{15}, $r = 0.021$) is caused by the fact that among the analyzed indexes of ecological competitiveness only that one (but for the humus balance, which is closely correlated with the already chosen factor), makes a positive effect on the resultant feature.

Thus, at the second stage, the researchers developed an econometric model of the SCI dependence on the chosen subindexes (Table 13.5).

The coefficient of multiple correlation for that model confirms availability of a very close relation between the factorial and resultant feature, and the coefficient of multiple determination demonstrates that variability of the SCI value is by 93.0% caused by the variation of five factors, included in the model. It is set that all subindexes made a positive impact on the SCI that argued the logical-economic assumption. However, the degree of that impact was various. Hence, the 1% rise of the subindex of competitiveness by the conditional yield contributed to the increase of the SCI by 0.319%, the gross added value – by 0.141%, the relative share of the market – 0.138%, the labor costs with tax-deductions – by 0.124%, the coefficient of conformity with the structure of cropping area – by 0.059%.

The analysis of the total volume of the SCI variation argues that it experienced the most significant impact by x_1 (32.5%), x_4 (22.3%), x_{11} (16.3%) and x_{18} (21.8%). The developed model is statistically adequate and credible that is proven by the calculated values of the criteria of Durbin-Watson, Fisher, and Student. Results of

Table 13.4 Description of variables for regression modeling of dependence of integral sustainable competitiveness index (SCI) of agricultural enterprises from subindexes, coefficient

Variable	Description of indicators
y – SCI	***Integral sustainable competitiveness index***
x_1	Competitiveness subindex by the conditional yield
x_2	Competitiveness subindex by the operating (production) expenses per 1 ha
ITC	***Technological competitiveness index***
x_3	Competitiveness subindex by the cash revenue per 1 ha
x_4	Competitiveness subindex by the gross value added per 1 ha
x_5	Competitiveness subindex by the payback (profitability)
x_6	Competitiveness subindex by the covering of production expenses with cash revenue
x_7	Competitiveness subindex by the profit per 1 ha
x_8	Competitiveness subindex by the marginal income per 1 ha
x_9	Competitiveness subindex by the expert monetary value of 1 ha of land
x_{10}	Competitiveness subindex by the conditional normative monetary value of 1 ha of land
IFC	***Financial competitiveness index***
x_{11}	Competitiveness subindex by the relative share of the market
x_{12}	Competitiveness subindex on the land rental market
IMC	***Marketing competitiveness index***
x_{13}	Competitiveness subindex by the balance of humus per 1 ha
x_{14}	Competitiveness subindex by the greenhouse gas emissions from fuel combustion per 1 ha
x_{15}	Competitiveness subindex by the coefficient of compliance with the structure of sown areas
x_{16}	Competitiveness subindex by the coefficient of erosion risk (hazard)
IEcolC	***Ecological competitiveness index***
x_{17}	Competitiveness subindex by the annual average number of persons employed in agriculture per 1000 ha of agricultural land
x_{18}	Competitiveness subindex by the labor costs with tax-deductions per 1 ha of agricultural land
x_{19}	Competitiveness subindex by the share of labor costs with tax-deductions in the structure of production costs
x_{20}	Competitiveness subindex by the average wage of 1 employee
ISC	***Social competitiveness index***

Source: formed by the authors

the analysis of the standard errors, *t*-statistics, and *P*-value argue statistical significance of all regressors and a free member under the level of significance making 0.95. Thus, the developed model is statistically qualitative and credible and can be used for (1) detection of the internal factors and reserves of growth of the sustainable competitiveness of enterprises, (2) assessment of the relative efficiency of management of SCI formation, (3) managerial decisions making and/or (4) the SCI forecasting.

Approbation of the developed econometric model secured quantitative evaluation of the impact of each factor on the SCI and examination of the reserves of its

Table 13.5 Parameters of the econometric model of dependence of the SCI of agricultural enterprises of Ukraine from the main subindexes, 2016 (n = 5597)

Statistical characteristics	Indicators and their meaning
Multiple linear regression model	$y = 0.199 + 0.338x_1 + 0.143x_4 + 0.136x_{11} + 0.079x_{15} + 0.131x_{18}$
Coefficient of multiple correlation (*R*)	*R* = 0.964 (very high correlation)
Coefficient of multiple determination (R^2)	$R^2 = 0.930$ (statistically significant, because significance $F < 0.05$)
Elasticity coefficients (*E*)	$E_1 = 0.319$; $E_4 = 0.141$; $E_{11} = 0.138$; $E_{15} = 0.059$; $E_{18} = 0.124$
Share of the factor in total variation (*C*), %	$C_1 = 32.5$; $C_4 = 22.3$; $C_{11} = 16.3$; $C_{15} = 0.2$; $C_{18} = 21.8$
Durbin-Watson test (*DW*)[a]	$DW_{fact} = 1.913$; $di = 1.892$, $du = 1.900$ – By 1% level of significance; $DW_{fact} > du$
Fisher's F-criterion	F_{fact} = 14,843; $F_{tabl} = 2.21$ – By 95% level of probability; $F_{fact} > F_{tabl}$
Student's t-criterion	$t_{fact} = 1029.7$; $t_{tabl} = 1.96$ – By 95% level of probability; $t_{fact} > t_{tabl}$

Note. [a]The critical value of DW is taken for 2000 observations
Source: authors' calculations based on the data of the form No. 50-s.g

Table 13.6 Forecast of increasing of the sustainable competitiveness of agricultural enterprises of Ukraine due to the use of internal reserves in the medium-term prospect (until 2025)

	Average values of variables by groups of enterprises, coef.			Difference of average values between:			Reserve of growth (or competitive advantage), coef.	
Variables	outsiders	leaders	TOP-100 leaders	leaders and outsiders	TOP-100 and leaders	Regressors, coef.	outsiders to leaders	leaders to TOP-100
x_1	0.749	1.303	1.859	0.554	0.556	0.338	0.187	0.188
x_4	0.678	1.403	2.279	0.725	0.876	0.143	0.104	0.125
x_{11}	0.748	1.314	1.979	0.566	0.665	0.136	0.077	0.090
x_{15}	0.522	0.539	0.560	0.017	0.021	0.079	0.001	0.002
x_{18}	0.637	1.453	2.154	0.816	0.701	0.131	0.107	0.092
y	0.749	1.241	1.876	0.492	0.635	–	0.476	0.497

Source: authors' calculations based on the econometric model (Table 13.5)

growth on the base of the comparative analysis of outsiders and leaders. It was used for the medium-term forecast (until 2025) of improvement of the competitiveness of agricultural enterprises by involving the internal reserves through rising of the average indexes of the outsiders (i.e. non-competitive enterprises) approaching the leaders, and the leaders – moving towards the TOP-100 of the best (Table 13.6).

Analyzing the obtained results, one can make the following conclusions, particularly, if the average value of the subindexes of competitiveness by the conditional yield of outsider-enterprises increases by 0.554, the gross added value per 1 ha – by 0.725, the relative share of the market – by 0.566, the coefficient of conformity with the structure of cropping areas – by 0.017, and the size of labor costs with tax-deductions per 1 ha of agricultural lands – by 0.816, they will be able to employ the reserve of the SCI growth by 0.187, 0.104, 0.077, 0.001 and 0.107 respectively. Thus, the total reserve of growth accounts for 0.476, i.e. the forecasted value of the SCI equals to 1.225. Hence, outsiders can be transformed into competitive enterprises. It is clear that such result can be achieved only under conditions of the adequate growth of all subindexes and the appropriate financial supply. That task can qualified as real one.

A similar conclusion can be made concerning the comparative economic analysis of leading enterprises and the TOP-100 of the best. Thus, growth of the subindex of competitiveness by the conditional yield at the leading enterprises by 0.556, and approaching the average index of the TOP-100 of the absolute leaders secures the opportunity to increase the SCI by 0.188. Moreover, the average value of the subindexes of competitiveness by the gross added value per 1 ha, relative share of the market, coefficient of conformity with the structure of cropping area, and labor costs with tax-deductions per 1 ha of agricultural lands, observed at the TOP-100, will increase the SCI by 0.125, 0.090, 0.002 and 0.092 respectively. Thus, the total reserve of growth of the competitiveness of leaders, in case they approach the TOP-100, accounts for 0.497, which can be used under conditions of a significant rise of all primary subindexes, included in the model. In that case, the forecasted level of the SCI accounts for 1.738, i.e. a high level of sustainable competitiveness. It is worth noting that in both cases, the most significant growth of the SCI can be achieved due to the increase of conditional yield that is argued by the author's assumption on one hand, and the necessity of a particular concern by the managers of agricultural enterprises on the other hand.

The degree of reach and opportunity of employment of the mentioned reserves is argued by the fact that 53.9%, i.e. 3019 agricultural enterprises, can achieve a higher level of sustainable competitiveness under the available competitive potential without attraction of extra financial resources due to improving the level of management and/or rational use of their resources (Table 13.7).

It is determined that 3019 agricultural enterprises can improve their SCI by 15.2% due to the appropriate level of management, and 148 agribusiness entities can do it by transforming into competitive ones. Approbation of the econometric model enabled quantitative assessment of the relative efficiency of management by forming the sustainable competitiveness. In particular, it is noted that at 46.1% of agricultural enterprises, the management is estimated as relatively efficient, whereas at the rest – it is relatively inefficient, comparing to the average level, and thus, there are reserves for its improvement. It is well-argued by the results of assessment of the relative efficiency of management by reaching the sustainable competitiveness on the example of the chosen 20 enterprises and reserves of its improvement by increasing the management level (Table 13.8). At 10 enterprises with relatively inefficient management, the reserve of the SCI growth by improvement of the management variated within 0.023–0.353. Thus, three of them can be transformed from outsiders into competitive ones. Following the same algorithm, one can forecast the increase of technological, financial, marketing, ecological and social competitiveness of agricultural enterprises by means of the reserves use, can explore those reserves,

Table 13.7 Results of the assessment of reserves of increase of sustainable competitiveness of agricultural enterprises of Ukraine by improving the level of management and/or use of resources, 2016

Groups of enterprises by reserve of growth of SCI, %		Number of enterprises in the group, units	Average reserve of growth of SCI, %	Share of competitive, %		Reserve of growth of share of competitive, %	SCI, coef.		Reserve of growth of SCI, coef.
				fact.	theoret.		fact.	theoret.	
I	To 5	1380	2.5	33.9	36.2	2.3	0.919	0.942	0,023
II	5.1–10.0	935	7.3	23.1	29.8	6.7	0.842	0.903	0,061
III	Over 10.0	704	15.2	13.8	21.4	7.6	0.729	0.840	0,111
On average		3019	6.9	25.9	30.8	4.9	0.851	0.910	0.059

Source: authors' calculations based on the econometric model (Table 13.5)

Table 13.8 Results of evaluation of relative efficiency of managing of formation of sustainable competitiveness on the example of specific agricultural enterprises and reserves of its increase due to improvement the level of management, 2016

No.	SCI, coef. fact.	SCI, coef. theoret. (normat.)	Efficiency of management of formation of SCI, coef.	Reserve of growth of SCI, coef.	Indicators for the econometric model x_1	x_4	x_{11}	x_{15}	x_{18}
Enterprises with relatively inefficient management of formation of SCI (coef. < 1.000)									
1	0.259	0.399	0.650	0.140	0.181	−0.098	0.395	1.000	0.151
2	0.817	1.170	0.698	0.353	1.305	1.439	1.532	1.000	0.279
3	0.317	0.447	0.710	0.130	0.508	0.145	0.085	0.409	0.087
4	0.485	0.613	0.791	0.128	0.691	0.320	0.412	0.627	0.226
5	0.434	0.545	0.796	0.111	0.543	0.281	0.401	0.462	0.241
6	0.511	0.631	0.809	0.120	0.781	0.229	0.516	0.359	0.284
7	0.856	1.051	0.815	0.195	1.480	0.469	0.135	1.000	1.427
8	0.864	0.933	0.926	0.069	1.202	0.688	0.958	0.765	0.297
9	1.090	1.154	0.945	0.064	1.021	0.700	0.971	0.532	2.560
10	0.701	0.724	0.968	0.023	0.746	0.484	0.642	0.640	0.501
Enterprises with relatively efficient management of formation of SCI (coef. ≥ 1.000)									
11	1.257	1.257	1.000	–	1.143	2.395	1.883	0.446	0.289
12	1.309	1.309	1.000	–	1.145	2.045	0.786	0.755	2.012
13	1.106	1.101	1.005	–	1.063	1.213	1.455	0.550	0.973
14	1.197	1.178	1.016	–	1.111	0.562	0.391	1.000	2.982
15	1.150	1.022	1.125	–	1.308	0.762	0.629	1.000	0.819
16	1.221	1.032	1.183	–	1.146	1.619	1.238	0.190	0.236
17	1.423	1.170	1.217	–	1.232	2.075	1.592	0.304	0.130
18	1.373	1.074	1.278	–	1.021	0.071	1.775	1.000	1.525
19	1.736	1.318	1.317	–	1.575	1.781	1.197	0.450	1.023
20	1.812	1.228	1.476	–	2.350	0.012	1.041	1.000	0.094

Source: authors' calculations based on the data of the form No. 50-s.g. with using the developed econometric model and assess the relative level of management efficiency of the competitiveness formation.

Results of the correlation analysis identified a high direct correlation relation of the SCI of agricultural enterprises at the market of winter wheat grain with the subindex of competitiveness by the yield ($r = 0.791$) and by the gross added value ($r = 0.753$); moderate direct correlation relation with the subindexes of the competitiveness at the market of land lease ($r = 0.412$), and by the labor costs with tax-deductions ($r = 0.463$), as well as a weak direct relation with the subindex of competitiveness by the greenhouse gas emission from fuel combustion ($r = 0.112$). The mentioned subindexes for the regression modeling are chosen according to the previously applied principle.

The developed econometric model (Table 13.9) has successfully passed all stages of testing in terms of credibility, statistical significance, and adequacy. The factors, included in the model, demonstrate a very high correlation relation with the SCI, while 89.1% of its variation is determined by the chosen factors. Economic interpretation of the model parameters substantiate that growth of the subindex of competitiveness by the yield per a unit causes rising of the SCI by 0.493, while its growth by the gross added value per a unit results in the SCI increase by 0.197; growth of the subindex of competitiveness at the market of land lease per a unit results in the increase of the SCI by 0.067 on average, by the greenhouse gas emission – by 0.090, by the labor costs with tax-deductions – by 0.092. Comparing the relative impact of the mentioned factors referring to the elasticity coefficient, one can conclude that growth of the factor x_{17} by 1% causes the SCI increase by 0.723%, i.e. it makes the greatest impact; increase of x_4 by 1% contributes to the SCI growth by 0.607%,

Table 13.9 Parameters of the econometric model of dependence of the SCI of agricultural enterprises of Ukraine on the winter wheat grain market from the main subindexes, 2016 *(n = 4264)*

Statistical characteristics	Indicators and their meaning
Multiple linear regression model	$y = 0.090 + 0.493x_1 + 0.197x_4 + 0.067x_{14} + 0.090x_{17} + 0.092x_{19}$
Coefficient of multiple correlation (R)	$R = 0.944$ (very high correlation)
Coefficient of multiple determination (R^2)	$R^2 = 0.891$ (statistically significant, because significance $F < 0.05$)
Elasticity coefficients (E)	$E_1 = 0.321$; $E_4 = 0.607$; $E_{14} = 0.476$; $E_{17} = 0.723$; $E_{19} = 0.496$
Share of the factor in total variation (C), %	$C_1 = 44.5$; $C_4 = 32.0$; $C_{14} = 4.7$; $C_{17} = 2.6$; $C_{19} = 5.3$
Durbin-Watson test (DW)[a]	$DW_{fact} = 1.897$; $di = 1.892$, $du = 1.900$ – By 1% level of significance; $DW_{fact} > du$
Fisher's F-criterion	$F_{fact} = 6947$; $F_{tabl} = 2.21$ – By 95% level of probability; $F_{fact} > F_{tabl}$
Student's t-criterion	$t_{fact} = 565.1$; $t_{tabl} = 1.96$ – By 95% level of probability; $t_{fact} > t_{tabl}$

Note. [a]The critical value of DW is taken for 2000 observations
Source: authors' calculations based on the data of the form No. 50-s.g

Table 13.10 Forecast of increasing of the sustainable competitiveness of agricultural enterprises of Ukraine on the winter wheat grain market due to the use of internal reserves in the medium-term prospect (until 2025)

Vari-ables	Average values of variables by groups of enterprises, coef.			Difference of average values between:		Regre-ssors, coef.	Reserve of growth (or competitive advantage), coef.	
	outsiders	leaders	TOP-100 leaders	leaders and outsiders	TOP-100 and leaders		outsiders to leaders	leaders to TOP-100
x_1	0.783	1.188	1.648	0.405	0.460	0.493	0.200	0.227
x_4	0.642	1.308	2.567	0.666	1.259	0.197	0.131	0.248
x_{14}	0.747	1.183	1.311	0.436	0.128	0.067	0.029	0.009
x_{17}	0.991	1.008	1.029	0.017	0.021	0.090	0.002	0.002
x_{19}	0.707	1.252	1.555	0.545	0.303	0.092	0.050	0.028
y	0.788	1.240	1.780	0.452	0.540	–	0.412	0.513

Source: authors' calculations based on econometric model (Table 13.9)

growth of x_{19} – by 0.496%, x_{14} – by 0.476%, and x_1 – by 0.321%. Hence, the subindex by the yield makes the greatest effect in absolute terms, confirming the author's hypothesis, but the least impact in relative terms.

Moreover, analysis of the total volume of variation demonstrates that the subindex of competitiveness by the yield explains the SCI variation – 44.5%; the second position is taken by the gross added value – 32.0%, other factors explain 2.6–5.3% of the SCI variation. Thus, main reserves of the SCI increase should be primarily related with the increase of the yield and gross added value. Comparison of the actual and table values of Fisher *F*-test and Student *t*-test argues statistical significance and credibility of the model. The values of the Durbin-Watson criterion manifest complete absence of autocorrelation of the residues of the first order with the probable deviation of 1%. All coefficients of the model are statistically credible, whereas calculation of β-factors show that maximum reserves of the SCI growth are primarily related with the factors x_{21} and x_1.

Testing of the developed econometric model along with the comparative analysis of outsiders, leaders, and the best achievements of the TOP-100 leaders secured the opportunity to make a forecast concerning improvement of the competitiveness of agricultural enterprises at the market of winter wheat grain due to the gradual approaching the best enterprises by worse business entities (Table 13.10).

Thus, in case the outsider-enterprises upgrade their operation and achieve the level of competitive enterprises by the yield, they will be able to utilize the reserve of the SCI growth by 0.200. If the average value of the subindex of competitiveness by the gross added value increases by 0.666, at the market of land lease – by 0.436, the greenhouse gas emission – by 0.017, and the labor costs with tax-deductions – by 0.545, the outsiders will be able to utilize the reserve of the SCI growth by 0.131,

0.029, 0.002 and 0.050 respectively. Thus, the total reserve of the SCI growth accounts for 0.412, whereas its expected level makes 1.200, i.e. outsider-enterprises can be transformed into competitive ones. It is clear that the expected SCI is a little lower than the level, reached by the leaders that is caused by the impact of other factors, not included in the model.

In case the leaders try to approach the TOP-100 of the best, their SCI value will increase by 0.513, also due to a growth of the subindex by the yield – by 0.227, the gross added value – by 0.248, at the market of land lease – by 0.009, the greenhouse gas emission – by 0.002, and the labor costs with tax-deductions – by 0.028. Thus, the expected value of the SCI makes 1.753 that is characterized as a high level of sustainable competitiveness. Achievement of the forecasted indexes is considered as an absolutely real goal.

Results of the correlation analysis manifest that the SCI of agricultural enterprises at the market of maize grain has a high direct correlation relation with the subindexes of competitiveness by the yield ($r = 0.778$) and by the gross added value ($r = 0.791$), a significant relation with the subindex by the labor costs with tax-deductions ($r = 0.572$), as well as a moderate direct relation with the subindex of competitiveness at the market of land lease ($r = 0.414$). They are chosen as the factors for development of the econometric model (Table 13.11). It is worth noting that the model does not include the indexes of ecological competitiveness, because one of them (the humus balance) closely correlates with the already-included subindex of the yield that is considered as autocorrelation, whereas the subindex of the greenhouse gas emission is characterized by a reverse relation that does not comply with the logical-economic assumption.

Table 13.11 Parameters of the econometric model of dependence of the SCI of agricultural enterprises of Ukraine on the maize grain market from the main subindexes, 2016 *($n = 3236$)*

Statistical characteristics	Indicators and their meaning
Multiple linear regression model	$y = 0.395 + 0.249x_1 + 0.192x_4 + 0.067x_{14} + 0.100x_{19}$
Coefficient of multiple correlation (R)	$R = 0.939$ (very high correlation)
Coefficient of multiple determination (R^2)	$R^2 = 0.882$ (statistically significant, because significance $F < 0.05$)
Elasticity coefficients (E)	$E_1 = 0.388$; $E_4 = 0.580$; $E_{14} = 0.468$; $E_{19} = 0.803$
Share of the factor in total variation (C), %	$C_1 = 29.8$; $C_4 = 35.0$; $C_{14} = 5.2$; $C_{19} = 18.2$
Durbin-Watson test (DW)[a]	$DW_{fact} = 1.924$; $di = 1.892$, $du = 1.900$ – By 1% level of significance; $DW_{fact} > du$
Fisher's F-criterion	$F_{fact} = 5960$; $F_{tabl} = 2.37$ – By 95% level of probability; $F_{fact} > F_{tabl}$
Student's t-criterion	$t_{fact} = 448.0$; $t_{tabl} = 1.96$ – By 95% level of probability; $t_{fact} > t_{tabl}$

Note. [a]The critical value of DW is taken for 2000 observations

Source: authors' calculations based on the data of the form No. 50-s.g

The composed model has successfully passed all stages of testing concerning statistical significance, credibility and adequacy. Thus, with the 95% probability, one can affirm that an increase of the factor x_1 per a unit causes the SCI growth by 0.249, x_4 – by 0.192, x_{14} – by 0.067, and x_{19} – by 0.100. The greatest absolute impact is made by the yield, arguing the presented hypothesis. Coefficients of elasticity show that the increase of the factor x_{19} by 1% causes the SCI increase by 0.803%, x_4 – by 0.580%, x_{14} – by 0.468%, and x_1 – by 0.388. It is determined that variation of the SCI depends on the factor x_1 – by 29.8%, x_4 – by 35.0%, x_{14} – by 5.2% and x_{19} – by 18.2%. All coefficients of the model are statistically significant. Calculation of the β-coefficients demonstrates that maximum reserves for growth of the SCI of the studied enterprises at the market of maize grain are related with the factor x_1, smaller ones – with x_{19} and x_{14}, and the smallest ones – with x_4. It is explained by differences in measuring the mentioned subindexes.

Considering the regressors of the econometric model as objective assessments of significance of some subindexes in formation of the SCI, the author have made a forecast of sustainable competitiveness of the enterprises at the market of maize grain (Table 13.12). If the average value of all subindexes of the outsider-enterprises, included in the model, achieves the average level of competitive business entities, the outsiders will be able to utilize the reserve of the SCI growth by 0.301, also due to the increase of the factor x_1 – by 0.109, growth of x_4 – by 0.157, and due to x_{14} and x_{19} – by 0.029 and 0.006 respectively. On a mid-term horizon, the leading enterprises should focus on the TOP-100 of the best ones. If they achieve the average indexes of those best enterprises, it will secure the growth of the SCI of the leaders by 0.231. Thus, the forecasted value of the SCI for the current outsiders accounts for 1.050, and for the leaders – 1.472. Enterprises of the TOP-100 group, having achieved the best indexes, have competitive advantages. They are integrally manifested by the SCI within 0.231, comparing to the leaders, and 0.532, comparing to the outsiders. In the future, it is necessary to maintain the advantages, and, if possible, to grow them in order to increase their share and competitive position at the European market.

Table 13.12 Forecast of increasing of the sustainable competitiveness of agricultural enterprises of Ukraine on the maize grain market due to the use of internal reserves in the medium-term prospect (until 2025)

Variables	Average values of variables by groups of enterprises, coef.			Difference of average values between:		Regressors, coef.	Reserve of growth (or competitive advantage), coef.	
	outsiders	leaders	TOP-100 leaders	leaders and outsiders	TOP-100 and leaders		outsiders to leaders	leaders to TOP-100
x_1	0.764	1.202	1.409	0.438	0.207	0.249	0.109	0.052
x_4	0.559	1.378	2.165	0.819	0.787	0.192	0.157	0.151
x_{14}	0.814	1.245	1.612	0.431	0.367	0.067	0.029	0.025
x_{19}	0.672	1.283	1.684	0.611	0.401	0.010	0.006	0.004
y	0.749	1.241	1.876	0.492	0.635	–	0.301	0.231

Source: authors' calculations based on the econometric model (Table 13.11)

13.4 Conclusions

In conclusion, we noted that this section presents empirical evidence on the zonal features of the formation of sustainable competitiveness of agricultural enterprises on the Ukrainian grain market. The provision on zonal features of formation of sustainable competitiveness of agricultural enterprises was further developed.

For the first time, a multi-factor linear econometric models of the dependence of SCI from the main subindexes were developed. The results of econometric modeling confirmed the hypothesis that there is a relationship between the subindexes value and the SCI level of agricultural enterprises, whereas the largest positive impact on the SCI is made by the subindex of competitiveness by the yield. The strategic direction of improving the competitiveness of agricultural enterprises involves gradually increasing the internal reserves through rising of the average indexes of the outsiders (i.e. non-competitive enterprises) approaching the leaders, and the leaders – moving towards the TOP-100 of the best. The degree of reach and opportunity of employment of the mentioned reserves is argued by the fact that 53.9%, i.e. 3019 agricultural enterprises, can achieve a higher level of sustainable competitiveness under the available competitive potential without attraction of extra financial resources due to improving the level of management and/or rational use of their resources. At the same time, the practical implementation of reserves in other enterprises requires additional financial resources.

The results of the study can be used for (i) identification of reserves for increasing sustainable competitiveness; (ii) express-estimation and forecasting the level of competitiveness use of agricultural enterprises; (iii) assessment of the relative level of effectiveness of the management of formation of sustainable competitiveness; (iv) making decisions on the advisability of increasing the level of competitiveness, as well as (v) other stakeholders (for example, investors, shareholders, state government and regional bodies, etc.) when making investment decisions, modeling potential SCI of new enterprises, designing events for improving competitiveness and financial support for sustainable development.

References

Atta Mills, E. F. E., Dong, J., Yiling, L., Baafi, M. A., Li, B., & Zeng, K. (2021). Towards sustainable competitiveness: how does financial development affect dynamic energy efficiency in belt & road economies? *Sustainable Production and Consumption, 27*, 587–601. https://doi.org/10.1016/j.spc.2021.01.027

Biazzin, C., Paiva, E. L., & De Figueiredo, J. C. B. (2021). Operational capabilities dissemination for sustainable competitiveness: towards an integrated framework. *International Journal of Services and Operations Management, 38*(3), 309–335. https://doi.org/10.1504/ijsom.2021.113593

Castro Oliveira, J., Lopes, J. M., Farinha, L., Silva, S., & Luízio, M. (2022). Orchestrating entrepreneurial ecosystems in circular economy: the new paradigm of sustainable competitiveness. *Management of Environmental Quality: An International Journal, 33*(1), 103–123. https://doi.org/10.1108/MEQ-11-2020-0271

Choudhary, A. (2022). Impact of relational bonding on environmental efficiency and Firm's sustainable competitiveness. *Indian Journal of Economics and Development, 18*(1), 169–174. https://doi.org/10.35716/IJED/21251

Danileviciene, I., & Lace, N. (2021). Assessment of the factors of sustainable competitiveness growth of the companies in Latvia and Lithuania. *International Journal of Learning and Change, 13*(4–5), 510–526. https://doi.org/10.1504/IJLC.2021.116678

Delgosha, M. S., Saheb, T., & Hajiheydari, N. (2021). Modelling the asymmetrical relationships between digitalisation and sustainable competitiveness: a cross-country configurational analysis. *Inf Syst Front, 23*(5), 1317–1337. https://doi.org/10.1007/s10796-020-10029-0

Dziembała, M. (2021). The enhancement of sustainable competitiveness of the Cee regions at the time of the covid-19 pandemic instability. *Sustainability, 13*(23). https://doi.org/10.3390/su132312958

Farinha, F., Bienvenido-Huertas, D., Pinheiro, M. D., Silva, E. M. J., Lança, R., Oliveira, M. J., & Batista, R. (2021). Sustainable competitiveness of tourism in the Algarve region. Critical stakeholders' perception of the supply sector. *Sustainability, 13*(11). https://doi.org/10.3390/su13116072

Fedulova, S., Dubnytskyi, V., Myachin, V., Yudina, O., & Kholod, O. (2021). Evaluating the impact of water resources on the economic growth of countries. *Agricultural and Resource Economics, 7*(4), 200–217. https://doi.org/10.51599/are.2021.07.04.11

Hategan, C., Pitorac, R., Hategan, V., & Imbrescu, C. M. (2021). Opportunities and challenges of companies from the Romanian e-commerce market for sustainable competitiveness. *Sustainability, 13*(23). https://doi.org/10.3390/su132313358

Karman, A., & Savanevičienė, A. (2021). Enhancing dynamic capabilities to improve sustainable competitiveness: insights from research on organisations of the Baltic region. *Balt J Manag, 16*(2), 318–341. https://doi.org/10.1108/BJM-08-2020-0287

Kucher, A. (2019). Zonal features of formation and reserves of increasing the sustainable competitiveness of agricultural enterprises. *Agricultural and Resource Economics, 5*(3), 77–105. https://doi.org/10.22004/ag.econ.293987

Kucher, A. (2020). Soil fertility, financial support, and sustainable competitiveness: evidence from Ukraine. *Agricultural and Resource Economics, 6*(2), 5–23. https://doi.org/10.51599/are.2020.06.02.01

Kucher, A., Kucher, L., Taratula, R., & Dudych, L. (2021). Formation of sustainable competitiveness of enterprises on soils of different quality. *International Journal of Information Systems in the Service Sector, 13*(3), 49–64. https://doi.org/10.4018/IJISSS.2021070104

Liu, C., & Wang, T. (2022). Effect of transformational leadership on environmental management and sustainable competitiveness in hi-tech industry. *Journal of the Balkan Tribological Association, 28*(1), 150–157.

Liu, X., Wang, Y., & Wang, L. (2022). Sustainable competitiveness evaluation of container liners based on granular computing and social network group decision making. *Int J Mach Learn Cybern, 13*(3), 751–764. https://doi.org/10.1007/s13042-021-01325-5

López-Fernández, A. M., Terán-Bustamante, A., & Martínez-Velasco, A. (2022). Machine learning sustainable competitiveness for global recovery. In A. M. López-Fernández & A. Terán-Bustamante (Eds.), *Business recovery in emerging markets. Palgrave studies in democracy, innovation, and entrepreneurship for growth*. Palgrave Macmillan. https://doi.org/10.1007/978-3-030-91532-2_13

Rajnoha, R., & Lesnikova, P. (2022). Sustainable competitiveness: how does global competitiveness index relate to economic performance accompanied by the sustainable development? *Journal of Competitiveness, 14*(1), 136–154. https://doi.org/10.7441/joc.2022.01.08

Shahbaz, M. S., Javaid, M., Alam Kazmi, S. H., & Abbas, Q. (2022). Marketing advantages and sustainable competitiveness through branding for the supply chain of islamic country. *Journal of Islamic Marketing, 13*(7), 1479–1491. https://doi.org/10.1108/JIMA-04-2020-0094

Skrypnyk, A., Klymenko, N., Tuzhyk, K., Galaieva, L., & Rohoza, K. (2021a). Prerequisites and prospects for sustainable development of grain production in Ukraine. *Agricultural and Resource Economics, 7*(3), 90–106. https://doi.org/10.51599/are.2021.07.03.06

Skrypnyk, A., Zhemoyda, O., Klymenko, N., Galaieva, L., & Koval, T. (2021b). Econometric analysis of the impact of climate change on the sustainability of agricultural production in Ukraine. *Journal of Ecological Engineering, 22*(3), 275–288. https://doi.org/10.12911/22998993/132945

Yahya, F., Abbas, G., Hussain, M., & Waqas, M. (2022). Financial development and sustainable competitiveness in arctic region: a dynamic panel data analysis. *Problemy Ekorozwoju, 17*(1), 267–278. https://doi.org/10.35784/pe.2022.1.24

Chapter 14
Economic Viability and Sustainability in Baking Industry/ Simeuna – Bread Producer with Own Grain Production

Aleksandra Figurek and Marinos Markou

14.1 Introduction

The key requirement for successful production planning is the availability of information related to resources, restrictions, and available production capacity, as well as the company's commercial and economic features. Nowadays, grain production confronts a variety of major issues, including rising demand, less cultivable land due to population increase, and a loss in agricultural productivity due to natural resource degradation. Slof and Argiles (2000) underline the importance of agriculture and the fact that scientists should pay greater attention to agriculture because of the value it has for the people.

The fast expansion of the economy, as well as the need of timely information for a specific activity, necessitate the establishment of an accounting information system in order to optimise company operations (Istvánné, 2000). Financial management is the lifeblood of every organisation (Santacoloma et al., 2009) and proofs necessity for its stability. If the producer regains control of their finances, it will have a good impact on the organisation of operations in the firm. This entails effective management, which necessitates the availability of high-quality production and economic data regarding company activities (Del'homme, 1997). Information obtained from accessible manufacturing processes may be viewed as a significant aspect in the interaction with other production variables such as land, labour, and capital, and can lead to more successful corporate operations. Furthermore, the

A. Figurek (✉)
Gnosis Mediterranean Institute for Management Science, School of Business, University of Nicosia, Nicosia, Cyprus

M. Markou
Department of Agric. Economics, Agricultural Research Institute, Nicosia, Cyprus

J. M. Ferreira da Rocha et al. (eds.), *Baking Business Sustainability Through Life Cycle Management*, https://doi.org/10.1007/978-3-031-25027-9_14

information distributed to producers through the media, educational institutions, and advisory services serve to enrich their knowledge, allowing them to make better decisions based on that information, take advantage of market opportunities, and continuously manage changes in their production activities.

It is crucial to analyse information flow and construct the system of economic viability while taking into account all of the aspects that may have a significant influence on the company's economic sustainability. This process begins with the producers, whose judgments and selection of information sources constitute the foundation of the entire system. If the producer concentrates on getting external information in addition to internal information, the prospects of strengthening organisational and management functions increase. Greater organisation in the execution of productive activities allowed producers to acquire insight into their production capacities as well as a better assessment of indications that reflect the success or failure of production activities (Sørensen et al., 2010).

This approach results in improved control of production processes, as well as increased prospects for higher product quality, manufacturing traceability, expanding market share. Control of operations, along with a database of historical company results, represents superior decision-making assistance. In this regard, an adequate system of organisation and process control can produce a stronger influence on the outputs of the manufacturing process. The producer operations schedule should be linked to the system for monitoring the consequences of executing production activities. The essential need for coordinating production activities with defined budgets is the recording of data on production processes (quantities and prices, income from sales).

Producers, as decision makers, must constantly work on problems which increase in a complex industrial sector that has always been dependent on numerous elements (Parker et al., 1997). If there are discrepancies between planned and actual activities in defined period, data recorded in the past can serve the purpose of identifying the critical points registered in previous production cycles, and it is necessary in the future to adjust the production plan of activities and the company's budget or financial means.

To create an adequate information base and ensure the economic viability and sustainability of novel products in the baking industry, producers must express their willingness to adopt new work practises such as continuous recording of production activity results, and, if necessary, pass additional training in learning new skills.

Having precise records (Heidhues & Patel, 2008) allows the producers to compare all expenditures (labor costs, etc.). Changes in income, raw material and final product prices, manufacturing volume are important for organizational process (Somiari, 2008).

Information on production related to quantity production, i.e. the yield, the dynamics of consumption inputs from the quantities of seeds, fertilisers, protective aids, time spent by paid and their own labour, is critical for the planning and organisation of the next production activities, with the goal of reducing the risk of getting a lower yield than expected. Crop yields are greatly influenced by meteorological or weather factors, which can greatly affect the achievement of lower yields, and it is required that producers in organisational terms, reviewing its past operations and economic results, achieve influence for the better results of the company's operations.

Information related to sales, marketing and market aspects, where they play a significant role in the realization of final products and adequate strategy in this regard, can play a decisive role in establishing economic viability in the production and sale of bread.

14.1.1 *Bread Producer with Their Own Grain Production – Simeuna*

By the heterogeneous companies (which produce their own inputs for bread production), due to the presence of several types of productional activities, there is a need for a more complex approach in the measurement and assessment of the results of their entire business. These businesses make a range of goods and employ a variety of inputs that must be tracked and correctly allocated. Such an approach presents a strong base for modelling information, allowing the construction of suitable information bases for decision-making.

Company Simeuna, by its example, showed positive experiences in bread production in BiH (Republic of Srpska entity) in the traditional way. With its production program, it has become unique on the BiH market, because it has completed the production cycle from sowing to table (Pictures 14.1 and 14.2)

Domestic food Simeuna is based on the healthy food, produced on traditional way, similar to organic system husbandry how it was from ancient times practice in homeland and which for centuries was given positive results – a health family (Picture 14.3).

Production of grain is realized without the use of pesticides and herbicides and production of bread and baked goods performances without additives and preservatives. In this way is reaching production of exceptionally high-quality food, which is together medicine and salvation for mankind. Food made in this way guarantees a healthy and long life, as Hippocrates said a long time ago: *"Let food be the medicine"*.

Picture 14.1 Traditional bread based on sourdough

Picture 14.2 Rye bread

Picture 14.3 From the field to the table

14.1.2 Importance of the Information for the Economic Viability and Sustainability

Information is vital for making business decisions. They provide guidelines and represent the direction and orientation of a company. Without timely information, the company is not able to make quality decisions, which are of great importance for successful business. Prior to determining the core business and product type, Simeuna collected all information on both the domestic and foreign markets. The business philosophy was to define the core business and produce a product that is completely unique in the market.

For the strategy of entering or introducing a new product on the market, it is very important to do quantitative and qualitative research. Strategy represent the way in which a company tries to differentiate itself from competitors, using its relative strength and to better meet customer needs.

When developing any business strategy, three main players must be taken into account: the company itself, the customer and the competition. The following factors are key in quantitative research:

1. consumption habits of the population
2. habits of buying specific brands of products
3. market share of individual product brands
4. product positioning in the minds of consumers
5. testing the effects of company propaganda
6. research before launching a new product
7. continuous monitoring of the effects of undertaken marketing actions
8. identification of target consumer groups according to demographic, economic and sociological indicators
9. assessment of the absolute potential (sales potential of all producers from that production branch) and the real market potential (sales potential of a specific producer)

Qualitative research should focus on groups and in-depth interviews and explore deeper factors of consumer behavior, such as: needs, motives, feelings.

14.1.3 Customer – Based Startegy

Customers are the most important link in the company's business chain. The strategy based on the customer should be based on getting to know them, their wishes and needs. That should be the ultimate goal of every business. Profit should never be the only goal the company strives for, but it is much more important to find a buyer, conquer the market, impose its name, become famous, and profit ultimately comes as a result of all that.

14.1.4 Competitor – Based Strategies

Competitor-based strategies can be built by looking for possible sources of differentiation in functions ranging from sales, design, and technique. Any difference between a particular company and a competitor must be related to one or more of the four elements that jointly determine profit, price, volume and cost. For example, achieving better prices due to better aesthetic design and product quality, or with more points, unlike a competitor.

In this particular case, the company Simeuna based its production program on uniqueness. The raw material (cereal cultivation) was produced in the traditional way without the use of pesticides and herbicides, and the final products (flour, flakes, bread, pastries, etc.) were produced without the use of additives and preservatives. That is how this company achieved its uniqueness.

14.1.5 Market Design of the Offer

Simeuna appeared on the market in accordance with modern market practice. It determined its brand, name and sign in accordance with tradition and became very recognizable on the market. It is a symbol of the identity of a nation, script and tradition (Picture 14.4).

Characteristics from the point of view of advertising are based on the next elements:

- Attracting attention – it attracts attention because it is unique.
- Understanding the message – it is a symbol of the identity of a nation, letters, traditions and cultural heritage.
- Emotional potential – it evokes intense positive emotions, tenderness, memory.
- Characteristics from the point of view of packaging – attracting attention.

The rule of successful business is that a company cannot stay on the market with the same production program for longer period without introducing new products or changes. Each company in bread industry should strive to introduce products into the

Picture 14.4 The logo of Simeuna

production program based on the knowledge of the wishes, needs, motives of customers, and not to come up with a new product and only then create a market for it.

14.1.6 The Price

As a significant economic indicator, the price reacts to all changes and trends in the world market. Factors influencing the price policy are: costs, demand, competitors, the state, etc. In a market economy, demand is the ceiling, i.e. the upper price level, and costs are the pathos, i.e. the lower limit. Decisions on prices are also regulated by the state by special laws. Price should be an instrument, not a goal in international marketing. It, in combination with other marketing instruments, should contribute to establishing a more stable position on the market in relation to other participants. The price also contributes to long-term trust with end customers and consumers. When determining prices, certain problems arise:

- problem of multinational and multicultural character,
- currency problem, which refers to the fluctuation of exchange rates,
- the problem of securing collection in less developed countries,
- administrative problem, which accompanies the transfer of goods across the border.

The company Simeuna has adjusted the prices of its products to the purchasing power of the population. Thus, it set a 70% lower price on the domestic market.

14.1.7 Sales Channels

Sales and distribution is the operational part through which the entire marketing policy of the company is realized. According to the praxis, it is set up in such a way that the goods are delivered to companies and customers in the right place at the right time, in appropriate quantities and under mutually acceptable conditions. Simeuna based the distribution of its products on large trading houses with whom it also concluded sales contracts.

This company has not organized its strategy of entering the foreign market, nor does it intend to apply it according to the principles of international marketing and established practice. It does not form indirect sales channels through trade intermediaries and distributors, nor does it use consortia or export groups and export management firms to promote exports.

Simeuna decided in a unique way to enter a foreign market and maintain its position in that market. Company organized also the distribution of its products in the form of cocktail delivery through the Chamber of Republika Srpska, for various events and receptions of foreign business sociaty visiting the Chamber of Republika Srpska, the Ministry of Agriculture, the Ministry of Trade and Tourism and scientific gatherings. It has business cooperation with the Banja Luka Fair (cocktails for the opening and closing of fairs) and where are also foreigners who are directly acquainted with the production program of the company (Picture 14.5).

The cocktails are produced in the traditional way, serves in earthenware bowls with wood cutlery, and places them on wicker wreaths of grain (Picture 14.6).

Picture 14.5 Cocktail prepared for foreign visitors

Picture 14.6 Cocktail in the nature

14.1.8 Promotion

The promotion of its products was also organized by Simeuna in a unique way. They uses all the events that the company monitors through the Chamber of Commerce for promotional activities in the form of sharing catalogs with foreign customers who are interested in this type of production. Thus, foreign partners have the opportunity to get acquainted with the products they are interested in through a visit to the Chamber of Commerce and through magazines that the chamber forwards to other chambers abroad. On that occasion, foreign customers get acquainted with the company's production program, and those interested establish direct contacts (Picture 14.7)

Considering that all the mentioned manifestations must be accompanied by the media, which regularly emphasize tradition and cultural heritage.

14.1.9 Fairs

Participation in fairs is a necessity for companies. The nature of the product itself sometimes requires a fair presentation, because then it provides direct and physical contact with the product itself.

Picture 14.7 Promotion of products

Picture 14.8 Participation in fairs

In its strategy, the company decided on another way to promote sales – participation in fairs, both in Republika Srpska and abroad. Company visited the largest European fair Bio-Fah in Nurnberg, the International Fair of Entrepreneurship in Belgrade, and exhibited her products in other European countries (Picture 14.8).

The company has won numerous awards:

- At the selection of the most successful companies in Republika Srpska for 2004, the company received a special award for the production of natural healthy food,
- Glas Srpske Award at HIP 2000,
- Recognition of the Association of Innovators of the City of Banja Luka,
- Recognition of the Association of Citizens of the Republic of Srpska,
- At the Fair of Education, Finance and Entrepreneurship for 2005,
- Award for Preservation of Tradition and Entrepreneurship,
- At the Selection of the Most Successful Companies of the Republic of Srpska for 2007, the company Simeuna received a special award for its contribution to the development of women's entrepreneurship.

14.2 The System of Economic Viability and Sustainability

Producers, as decision makers, must constantly work on problems that develop in a complex agricultural-manufacturing industry and which has always been dependent on numerous elements. Decision-making and management of production activities with efficient data processing and their transformation into useful information for companies, is an important aspect in terms of production management. Economic indicators on the production activities are the basis for planning their future business activities.

Adequate management of production processes provides an opportunity for companies to achieve significant market share. The efficiency of corporate decision-making under competitive situations is determined by the amount, timeliness, and quality of information presented. Florey et al. (2004) underline the need of providing producers with technical support in organising and keeping accounting information. According to Frye (2009), documenting financial occurrences is essential in order for businesses to plan production and general management. Arfini (1998) highlights the rationale for the system to collect data from production and emphasises its relevance, particularly for the goal of annually reviewing the entity's income.

According to Dalci and Tanis (2002), accounting is important in terms of giving quantitative information. The data acquired by the information system is analyzed in order to give users with the information they need to plan, control, and continue business activities. Along with the development of system comes the implementation of the JIT (just-in-time) idea. A production system that employs the JIT principle may be competitive and continuously ready to satisfy market demands. In this regard, the utilisation of information technology is critical for manufacturers, since the JIT system provides considerable benefits.

Understanding the company's business allows to identify all of the specifics associated to cereal and bread production, as well as the procedures that are inherent. Providing information on firm activities for the purposes of the management process necessitates the establishment of an economic viability system. The system should be designed to help with the following:

- Decision support,
- Achieving planned economic outcomes,
- Direct control over ongoing company operations.

The process of creating the economic viability system for the aim of enhancing production, which includes the mutual agreement and operation of the accounting information system and corporate decision-making. Accounting informing offers the information required for strategic, tactical, and operational planning, documenting, and accounting analysis for decision-making.

The objective of the system's existence is to measure the conversion of inputs into outputs. Finding critical areas in the production and organisational processes is tough and confronts numerous hurdles due to a lack of appropriate financial information by organisations. It is important to create a comprehensive framework that includes the following elements:

- Monitoring grain and bread production operations,
- Monitoring production and financial outcomes,
- Information processing,
- The creation of a database of all active production activities.

It is essential that data on company operations be recorded using a consistent manner based on trustworthy and systematised documentation. The construction of suitable records within the information system is anticipated to include information on the state and flow of resources and manufacturing processes in the sector. The records of business operations offer systematic documenting of business actions that occur on the filed during a specific time period. To achieve such requirements, the following categories of records should be created:

- Records of the available capacities,
- Records of costs for cereal production,
- Records of costs for bread production,
- Records of the income, obligations, payments, etc.

Based on these records, it is able to generate reports on certain cost categories, such as implemented operations, working hours spent for each type of production, suppliers, consumers, and so on (Diagram 14.1).

These records serve as the foundation for creating summary reports (reports for production, monthly reviews of inputs, work hours, achieved sales of the products, etc.). The records should give insight into the flow of production resources as well as their purpose, accurate, and reasonable utilisation. All additional actions that impact operational performance should be included in this regard. The efficiency of all components involved, as well as the cost-effectiveness of expenditures based on objective or subjective impacts, can be analysed from the records.

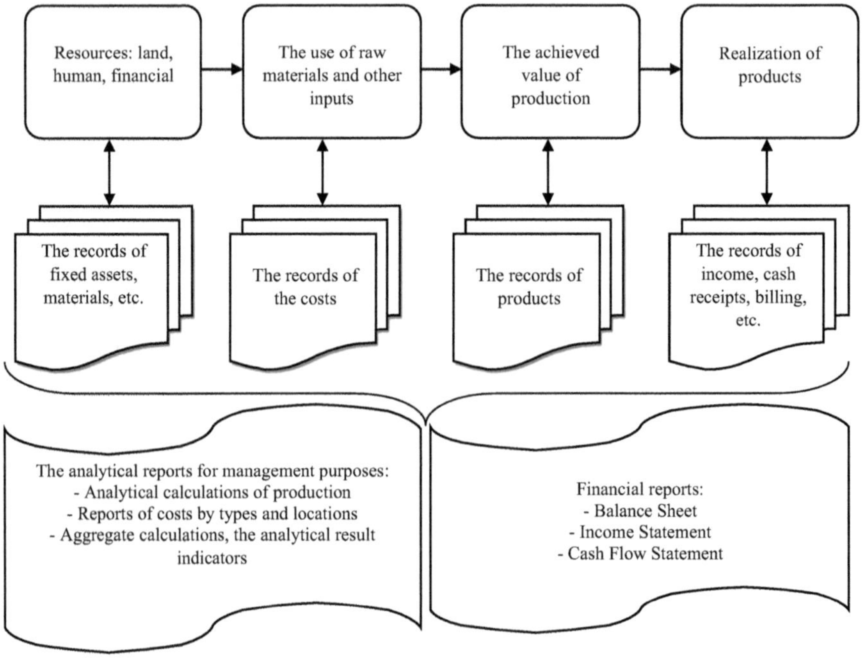

Diagram 14.1 The system of records

Such an information base, combined with production factors, can have a substantial influence on long-term company performance improvement. Accounting data enable the performance of credible macroeconomic indicators and, on the basis of their analysis, the adoption of suitable policies in a variety of fields (credit, monetary policy, etc.). At the micro-level, high-quality information serves as the foundation for making better decisions and optimising business operations.

14.3 Conclusions

The proper utilisation of resources is necessary for the survival and economic success of organisations engaged in cereal production. This entails competent producer management and adequate system of economic viability, which necessitates the availability of high-quality production-economic information regarding company activities. Information derived from accessible manufacturing processes may be seen as a significant component that interacts with other production variables such as land, labour, and money to contribute to a more successful corporate operation. By obtaining greater coordination in the implementation of production activities, producers gained insight into their production capacities and a better evaluation of indications that present the success or failure of production activities. This results in improved management of production processes, as well as increased prospects for superior product quality, manufacturing traceability, and increased market share. Control of operations, paired with a database of prior results of operations, gives quality decision-making assistance.

To plan and regulate production activities, following decisions that should be made with great care and on the basis of well-planned future actions. As technology advanced and improved, certain requirements for efficiency and integration of production planning and control operations should be imposed for the bread producers which has their own grain production. Prior to starting the planned production activities, it is vital to investigate the variables that might lower the predicted output or outcomes of operations. The essential part of production activity planning and implementation is gathering information about the companies current and future status in various fields (credit, marketing, sale circumstances, etc.). Monitoring the production processes and specific operations, as well as the expenses that result from the production processes, can help with the control or management.

References

Arfini, F. (1998). *The EEC accounting information network: Methods, data and problems*. Universita di Parma, Facolta di economia - Instituto di economia agraria eforestale, Parma..

Dalci, I., & Tanis, V. N. (2002). Benefits of computerized accounting information systems on JIT production system. *Review of Social, Economic and Business Studies, 2*, 45–64.

Del'homme B. (1997). Information and decision in agriculture: How relevance is management diagnosis is improved by an expert system - proceedings of the first European conference for information Technology in Agriculture, first conference, Copenhagen, EFITA, 221–227.

Florey, B., Adams J., & Robinson, M. (2004). Figures for a farming future, getting started in farm management accounting, using the farm accounts to point the way, Department for Environment, Food and Rural Affairs edited by farm and animal health economics division, Defra, London.

Frye, D. (2009). Finance, Insurance & Real Estate, Agriculturist. *Wisconsin.*

Heidhues, E., & Patel, C. (2008). The *role of accounting information in decision-making processes in a German dairy cooperative*. In *10th international conference on Accounting & Business.*

Parker, C. G., Campion, S., & Kure, H. (1997). *Improving the uptake of decision support systems in agriculture, proceedings of the first European conference for information Technology in Agriculture* (pp. 129–134). EFITA.

Santacoloma P., Röttger A., Tartanac F. (2009). Business management for small-scale agro-industries, FAO agriculture management, marketing and finance service rural infrastructure and agro-industries division, food and agriculture organization of the united nations.

Slof, E. J., & Argiles, J. M. (2000). New opportunities for farm accounting, the European, accounting review 10 (2). *Barcelona*, 361–363.

Somiari, F. R. (2008). *Farm menagement and record keeping monitoring and evaluation officer, third national fadama development programme*. Rivers state coordinating office, ADP premises, Port Harcourt.

Sørensen, C. G., Fountas, S., Nash, E., Pesonen, L., Bochtis, D., Pedersen, S. M., Basso, B., & Blackmore, S. B. (2010). Conceptual model of a future farm management information system. *Computers and Electronics in Agriculture, 72*, 37–47.

Vajna Istvánné Tangl Anita. (2000). *Comparison the informations from the hungarian accounting system to the reqiremets of the european union's farm accontancy data network, Ph.D. thesis.* Szent István University - Gödöllõ.

Part IV
Innovations in the Bakery Industry

Chapter 15
Impact of Bakery Innovation on Business Resilience Growth

Elena I. Semenova and Aleksandr V. Semenov

15.1 Innovation and Consumer Preferences

Innovations in business represent the materialization of scientific knowledge in the process of investment, creation, and distribution of a new (innovative) product to meet production and social needs, provide beneficial effects on the development of economic entities, and achieve the necessary socio-economic effect.

The importance of innovation in bakery production is due to the importance of this product in the consumers' diet.

An innovative product with different consumer properties is created to meet the variety of tastes and preferences of consumers.

Nowadays, the following bakery products can be distinguished:

- Traditional and new products;
- Functional products (healthy eating);
- Therapeutic products;
- Products for children, nursing mothers, pregnant women, and pensioners;
- Special purpose products for certain groups of the population: special contingents, workers of extreme working conditions.

The traditional bakery products in Russia are as follows:

- Rye bread (weight or piece) made of molded or raised flour;
- Improved products with the addition of wheat flour, malt, coriander, cumin, etc.;
- Rye bread made from crimped and wheat flour;
- Rye-wheat and wheat-rye bread;
- Wheat bread;
- Rolls and buns.

E. I. Semenova (✉) · A. V. Semenov
Federal Research Center of Agrarian Economy and Social Development of Rural Areas – All Russian Research Institute of Agricultural Economics, Moscow, Russia

J. M. Ferreira da Rocha et al. (eds.), *Baking Business Sustainability Through Life Cycle Management*, https://doi.org/10.1007/978-3-031-25027-9_15

Functional products include products enriched with functional ingredients, gluten-free and lactose-free products, low-fat products, low-carbohydrate products, etc.

Dietary bakery products can be salt-free, low acid, low carbohydrate, low protein, and high fiber, with added lecithin, high iodine content, etc.

Products with specified properties are instant products that do not require boiling: cereals from different types of legumes – rice, buckwheat, oats, rye, barley, millet, wheat, peas, beans, and lentils.

In response to changing consumer preferences, innovation drives change and value growth in the market for baked goods.

Today's consumers lead active lives and follow diets. Thus, breakfast is increasingly often replaced with a snack. The global bakery industry faces increasing competition from the "fast food on the go" segment.

The so-called Mealtime 2.0 (Mealtime 2.0 meal planning and accounting) is entering everyday life. Therefore, innovations are required to interact with existing and new potential customers.

The growing popularity of low-calorie products is associated with the growing concern of consumers about health and the desire for a balanced diet. The trend of naturalness is closely related to this trend, i.e., the abandonment of artificial additives in favor of natural ingredients in the manufacture of products, including bakery products.

Production of bread for fast food organizations increases because consumers switch from unpackaged to packaged portioned meals.

Bakery brands develop these new food formats. In France, Jacquet bread is produced specifically for making sandwiches for light meals on the go. Australia makes Tip Top Grab'n Go Fruit & Fiber Breakfast Buns, which require no oil, jams, or toasting. The manufacturer positions such bread as snack rolls, which already contain everything necessary; one only needs to heat them in the microwave. Italian Focacelle is marketed as "the first tasty snack of bread consumed anywhere and everywhere," thanks to its practical and individually packaged portion packs. The product comes with several toppings: original extra virgin olives, black olives with cherry tomatoes, and new rustic and rosemary varieties. The range of Focacelle has become the best-selling in Italy in recent years in its segment.

Another innovative trend in bakery production is baking hybrids. In Japan, Caramel&Almond Croissant Doughnut has produced crispy cookies covered in caramel, chocolate, and almonds. What makes these cookies special is that they have a unique texture because they combine a croissant and a donut. The donut and bun are also a combination of Tesco's Raspberry Loafs, launched in the UK because they contain donut-buns coated in sugar and filled with raspberries.

Dawn Food Products (USA) is in the gluten-free baking business. Cake Mix offered new gluten-free fruit cupcakes.

To meet the demand of consumers (Russia), multi-colored bread was developed with the use of natural food dyes. The raw materials are water, flour, sugar, salt, yeast, and additives (dyes). Each dry food coloring contains maltodextrin, which is easily absorbed by the body and, unlike glucose, does not cause blood sugar spikes. Other substances were added to the dyes, depending on the color: curcumin (orange

Table 15.1 Expenses for innovations in the organizations-leaders of the 2020 rating by type of activity 10.71.1 "Production of bread and bakery products of short-term storage" depending on the scale of activity

Category of organizations	Number of organizations – leaders in revenue in 2020	Revenue per organization, million rubles		Payments for the acquisition, creation, modernization, reconstruction, and preparation for the use of non-current assets per one organization, million rubles	
		2019	2020	2019	2020
Large	2	4774.75	5215.00	481.22	1573.72
Medium	2	1157.75	1147.50	28.41	16.57
Small	2	639.98	635.53	199.04	39.03
Microenterprises	2	234.89	186.82	2.27	0.03
Small-scaled	2	10.66	9.80	0	0

Source: Compiled by the authors

or yellow), chlorophyll (green pigment), betanin (beetroot red), and carmine (bright red).

Natural food dyes are not harmful to human health, do not affect the taste, smell, and quality characteristics of the bread, but only give an unusual attractive color to the bread. The profitability of such bread is 20–21%.

In the promotion of bakery products on the market, the emphasis is placed on the appearance of the product; the main slogans are "attention to detail," "bread as a source of pleasure," or "pleasure in every bite."

Branding of bakery production "goes into the packaging"; the customer must leave the store with the brand name, and this name is guaranteed to be seen by all.

The ability to innovate is higher for large and medium-sized enterprises. Line 4221 in the accounting statement of cash flows reflects payments in connection with the acquisition, creation, modernization, reconstruction, and preparation for the use of non-current assets, i.e., costs associated with innovations in organizations. Among the organizations within the type of activity 10.71.1 "Production of bread and bakery products of short-term storage," this item was analyzed according to the reports of organizations leading in the sector (Table 15.1).

As shown in Table 15.1, large, medium, and small enterprises make significant investments in innovation, which provides them with the corresponding revenue. Small-scaled businesses do not have the financial capacity to innovate.

15.2 Implementation of Innovations by Stages of Bread Production

The implementation of innovations in all stages of bread production is presented in Fig. 15.1.

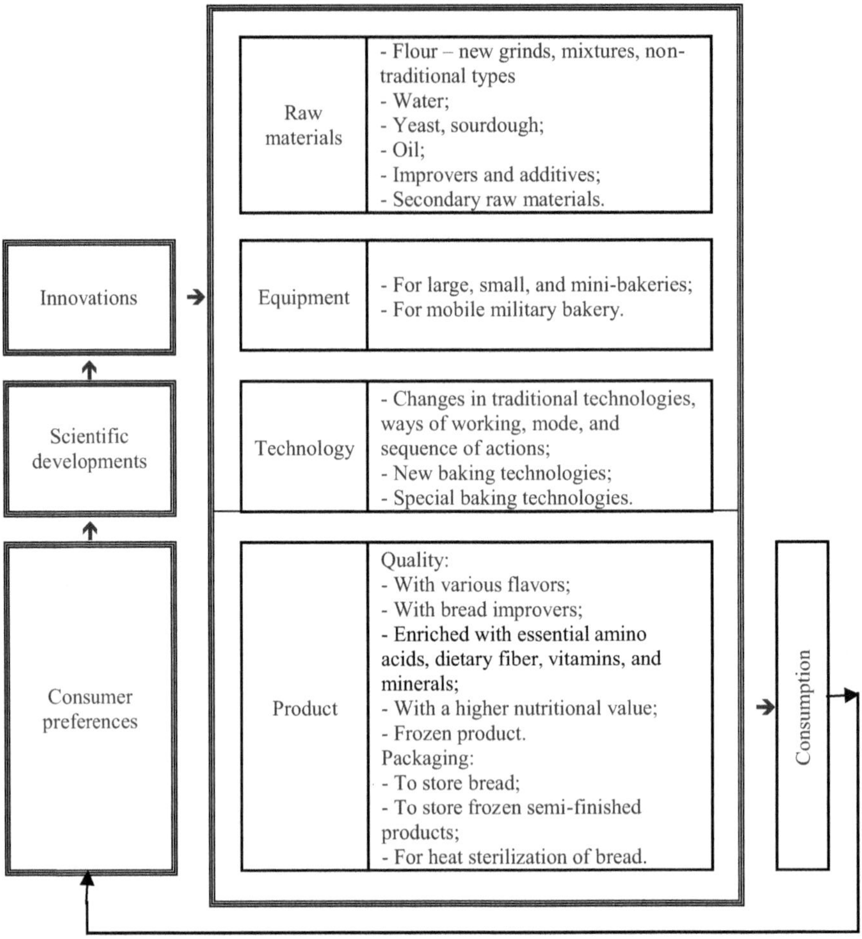

Fig. 15.1 Innovations in bakery production. (*Source*: Compiled by the authors)

Ensuring the stable quality of finished products is based on the optimal way of conducting the technological process, considering the quality indicators of raw materials, on directed and reasonable use of micro-ingredients, including complex additives with multifunctional action, and on the introduction of current methods of technical and chemical quality control of raw materials, semi-finished products, and finished products. Today's food technology uses more than 600 raw materials and ingredients, including food additives. To ensure consistent quality of finished products, it is necessary to implement and maintain procedures based on Hazard Analysis and Critical Control Points (HACCP) principles.

Let us consider innovation at each production stage (according to the diagram shown in Fig. 15.1).

15.2.1 Raw Materials

The main raw material of the baking industry is flour, the quality of which largely determines the quality of bread products. The global trend is a decrease in the level of protein in grain and, consequently, in flour. For example, flour with a gluten content of 30–42% was used in the USSR in the 1930s, 28–37% in the 1970s, and 24–33% now. Additionally, deterioration of microbiological indicators and significant fluctuations in the activity of hydrolytic enzymes in different batches of flour were noted.

To obtain more chemically balanced bakery products, composite mixtures of flour with different compositions based on mixtures of cereals and cereal products are used, mainly in the form of flour from different crops.

It is possible to use oat flour and rice flour combined with wheat flour. The addition of rice flour and oat flour of up to 10% to the mass of wheat flour increases the volume of baked goods, improves porosity structure, and strengthens the gluten. The addition of rice flour to the first and second grades of wheat flour produces products with lighter colors.

The simultaneous use of different flours from cereals and grains can give a more complete effect of bread enrichment. The optimum ratio of cereals included in the mixture (48.4% polished millet, 35.5% buckwheat, and 16.1% rice groats) was established. Bread with a mixture of cereals has an attractive and unusual appearance, pleasant taste and flavor, and a high nutritional value.

The use of legume flour in composite mixtures has a significant effect on enriching products with proteins. Soy flour is used more often, which contains active enzymes needed to prepare yeast bread, increases water absorption capacity in the dough, and reduces the absorption of fats. However, most soybean varieties are the product of genetic modification (Rusina et al., 2012).

Legumes have good dietary properties and soothing effects and are useful for dental health. Beans and peas are essential foods in the diet of patients with diabetes. The introduction of pea and bean flour in the amounts of 10% and 15%, respectively, to the mass of wheat flour of the highest grade has no negative impact on the main technological quality indicators of the obtained composite mixtures. The dough method is the best way to bake products based on composite mixtures in technological and economic terms.

The nutritional value of buckwheat flour is as follows: protein – 12%–35.2%, fat – 1%–8.3%, and carbohydrates 37.1%–72%. It is reasonable to replace 20% of rye flour with buckwheat flour. Rye-wheat bread based on a flour mixture consisting of rye-wheat flour with the addition of buckwheat flour in a ratio of 40:40:20 has a more pleasant taste and pronounced aroma of buckwheat, increasing the yield of bread.

Flax flour, a gluten-free product with high fiber content, is used as an additive to cereal flour in producing baked goods. Non-fat linseed flour (up to 10% of residual oil) is usually introduced into bakery products, containing 26–28% of protein, 30–40% of dietary fiber, up to 30% of crude carbohydrates, and 7.8 mg of minerals per 100 g. In the production of baked goods, flaxseed polysaccharides act as water-holding agents, texturizers, and binders, while having a protective effect on the digestive system. Flaxseed proteins have a high nutritional and biological value because they have a fairly balanced amino acid composition. The use of flax flour in the human diet normalizes the work of the gastrointestinal tract. Flaxseed flour is a promising source of biologically active substances that can be used in baking to create functional bakery products of high nutritional and biological value.

Soy flour is used to increase the protein content in bread. The water-absorbing function of soy proteins allows for drier doughs suitable for machine processing, as well as bread that stale slowly due to better moisture binding. As a result, there is an opportunity to increase the yield of bread. If more than 15% of soy is added to bakery products, they should be classified as dietary products. In this case, it is necessary to use additives that strengthen the gluten and can retain an acceptable specific volume.

The gas-forming ability of flour is of great technological importance in producing bread or bakery products, the recipe for which does not provide for the introduction of sugar in the dough (Spirichev et al., 2005). The gas-forming ability of flour is determined by the content of its "own" sugars and its sugar-forming ability. The sugar-forming ability of flour is related to the action of amylolytic enzymes contained in it on the starch, as a result of hydrolysis of which sugars are formed in the dough. The use of composite mixtures contributes to increasing the gas-forming ability of the dough and improving the quality of the gluten.

The ratio of water to flour in the dough is important in baking technology. It determines the structure of the dough, the properties of the dough when processed by dough dividers, proofing and baking times, and the yield of bread and bakery products. The ratio of flour to water depends on the moisture content and properties of the flour, its yield, the type of cereal from which the flour is obtained, the type of bakery product, the amount of sugar, fat, and other ingredients in the dough, the method of making dough, etc. For flour products, yeast dough is prepared at a water-flour ratio of 1:0.45–0.55. Adding flakes in the amount of 5–20% to the mass of the flour reduces the mass and the amount of raw gluten. It is optimal to add 15% flakes to the mass of flour.

Magnetized water affects the lifting power of the yeast, which affects the period of fermentation and proofing of the dough. Reducing the fermentation and proofing period allows for intensifying the technological process and increasing the productivity of bakeries. Magnetized water positively affects baking indices of flour quality and bread quality.

Nowadays, the basis of baking is baking yeast, which is a biomass of the monoculture *Saccharomyces cerevisiae*. Baking yeast can be replaced by symbiotic fermentation starter derived from the natural microflora of flour; kefir fungus-based fermentation starter, or leaven of microbiological cultures obtained from the processing of some food waste (curd whey, whey powder, beer pellets, etc.).

There are three main actions of fermentation starter regarding the improvement of the nutritional value of bread:

- Reduction of phytic acid content and, accordingly, an increase in the bioavailability of mineral salts;
- Isolation of exopolysaccharides that act as prebiotics;
- Hydrolysis of the prolamine fraction of the protein, which increases the availability of the product for people suffering from gluten enteropathy.

According to the method of production and application, fermentation starters can be classified into three types:

- Traditional renewable fermentation starters;
- Industrial starters prepared on specially selected strains of microorganisms;
- Dry starters made by different technologies: tumble drying, spray drying, fluid-bed drying, and freeze-drying.

There are more than 50 species of lactic acid bacteria and more than 25 species of yeast in fermentation starters. Sourdough bacteria affect the flavor of bread by releasing organic acids, alcohols, ketones, aldehydes, esters, and sulfur-containing compounds, including those not released by conventional yeast.

Replacing baker's yeast in the recipe of bakery products with fermentation starters with minor adjustments in the technological process can increase the nutritional value, improve organoleptic characteristics, and extend the product's shelf life.

Hop extracts, liquid yeasts, and hop leavening with stable, predictable biotechnological properties are used in bakery technology.

In recent years, traditional technology of producing bakery products has been increasingly subjected to profound changes to obtain products of high biological value and high organoleptic characteristics, possessing dietary, therapeutic, and preventive properties.

The leading trend in the technology of producing bakery products is the inclusion of compounds of various natures referred to as food additives in the recipes of these products.

According to the FAO international classification, food additives used to improve the efficiency of food technology include process enhancers, including enzyme preparations, myoglobin fixers, flour whitening agents, bread improvers, polishing agents for confectionery, and clarifiers for beverages.

The need for additives and non-traditional raw materials in baking is due to widespread nutritional disorders – widespread and profound deficiency of vitamins, minerals, and protein. According to the data of the Institute of Nutrition of the Russian Academy of Medical Sciences, there is a deficiency of polyunsaturated fatty acids, proteins, and vitamins (ascorbic acid – in 70–100% of the population; thiamine, riboflavin, and folic acid – in 60% of the population; β-carotene – in 40–60% of the population) and a variety of minerals (calcium, iron, iodine, fluorine, selenium, and zinc) in the diet of the Russian population, which is observed throughout the year in the nutrition of all age and professional groups. Protein deficiency in the diet of the population averages 26%, which is the reason for the development and use of new technologies in baking with the use of protein-containing products of plant origin. Bread has a low protein content and the reduced biological efficiency of proteins due to insufficient balance of the amino acid composition of proteins in quantity and quality.

In practice, bakery production uses food additives, divided into two groups, to improve the quality of bread and bakery products and regulate the parameters of technological processes (Buldakov, 1996):

- Derived from plant and animal raw materials (legume seed products, fruit powders, dairy products, etc.);
- Produced chemically (oxidants, synthetic vitamins, etc.), by microbiological synthesis (enzyme preparations), or by processing natural materials (modified starch).

Functional ingredients have special requirements: no ability to reduce the nutritional value of the food product and safety in terms of a balanced diet and naturalness. Functional additives include high-protein products (protein concentrates from soybean and pea seeds, high-protein lupine flour) and special protein-containing additives based on mushrooms, amaranth, etc., dry gluten; new sweeteners; additives for flour with increased autolytic activity (rye flour); additives to protect bread from microbes and pathogens (propionic acid).

Nutritional enhancers used in the baking industry to regulate the technological properties of dough are divided into the following groups:

- Oxidative enhancers – potassium bromate, ascorbic acid, azodicarbonamide, calcium peroxide, and benzoyl;
- Enhancers based on modified starches derived from various types of vegetable raw materials – extrusion, oxidized, swelling, phosphate, carboxymethylated starches, corn starches oxidized with calcium hypochlorite, potassium bromate, or potassium permanganate;
- Surface-active enhancers;
- Enhancers of enzymatic nature and mineral enhancers to increase yeast activity – salts containing nitrogen and phosphorus, ammonium salts of phosphoric acid, and other phosphates, most often sodium and potassium salts.

Almost all known complex enhancers include enzyme preparations and oxidizing agents. Most complex enhancers contain amylolytic enzyme preparations, their mixtures with proteolytic enzyme preparations, enzyme-active products of plant origin – malt, malt flour, soybean meal or preparations based on them, surfactants, mineral salts, oxidants, as well as dietary and preventive supplements.

The use of enhancers affects the protein and carbohydrate complex of flour, vital activity of yeast, rheological properties of dough, and the process of gas formation during fermentation, which can improve the quality of bakery products: increase the specific volume by 20–25%, porosity, and structural and mechanical properties of crumb, form stability of bread, and delay its staling during storage (Puchkova et al., 1988).

Multigrain, functional, and enriched bread involve a number of different ingredients: multigrain mixes, sprinkles, enhancers, etc. Thus, in the production of enriched and functional bakery products, it is advisable to use a complex mixture rather than individual nutrients.

When adding additives, the main thing is to develop a recipe and evaluate the consumer properties of the bakery product.

The use of peanuts can increase the nutritional value and expand the range of baked goods, giving them functional properties. In France, there is a patented method of making high nutritional value bread from a mixture of wheat and rye flour with the addition of walnut kernels and nut oil.

The addition of vegetable oil with a high content of polyunsaturated fatty acids and phospholipids to the sourdough contributes to the formation of more stable complexes with proteins and carbohydrates of flour. As a result, an increased volume of bread is formed with well-developed porosity and elastic dry crumb. In the production of bread from wheat baking flour of the highest grade, hemp oil can be used instead of sunflower oil in an amount up to 2.0% when it is added to the sourdough. The use of hemp oil in the production of wheat bread will produce a more valuable product of dietary and therapeutic nature due to up to 75% of polyunsaturated fatty acids contained in hemp oil (Zhuravleva et al., 2012).

An important direction in baking is the use of local raw vegetable materials. Such raw materials are dried banana flour (Latin America), dried coconut (Philippines), cassava and potato fruit flour (India), and citrus fruit pulp (the USA).

In Russia, non-traditional raw materials and by-products of the food industry are used: fresh leaves of celery, green onion, plantain, and nettle, added in an amount of up to 0.5% of the weight of the flour. Tomato seeds and powder from grape squeeze and grape seeds are also used.

Among the non-traditional species for baking, a special place belongs to the products of amaranth grain grinding. Amaranth grain is subjected to conditioning (yield 95–98%), which is then divided into native amaranth flakes (53–56%) and native germ coarse (28–35%). Then, amaranth variety flour native (82–88%) is formed from the flakes, and oil is extracted from the coarse (6.5–7.5%); a by-product – semi-strained germ flour is formed (28–35%), which contains, respectively, the semi-skimmed protein bran (18–25%) and semi-skimmed protein flour (75–82%).

In terms of nutritional aspects, protein semi-strained flour, graded native flour, and protein semi-strained bran are the most valuable among amaranth grain milling products. Amaranth protein flour is an effective protein enrichment and technological enhancer in producing bakery products from wheat flour. Flour of native variety is an effective quality enhancer for baked goods made from a mixture of rye and wheat flour. It is advisable to use semi-skimmed protein bran as a source of dietary fiber in the development of bakery products for functional, dietary or therapeutic, and prophylactic purposes.

The most widespread food additives in the baking industry are products of processing vegetables, fruits, and waste juice production – juices, purees, boils, and vegetable and fruit powders from whole fruits or their squeeze. The composition of these products has a significant content of sugars (40–60%) – sucrose, glucose, and fructose, as well as pectin substances (7–15% of pectin and 2–4% of nitrogenous substances), which allows their use instead of sugar in the manufacture of confectionery products.

For bakery products made of rye and wheat flour, a technology was developed based on semi-finished products using:

- Beet powder – a suspension (beet powder, yeast, and water), after standing of which for a certain time, the dough is kneaded on a thick rye fermentation starter; or
- Carrot or pumpkin powders – leavened sourdough, based on which dough is kneaded.

Dough for wheat bakery products is prepared by using the evaporation method. Powders are added when kneading dough, except for products with pumpkin and potato powder, which are added when kneading sourdough. These technological solutions help improve the consumer properties and microbiological safety of the developed assortment (Shlelenko et al., 2014).

Adding pumpkin puree to the dough leads to improved organoleptic characteristics, namely, the improved color of buns, porosity, the elasticity of the crumb, and the shape of the finished products. The most optimal dosage in the production of bakery products should be considered 15% of the flour mass, which allows one to get products enriched with pectin and carotene without losing quality and even improving it.

A convenient form of plant raw materials for use in the baking industry is its use in the form of powders, which are well soluble in water and provide a homogeneous color of products.

For example, the addition of pumpkin and carrot powders to the bread recipe positively affects the crust coloring, crumb elasticity, and taste and aroma of buns. The addition of 10% of pumpkin powders increases the bread acidity by 1.4 °T and delays the formation of mold in buns by six days. The addition of 10% of carrot powder increases the carotene content of the buns by 0.64 mg%. The addition of 5% of pumpkin and carrot powder each increases the carotene content by 0.66 mg%.

In Germany, a method of making bread from wholemeal flour, medicinal herbs, and various fruits, as well as dried green algae powder, is used.

The addition of 5–10% of quince fruit powder improves the porosity and specific volume of bread. However, it has almost no effect on the moisture and acidity values of the wheat bread.

The introduction of topinambur powder significantly affects the main physicochemical parameters of wheat bread – improvement of porosity and specific volume. The best porosity and the highest specific volume were obtained by adding 5% of topinambur powder (363 cm^3 per 100 g of bread).

The use of burdock root powder in an amount of 10% changed the crumb's color and structure. The baked goods were characterized by an intensely colored crust and a more pronounced flavor.

Plant resources, namely cultivated mushrooms (mushrooms cultivated two-spore champignons and oyster mushrooms), are used in baking due to their unique chemical composition – a significant content of proteins and the peculiarity of the carbohydrate complex, biologically active, and aromatic substances. Unlike wheat and rye flour, mushroom powder contains glycogen, which is not found in plant organisms. Mushroom powder from champignons and oyster mushrooms is characterized by a fairly high content of fiber, which acts as dietary fiber, providing the body with a positive physiological effect. The addition of 5% of mushroom powder from champignons and oyster mushrooms in the dough improves the quality of bread in terms of physical and chemical parameters: porosity, moisture, and acidity.

Bread products are enriched with biologically active substances, such as garlic preparations (powders, homogenates, pastes, extracts, and chemically purified components) into the dough. Garlic additives are pre-treated at temperatures above 100 °C separately or mixed with water, milk, and additives from bran, vegetables, and fruits before being added to the dough. The use of black chokeberry (*Aroniya Melanocarpa*) as a biologically active additive in baking allows for improving the quality of finished products. Bread with the addition of 10% of chokeberry juice from the total amount of liquid in the dough by the unbleached method contains up to 0.00612 g of flavonoids in 100 g of the product.

The addition of puree from wild apples, cranberries, and rowanberries intensifies microbiological and biochemical processes in the dough and improves the properties of gluten, dough structure, form-holding capacity, and organoleptic characteristics.

Cranberries contain huge amounts of antioxidants and other phytonutrients. Dried cranberry from the USA is actively used in baking and confectionery: bread, various kinds of cookies, fruit bars, candy, nougat, and snacks with dried cranberries. Dried cranberry is easy to use, does not require special storage conditions (in a cool, dry place, the temperature should not exceed +18 ° C), has a high dietary value, and is available year-round. The use of dried cranberries allows bread production plants, bakeries, and confectionary shops to expand the range of products with enhanced consumer properties and strengthen customer demand. Dried cranberries are added to the dough in the process of kneading or 2–3 min before its end. Finely sliced cranberries (1–3 or 3–5 mm) are added to sieves, puffs, sugar cookies, and crackers.

According to the concept of healthy eating, plants containing polysaccharides belong to the group of physiologically functional ingredients. This group can have a favorable effect on one or more metabolic reactions of the human body. Chicory, echinacea, licorice, and stevia are used as sources of dietary fiber, polysaccharides, inulin, and natural sweeteners.

The use of inulin-containing raw materials in the technology of flour products will expand the range of preventive anti-diabetic products. The addition of chicory products to wheat flour leads to an increase in the elastic properties of gluten. The addition of chicory promotes the most active growth and vitality of yeast. Inulin and oligofructose can be included in the formulation of all known baked products – semi-sweet, shortbread, and oatmeal cookies, crackers, biscuits, rolls, waffle sheets, muffins, gingerbread, bread rings, etc.

Stevia can boost the body's immune system, making it resistant to disease. Stevia can relieve fatigue after a hard day's work and fight stress, depression, and nervous tension. The use of extracts of this plant is especially relevant for people who complain of chronic or rapid fatigue. Licorice root extract is a unique and versatile product that contains an extensive complex of biologically active substances. It has a bacteriostatic effect on spore microorganisms, inhibiting their development and the contamination of bread with potato disease (Skorbina, 2015).

The use of *Echinacea purpurea* in baking technology has a positive effect on the quality of raw materials and increases the amount of crude gluten and its extensibility. The yield of the finished product can be increased by 4–6% by adding plant polysaccharides. They improve the baking properties of flour, increase moisture absorption capacity, etc.

Of additives of animal origin, the most common are those based on milk and milk products – whey, cottage cheese, whey powder (Poland), and protein whey concentrates (USA).

Bread is enriched by introducing dietary fiber – additives with a high content of hard-to-digest polysaccharides, including synthetic ones such as methylcellulose.

Cereal bran (both coarse and finely dispersed) and other similar components from pea seeds (cellulose), sunflower, wheat germ, oat flakes, banana and apple powders, sea buckthorn meal, etc. are used for this purpose.

Improving the range and technology of production of new food products should be based on the integrated use of secondary raw materials and low-waste and resource-saving technologies.

One of the current directions of enrichment of bakery products is the use of secondary raw materials from the processing of raw vegetable materials and non-traditional plant components (e.g., fruit and berry crops, mushroom mycelium powders, starch products, flours from Sudan grass grain, amaranth, lentils, etc.).

The additives used are residual brewing products: soaked grains, beer pellets, and residual beer yeast. From 6% to 25% of beer pellets can be added to the dough to prepare dietary products. Beer pellet flour contains about 40% protein and 10% fiber. In the USA, beer pellets are pre-dried to a moisture content of 7–16% and milled to a certain degree of dispersion.

One of the possible sources of valuable nutrients is grape pomace, which is practically not processed at enterprises of primary winemaking and is exported as an organic fertilizer to the fields. Grape pomace accounts for up to 30% of the volume of grapes processed and is an essential raw material for obtaining valuable products such as dried grape skins and grape oil (Vershinina & Tezbieva, 2014).

The introduction of grape skin powder into the dough improves bread quality, slows down staling processes, and increases the content of minerals and pectin, enriching it with dietary fiber, which has a positive physiological effect on the human body. The best bread quality is observed when 5% powder is added and when the ratio of rye flour to first-grade wheat flour is 50:50.

The main component of additives, which are the waste products of the primary winemaking, is grape pomace; it can be used as an activator of dough fermentation.

The result of pressing the soy mass on a filter press in the production of soy milk is soy okara – insoluble soy powder of light-yellow color, neutral taste, with a characteristic crumbly consistency. It is a source of divalent iron and fiber and can contain up to 5% of protein and 4% of fat. It is added to the first and second courses, meat and vegetable cutlets (up to 30%). It is used to prepare baked goods, cookies, gravy, sauces, and as an additive to pancakes. The addition of soy okara in the amount of 5% of the weight of wheat flour of different qualities proved effective in increasing the nutritional value of the bread.

Deep-milled rice flour is a recycled product of rice grains. A distinctive feature of the amino acid composition of deep-milled rice flour is the high content of arginine and leucine. It is rich in lipids. The gluten content in rice flour was less than 2 mg/kg, which gives grounds to recommend it as a valuable raw material for the production of gluten-free products. For bakery products, for the purpose of enrichment, it is advisable to make a dosage of 15% rice flour in the dough.

Deep-milled oat flour, a by-product of cereal production, is quite high in protein (25%), fat (15%), and fiber (18%). The protein complex is balanced in amino acid composition and more complete than the protein complex of whole grain. The introduction of 25% of deep-milled oat flour instead of wheat flour leads to an increase of protein content in bread by 1.2 times, vitamin B_1 by 1.2 times, vitamin B_6 by 1.8 times, and E by 1.2 times. This bread is recommended for inclusion in the diet of persons with diabetes and for mass consumption.

It is possible to use the products of huskless oats processing in baking, namely whole-milled grain and seed flour mixed with high-quality wheat flour. Wheat-oat bread baked by dough method with the addition of 10%–20% of seed flour or 10% of wholemeal grain has a pleasant aroma and taste and light, quickly recovering crumb.

In the baking industry, it is recommended to use the following vegetable by-products in the mass of wheat flour: microcrystalline cellulose and pine nut, sesame, and pumpkin pomace in amounts of up to 3, 15, 10, and 14%, respectively; the combined use of vegetable additives in the following studied amounts 2% of microcrystalline cellulose, 10% of pine nut pomace, and 3.5% of pumpkin pomace. The most effective experiment was the use of a 15% plant mixture.

The implementation of innovations in the technological process allows us to improve the technical and economic parameters of technologies – to increase resource efficiency, reduce the duration of technological processes, reduce the cost of consumables, reduce rejects, increase the share of high quality products in total production, and improve product quality and expand its range. This affects the effectiveness of the organization and its sustainable development.

15.2.2 Equipment and Technology

Multi-profile holding Group of Companies "NHL" (Russia) is a dealer of manufacturers of equipment for bakery production. The company has its own production of food ingredients for baking and confectionery mixes, spare parts, and equipment – band saws and blades, clipping machines, etc.

"CLIMAT" climatic units for proofing chambers provide automatic heating, humidification, and air distribution in the proofing chamber. The fan systems of the unit generate air flows that ensure uniform proofing of dough pieces anywhere in the proofing chamber. When operating, traditional steam generators produce hot steam; it is not always possible to achieve the desired temperature and humidity conditions, especially in the summer. In turn, "CLIMAT" climate units can provide this. The technology used allows us to recover excess heat and maintain a relative humidity level of up to 100% at all proofing chamber temperatures.

Clipmashine "SPUTNIK" is designed for closing various soft packing (bags, nets, sacks, etc.) by placing a color metal-plastic clip on the neck of the package; the date of the product packaging is printed on this clip using thermal transfer printing (Markova, 2018).

In the Aquadoz WDM-15 T water dispenser mixer, produced in Russia, the electronic dispensing flow-mixing device based on an electronic controller with a touch panel allows us to mix hot and cold water to get the desired output volume of liquid at the desired temperature.

Another innovative area is the depositing and glazing machines, transport systems, and grain dryers. The company supplies cyclothermic ovens, gas and electric direct heated ovens, convection and hybrid ovens, and proofing and baking units. Pads for tunnel ovens can be made of a mesh of different densities and weave, plate strip, using granite or cast-iron tiles, as well as with Teflon coating. For baking pastries, such as pies in paper forms, the company offers special articulated trays. J4 ovens include the following new features:

- Optimization of the oven heating process, including draught and vacuum control and automatic temperature maintenance in individual zones;
- Complete monitoring of the baking process (monitoring the temperature curve, control and regulation of steam pressure in the steam zone, measurement of temperature in the center of the product during baking, etc.);
- Closed flame burners;
- Lengthening the inlet and outlet parts of the oven, etc.

J4 ovens can run on all types of fuel, including biofuel. New design elements, higher-quality insulation materials, and waste gas recovery technology are being introduced, considering the need for energy-saving technologies to reduce losses and ensure heat reuse. According to the experience of practical use, the proposed heat-exchange units give an average yield from one chimney from 10 to 55 kW; according to statistics, the practical gas savings reach 150 thousand m^3 per year.

Hasborg's baking machines can produce a wide range of cookies, such as two-color with two kinds of filling, two-color with filling in the middle and decorating on top, three-color, biscuit cakes, products produced with the "string cutting," "guillotine," and "diaphragm" options, and many more. "LUXOR" icing machines are used for icing cookies, gingerbread, bars, and sandwiches. Cooling tunnels and "ROYAL ROLLS" premium industrial machines are designed to work on the production line for the production of shortbread cookies or gingerbread pressed from the roller.

Turkish bakery equipment produced by the company "Fimak" (rotary and multi-level deck ovens, automatic spiral kneading and sheeting machines, dough dividers, conical dough sheeters, intermediate proofing cabinets, and dough rolling machines) has good technical characteristics and a good price-performance ratio (Markova, 2018).

A large amount of heat and electricity is used to produce bakery products. The main percentage of energy consumption is accounted for by ovens (Auerman, 2003). To reduce energy costs, the company "Spooner" (UK) offers energy-efficient ovens for bakery production. The energy savings come from shorter baking times at lower temperatures. Even the traditional lines of older models can be made just as energy-efficient through upgrades and modifications offered by the company "Spooner Plus." For the same performance, this brand's energy-efficient ovens are smaller than traditional ones. Cable lugs are integrated into the units, which also contributes to energy savings.

Electric contact heating and baking are used in baking to produce crustless bread. Electric contact heating allows for fast and even heating of the dough in the entire volume of the dough piece; the heating temperature does not exceed 100 ° C, which ensures the absence of crust in the finished products. The raw material is placed between two metal plates – electrodes connected to alternating current. Due to electrical conductivity, during the passage of current, there is a release of heat according to the Joule-Lenz law. In most cases, an AC of 220 V and 50 Hz is used for electric contact baking.

One of the promising and highly efficient technologies that combine thermal, hydraulic, and mechanical processing of raw materials is extrusion technology. It allows obtaining a new generation of products with predetermined properties by changing the initial composition of the extruded mixture and controlling the mechanism of physical, chemical, mechanical, biochemical, and microbiological processes occurring during the extrusion of food masses.

The method of extrusion processing makes it possible to intensify the production process, increase the utilization of raw materials, obtain ready-to-use food products or create components for them that have high water- and fat-holding capacity, reduce production and labor costs, expand the range of food products, reduce their microbiological contamination, increase digestibility, and reduce pollution of the environment. In the extrusion process, the product is gripped by the screw, moved along the housing, passes the zones of compression and heating due to the friction of the product on the surface of the rotating screw and housing, homogenization, and shear deformations in the product itself, as well as the extrusion zone and unloading zone. The duration of processing is 1–2 min, and the pressure and temperature increase and reach 50 MPa and 180 °C.

The extrusion process is carried out on cereals (wheat, barley, rice, corn, etc.), whose main component is starch, which undergoes significant changes during the extrusion, which leads to molecular disorganization (Shmalko et al., 2007).

The use of extruded products in baking contributes to the nutritional and biological value of baked goods.

Thus, it is advisable to use wheat extrusion bran in the production of wheat bread. When they are added to wheat flour, autolytic activity, gas-forming ability, and gluten quality increase. The optimal amount of additive introduced into the recipe of products is 5–15% of the flour's weight. In the production of bakery products, it is possible to use extrudates (up to 25% of flour according to the recipe) produced from whole barley grains in an extruder at a temperature of 125–195 °C for 30–40 seconds with the screw rotation speed of $38.2 \pm 2\ s^{-1}$ and the diameter of the outlet hole matrix 8 mm. The consumption of bread with extrudate contributes to the satisfaction of more than 30% of the daily need for protein.

For enrichment of bakery products, a composition is offered, consisting of 25% of amaranth carbohydrate-protein fraction (a by-product of *Amaranthus caudatus* grain oil production), 65% of barley grits, and 10% of pea grits obtained in the extruder at the following parameters: temperature inside screw chamber 150–160 °C, output 120–125 °C, outlet diameter of matrix orifices – 11 mm. The obtained extruded mass was cooled, milled, and added at the dough kneading stage in dosages of 5%–11%, which resulted in a significant slowdown of staling of the finished products due to the increased water absorption capacity of the starch added as part of the textured composition.

The new electrophysical technology includes magnetic treatment of water, ozonation of flour, and kneading of dough under the influence of a negatively charged electric field, which significantly improves the organoleptic properties of bread and accelerates the fermentation process on average by 37–40%. Bread made using electrophysical influences on the technological process will not get potato disease within eight days; it is different in appearance, volume, and porosity.

A technical device to intensify the cooling of bread and bakery products is proposed to improve the quality indicators to increase the shelf life of freshly baked bread in difficult natural and climatic conditions. The technical solution is implemented through the joint application of the effects of low vacuum and ultrasound in pulsed mode, providing intensification of the volumetric process of cooling bread and bakery products. The synergy of the fields allows reducing the cost of creating a vacuum in the cooling chamber by 2–3 times and making better use of the ultrasonic capillary effect, which provides an instantaneous evaporation process of moisture with thermal energy on the product's surface. The device can be used during the production of bread in the field, which will reduce the maintenance of the production area and accelerate the delivery of freshly baked bread to consumers.

The use of cryogenic technology in baking allows for preserving bakery products without changing the structure and chemical composition. The technology is based on the properties of a freezing liquid, whose molecules crystallize when the temperature drops. When using cryogenic technology, the formed ice crystals have the same size as the water molecules, which allows for keeping the product's structure after defrosting and retaining the valuable properties of the product.

Although increasing the cost of freezing and storage of products from 15% to 30%, cryogenic technology in bakery provides fast baking or simple defrosting of the product, extends shelf life, and allows transporting the product over any distance in frozen form without loss of quality characteristics, managing demand fluctuations through long-term storage stock, reducing costs for the return of unsold products, and decreasing the need for space for production and storage.

There are three main methods of freezing baked goods: freezing dough pieces, freezing partially baked products, and freezing finished products.

The main technologies of blast chilling used by most manufacturers are as follows:

1. Ready to mold. With this technology, pieces weighing from 100 g to 3 kg are formed as flat as possible right after kneading the dough. The formed pieces are blast-frozen and stored at −18 ° C. Before use, the dough pieces are thawed for 10–20 h at 4 °C, after which the technological process is resumed. The advantage of this technology is the ability to make products in different shapes with the addition of various starters and steams in the product's composition. The disadvantage lies in need to use gluten enhancers because it is necessary to reduce the moisture content of the dough to obtain a firmer consistency, which leads to a lower yield of dough.
2. Freezing after forming dough pieces – Crusurgele technology. The main advantages include baking semi-finished dough products (billets) at the point (bakery) without the forming stage; unbaked billets are less sensitive to storage temperature fluctuations and take up less space. However, this technology is energy-intensive because it requires a defrosting cabinet for defrosting.
3. Ready for baking. This is a blast chilling technology after the final proofing of dough pieces. The short shelf life of finished products obtained by this method is its main disadvantage (Gerasimova & Labutina, 2019).

In the production of frozen bread, enterprises use different types of freezing equipment: cabinets, cameras, and tunnels.

There are various options for storing frozen semi-finished dough products:

- Oriented polypropylene (OPP) is used to store frozen semi-finished dough products. Compared with low-density polyethylene (LDPE), OPP has almost double the tensile strength; it is less permeable to oxygen and more resistant to heat.
- Linear LDPE (LLDPE) has higher tensile, puncture strength (perforation resistance), and abrasion resistance. However, it is less transparent and more expensive.
- Coextrusion films, such as ethylene-vinyl acetate (EVA) and LDPE or LLDPE and HDPE (high-strength polyethylene).

Belarus has developed technology for thermal sterilization of bakery products – preservation of bakery products by suppressing the vital activity of microorganisms in them. Thermal sterilization of bakery products is the heat treatment of packaged bakery products. One-stage and two-stage thermal sterilization is used. For thermal sterilization, cooled bakery products are packed in film or polypropylene bags with a thickness of 25–72 microns, which guarantee the safety, quality, and security of products during transportation, storage, and sale. Thermal sterilization of packaged bakery products is carried out in baking ovens of various types and designs at temperatures up to 110 °C (Laptsenak & Sevastsei, 2017). After the thermal sterilization, the baked goods are placed on the trays of trolleys, containers, or in coolers to cool for 2–4 h at 20 ± 5 °C and relative humidity of no more than 75%.

The military bakery organizes the baking of bread at mobile bakeries. Bakeries with equipment placed in the truck bed have sufficient capacity and can use any transport base. Among the equipment used in mobile bakeries, there are ovens running on any type of energy carrier, where a diesel burner can be replaced by electric heating. However, they are not adapted for operation in low ambient air temperatures (Sharonov et al., 2015).

In the U.S. Army, bread is included only in standard and hospital rations. Bakery Container OZTI, with a capacity of 7.71 t/day, is designed to make bread from different types of flour in the field and supply it to military units. The bakery consists of three compartments – a dough mixing trailer with a monorail frame, dough dividing and rounding machines; two baking ovens mounted on trailers; two power plants; a proofing cabinet; two racks for cooling and storing finished bread; two collapsible conveyors to feed bread in forms to the racks and return forms to ovens; inventory and tools. The dough mixing, dough dividing, and rounding machines, as well as the lighting, are driven by two gasoline power plants of 25 kW each mounted on trailers with a weight of one ton. Baking is carried out at 190–220 °C for 70–75 min. The duration of setting up is two hours, tear-down –1.3 h.

SERT ELC 500 Field Kitchen (France) is designed to provide hot food and bread in the field for up to 800 people. The bakery kitchen is based on two 20-foot standard ISO containers with twist locks. Containers are built into DROPS platforms, allowing them to be easily transported on standard freight platforms. The bakery kitchen is equipped with four generators with a six-cylinder diesel turbo engine, making it fully autonomous and independent of external sources. Equipment composition includes baking ovens, heating plates, deep-fryer tubs, built-in frying pans, refrigerators, built-in boilers for first courses, water tanks, etc. (Sharonov & Sharonov, 2020).

A mobile bakery developed by "Pavalie" (France) is designed to supply bread and confectionery products for the military and the population (from 200 to 5000 people) of the army compounds. The mobile bakery can operate independently from its own electric unit, or it can be connected to the stationary electricity and water supply network. It is transported by a 100–150 hp. tractor on good roads at a speed of up to 70 km/h. The bakery is operated by two workers.

The Marshalls MFBS mobile self-contained bakery for the UK Armed Forces is designed to cook 5000 units of bakery products for 24 h autonomously. It includes two transformable ISO containers of 20 ft. with the necessary equipment and a 200 KW diesel generator. The bakery is equipped with all necessary equipment for the automated production of bakery products. The temperature range of the bakery varies from −50 °C to +50 °C.

The autonomous mobile mini bakery Bassanina Mobile 2500 of the Italian armed forces is mounted on a high cross-country vehicle. It allows producing 865 kg of products in eight hours of work and 2595 kg in a three-shift mode of operation (Polandova et al., 1986).

Promising developments include methods of baking without the use of yeast and fermentation starters, which can reduce the time necessary for dough preparation from several hours to 5–10 min (Chervyakov et al., 2011). This is provided by special equipment – foaming dough (preparing dough with pressured air). Depending on the type of dough, the foaming machine can produce 60–100 pieces per hour. The expected capacity of such a complex could be 1000 baked goods in a 12-hour shift or even more. It will allow producing a wide range of bread: gluten-free bread, bread from flour with low gluten content, and bread with a uniquely low glycemic index.

References

Auerman, L. Y. (2003). *Baking technology: a textbook* (9th ed.). Professiya.

Buldakov, A. (1996). *Food additives: handbook*. St. Petersburg, Russia: Ut

Chervyakov, O. M., Sharonov, A. N., Polischuk, A. P., & Avjyan, A. S. (2011). *Patent for invention RU 2423843 C1 "Method for accelerated graded wheat flour bread baking in field environment"* (July 20, 2011; Application No. 2009145799/13 December 9, 2009). Moscow, Russia: Rospatent

Gerasimova, E. O., & Labutina, N. V. (2019). Cryogenic technologies in bakery. *Izvestiya Vuzov. Food Technology, 1*(367), 6–9. https://doi.org/10.26297/0579-3009.2019.1.1

Laptsenak, N. S., & Sevastsei, L. I. (2017). Innovative technologies in the bakery prolonging the life of food. *Food Industry: Science and Technology, 2*(36), 20–28.

Markova, M. (2018). Conference "Innovations in Baking." Khleboproducty, 7, 16–18

Polandova, R. D., Shkvarkina, T. I., Bystrova, A. I., et al. (1986). *The use of complex baking improvers: a review*. Research Institute of Information and Technical and Economic Research in the Food Industry.

Puchkova, L. I., Matveeva, I. V., Sidorova, O. G., et al. (1988). *The use of non-traditional raw materials in the production of improved and dietary bread from rye and wheat flour: a review*. Central Research Institute of the Ministry of Grain Products of the USSR.

Rusina, I. M., Makarchikov, A. F., Trotskaya, T. P., Mistsiuk, Y. V., & Kavaleuskaja, S. S. (2012). Possibilities of kidney beans flour and pea flour using for bread production. *Food Industry: Science and Technology, 4*(18), 22–27.

Sharonov, A. N., & Sharonov, E. A. (2020). Means of field baking of the armed forces of NATO. *Current Problems of Military Scientific Research, 7*(8), 290–306.

Sharonov, A. N., Vostryakov, I. V., & Sivakov, A. S. (2015). Techniques development trends for military food services (leading foreign countries). *Bulletin of the Military Academy of Material and Technical Support named after Army General A V Khrulev, 1*, 45–50.

Shlelenko, L. A., Tyurina, O. E., Borisova, A. E., Nevskaya, E. V., & Dobriyan, E. I. (2014). The use of vegetable and fruit powders in baking. *Khleboproducty, 7*, 42–43.

Shmalko, N. A., Belikova, A. V., & Roslyakov, Y. F. (2007). The use of extruded products in baking. *Fundamental Research, 7*, 90–92.

Skorbina, E. A. (2015). Development of functional and specialized baking in the Stavropol territory. *Food Industry, 4*(26), 50–51.

Spirichev, V. B., Shatnyuk, L. N., & Poznyakovsky, V. M. (2005). *Food enrichment with vitamins and minerals: science and technology* (2nd ed.). Siberian University Publishing.

Vershinina, O. L., & Tezbieva, M. K. (2014). The use of grape skin powder in baking. *Khleboproducty, 2*, 48–50.

Zhuravleva, L. A., Zhuravlev, A. P., & Terekhov, M. B. (2012). Hemp oil and its use in baking. *Bulletin of the Altai State Agricultural University, 4*(90), 66–69.

Chapter 16
Innovation in the Commercial System of the Small Bakery Industry

Adriano Fidalgo and Jorge Miranda

16.1 Introduction

Based on the latest study by the European Commission, published in 2010, the bread and pastry market, at a European level, exceeds 32 million tons sold, with the bread and pastry industry being one of the oldest industries. Workers who built the pyramids in Egypt were paid in bread.

Although the long history of the industry, the habit of daily intake of bread in several countries and its high consumption, according to the Federation of Bakers (FOB) it amounts to 50 kg per year, showing a slight decline in demand because certain bread and pastry products are associated with a less healthy diet.

However, new business segments and product typologies have emerged, seeking to meet the demand for healthier and more natural bakery and pastry foods, movements supported by advances in food science.

There is at the same time a paradigm Shift in the profile of bakery and patisserie owners, from people with little education and culinary skills, to people with higher education in the most diverse areas, oriented towards business, innovation and the development of products aimed at changing consumption habits.

In view of changes in consumption patterns and developments and innovations in marketing and commercial management in other industries, the bakery and pastry industry has been adapting to this new reality, characterized by the presence of the online business in search of food healthier.

A. Fidalgo
Astrolábio – Orientação e Estratégia S.A., Porto, Portugal
e-mail: adrianofidalgo@astrolabio.com.pt

J. Miranda (✉)
InCubo, Arcos de Valdevez, Portugal
e-mail: jorgemiranda@incubo.eu

J. M. Ferreira da Rocha et al. (eds.), *Baking Business Sustainability Through Life Cycle Management*, https://doi.org/10.1007/978-3-031-25027-9_16

This chapter addresses innovative management practices in the industry in terms of new commercial and marketing management approaches. In terms of methodology, three methodological practices were used:

1. Collection of identifying elements and evaluation of digital presence: using the Orbis platform, owned by Bureau Van Dikj, a Boolean survey was carried out based on the following structure:

First, all companies were evaluated with NACE Rev. 2107 – "Manufacture of bakery and farinaceous products" present on the platform, with a sample of 182,183 companies;

The search criteria "company with an available website" was added, and from the above, there were a total of 24,943 companies in the sector, headquartered in the European Union, with an active website, representing a total of 13.19% of the whole the companies present on the platform;

Then, based on the 24,943 companies, the sample was constructed, using the number of residents as the standard for distributing of the sample by country.

Thus, the following representative sample was obtained in a total of 108 companies:

After identifying the sample, the various websites were consulted in order to assess the company's digital presence, responding affirmatively or negatively to the existence of the following elements: online store, online orders, home deliveries, specialized products, contact information and social networks, exhibition photos of the products, and a quick conversation window.

2. Interview conducted with the President of the Association of the Bakery, Pastry and Similar Industry (AIPAN-PT)

To collect qualitative information, a semi-structured interview was carried out with the President of AIPAN (the main association of companies in the bakery and pastry sector in Portugal). The questions were asked to obtain information identifying the practices and processes of marketing and commercial management adapted by the industry in Portugal.

3. Survey of bakeries and patisseries in the North of Portugal

Through collaboration with AIPAN, a survey was carried out with open and closed responses, having obtained a total of 42 respondents that represent the sample that allows a sensitivity analysis. The following characterization was observed:

Regarding the number of companies indicated in Fig. 16.1, we evaluated a set of requirements related to their digital presence, ensuring the consistency of the data, obtaining a representative sample of the bakeries in Portugal, both in terms of business dimension and bakeries typology (Figs. 16.2 and 16.3).

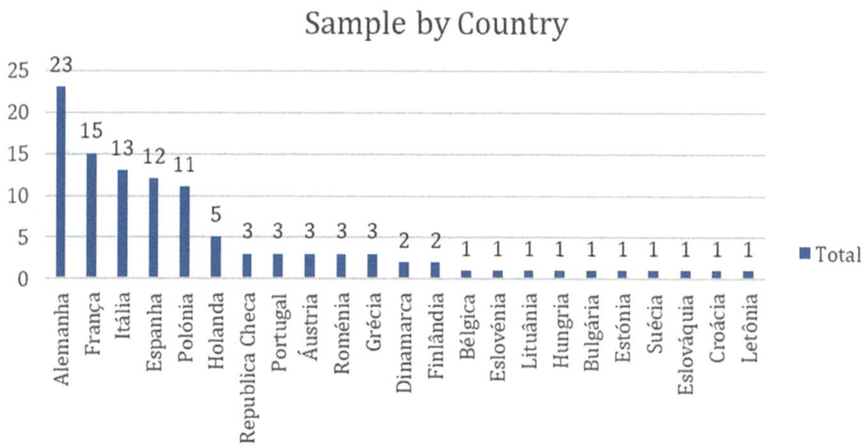

Fig. 16.1 Sample Companies with website. (Source: Orbis 2021)

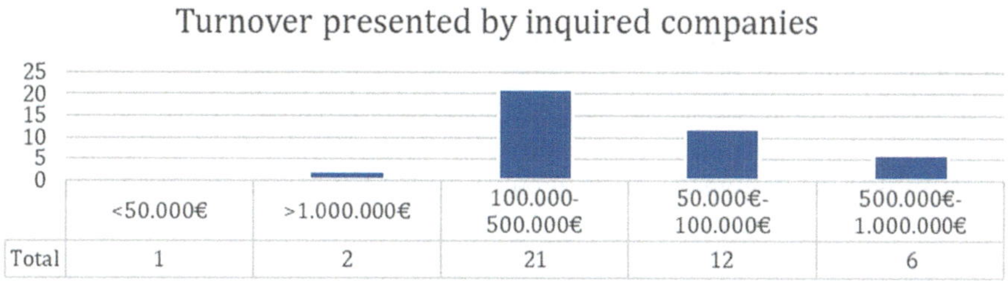

Fig. 16.2 Distribution of inquired companies by turnover presented. (Source: AIPAN Membership Questionnaire 2021)

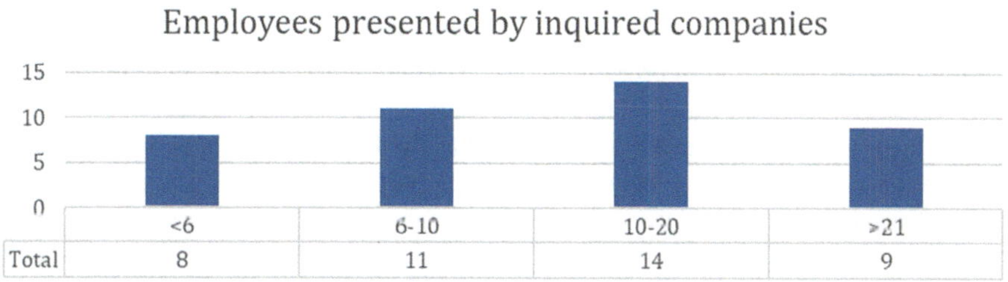

Fig. 16.3 Distribution of inquired companies by number of employees. (Source: AIPAN Membership Questionnaire 2021)

16.2 Marketing and Commercial Innovation

According to the Oslo Manual, a document guiding the main definitions and processes adopted by the OECD about innovation, defining it as a "new or improved product or process (or a combination of both) that differs significantly from previous products or processes" (OECD, 2018). The same states that there are innovations at four different levels of innovation, namely product, processes, marketing and organizational.

Concerning product innovations, traditional bakeries and pastry shops are betting on their own manufacturing as an innovation factor and trying to innovate the base recipe with the introduction of new ingredients, formats and finishes. Thus, the products manufactured in-house can meet the needs of consumers, finding solutions for meals throughout the day. In the case of bread production, this has undergone many changes, the manufacture evolved with the introduction of new cereals, vegetables and even spices such as turmeric or beetroot, influencing and bringing benefits to the adoption of a healthy diet.

Concerning productive space, it is also necessary that the establishment adopts a convenience store-style and offers comfort. Concerning sales management, the layout must allow the customer to move freely, and perceive all the products' aromas, textures, and colors.

Innovation in the marketing and commercial area can be defined as a redesign of the product's design, its positioning and packaging, the way it is promoted and the method used to define the price to make the product more attractive and salable. However, the pastry and bakery sector has very specific characteristics, such as irregular production of products through perceived demand, direct production orders from customers and even high perishability of products.

Thus, in the development of proximity businesses, it is necessary to have a clear vision of who the customers are, the products they consume the most, and the products they might like to consume.

To understand innovative practices that pastry shops and bakeries can adopt in terms of marketing and commercial processes, it is necessary to understand the concept of proximity markets and its implications.

According to the President of Aipan, the intervention of classic marketing in Portuguese markets is reduced (except for the new generations more adapted to new telecommunications). Management model adapted to small business models, whose product asserts itself and in proximity, which, when aimed at moving to other markets, must be planned through a marketing plan.

The new generations have a profile closer to the food area with technology integration, and attend training centers in various parts of Portugal, such as Lisbon, Porto, Coimbra and Algarve.

16.3 Proximity Markets

The commercial activity has been following the development of society in different geographical realities. We witness the appearance of fairs in the twelfth century, which took place inside the castle walls due to factors such as the influx of people, living in the vicinity, and factors associated with the security of transactions. Around these cores developed the first population agglomerates, giving rise to larger settlements.

Thus, the concept of distribution emerged, associated with the commercialization of generality of goods.

The perspective that distribution consisted of a product outflow channel appears centered on a backwards production perspective, currently called "push" strategy, where the distributor acts only as a passive agent in the product delivery process. This modus operandi had great relevance in times when demand was greater than supply.

In recent decades, in most developed countries, the scarcity season has been replaced by abundance, which is characterized by most consumers seeing their basic needs fully satisfied.

In these markets characterized as mature, supply exceeds demand, which makes the push approach unfeasible, since the consumer's decision does not fall on existing products. Hus, a pull approach becomes essential, so that supply responds to demand, producing and delivering the product when, how and where the customer wants it.

It is from this perspective of approximation and focus of the agents responsible for the offer, that the companies, the economic agents that are part of the demand, and the customers, that the proximity trade emerges.

To understand the concept of proximity trade (short market), a good understanding of the concept of trade is essential. There are numerous definitions of commerce and it should be understood as any activity that is carried out with a professional character, through the intermediation of goods and services.

16.3.1 Factors That Influence the Place of Establishment/ Operation of Bakeries and Pastries

Similar to the more traditional sectors of activity, the companies in the sector are located close to a specific population center/agglomerate or site with high daily affluence. Thus, when choosing the location of the establishment, companies can be affected by two sets of variables in the environment, such as the competitive environment and the preferences and habits of the local consumer.

In the competitive environment, being a proximity business, they will face direct competition, often side-by-side, where each party tries to capture the customer, whether through a differentiation strategy or a cost strategy.

Likewise, the range of pastry and bakery products from different countries, regions, cities and towns is diversified. Many are local, spread across Europe that have specific products or standardized products with specific characteristics.

More than in any other sector, pastry and bakery companies must know the customers, the products in the region, the times of the year when a certain product is sold and maintain constant feedback. With customers. This relationship with clients helps to assess whether their products meet their needs, whether the quality is comparable to that of competitors and what new products and which variants the customer are looking for.

According to the President of AIPAN, the factors that directly affect the decisions on the location of the establishment and the performance of bakeries are related to population density, services adjusted to the needs of the region, proximity to socio-economic agents and other facilitators. It is visible that it is more difficult to survive outside urban centers, and small industries in the interior have more economic difficulty.

According to the study carried out through a survey of 42 bakeries/pastries, 42.86% referred that population density is a decisive factor in choosing the location for opening a new establishment, and 11.9% of respondents indicated that the low competitive density also helps in decision making.

16.3.2 Size and Typology of Establishments

The opening of a bakery or pastry shop, introducing new products, or changing facilities bring many challenges and dependencies. The problem of defining plans and implementing marketing actions and commercial techniques occurs in companies that produce and provide services to the final consumer. Thus, depending on the initial investment to be made, the available space and the range of products they intend to sell, various types of business models can be developed within the sector:

Coffee and Bakery It is a commercial space, characterized by selling a wide range of cakes, breads, drinks, cafeteria products and desserts. This kind of establishment has area for customers to consume the products purchased, and it is necessary to incur higher costs in terms of human resources, acquisition and rental of buildings, due to the space required.

Counter Service This kind of establishment is most often found in densely populated population centers, shopping centers and places where building prices and rents are high.

Online Store or Marketplaces It is a business that can only take place online or online with a physical store. It allows reaching a more significant number of customers; however, despite the reduction in indirect costs typical to physical stores, there are additional costs related to the website, content production and maintenance of order platforms.

With the evolution of times and the fight against the Covid-19 pandemic, many bakeries and pastry shops witnessed the movement of marketplaces such as Uber, Glovo, which started selling products with the delivery service, consolidating themselves in an increasingly attractive market to explore in the bakery and pastry sector. This kind of mobile establishment covers a larger territory and is commonly found in areas with insufficient customers to make the establishment of a fixed store profitable. This kind of business needs special attention given the specific rules for food transport vehicles.

Specialized Bakeries/Pastries This kind of business does not have a typology of its own establishment. They focus on meeting the client needs and do unique cakes and bread. This business presents a demand with a lot of fluctuations since it deals with products that are not purchased recurrently.

Considering the survey carried out, 73.1% of respondents have an establishment of the type of cafe or bakery with table service and 19.5% have an establishment with only counter service and with an equal percentage they have a specialized bakery/pastry shop, which it only sells products to order.

In a result of the survey, it is observed that half of the respondents tend to own a bakery or pastry shop with a turnover between €100,000 and €500,000, and only a small percentage, 2.3% and 4.7% achieve a lower turnover to €50,000 and over €1,000,000 respectively.

16.4 Local Valuation and Territorial Strategy

As bakeries and patisseries are businesses that traditionally fall within the proximity trade, the products must meet the surrounding area's needs, tastes, preferences and consumption patterns. Thus, this range of products in the bakery area, usually consumed by those who live and frequent the surroundings of the establishment, includes various types, including traditional products, innovative products, products that meet a niche demand and basic products.

According to Barberis (1992), the traditional products are those that are identified for being unique, for the raw material and applied knowledge, distribution and consumption practices, and for their geographic origin, production processes and intrinsic qualities of the region. Resulting from the differentiation between traditional products and products considered normal, the EU created in 1992 the system of protection and enhancement, featuring three distinctions from traditional products, namely DOP - Protected Designation of Origin, IGP - Protected Geographical Indication and ETG - Specialty Traditional Guaranteed, which cover a wide range of products including baked goods and pastries.

In this way, bakeries and patisseries find a set of opportunities and threats in the tradition of the areas in which they are based. By marketing products recognized as traditional, bakeries and pastry shops guarantee the offer of a product with an established demand, transversal to different age groups and which has a high potential for acceptance by tourists visiting the region. However, to guarantee quality, comply with tradition and maintain the identity associated with traditional products, bakeries and patisseries must abide by a set of specific production processes, use raw materials from the region and ensure that the format, packaging and appearance of the products are in accordance with the standards. In this way, traditional bakery and pastry products contribute to the development of the region and should be included in the marketing plans of localities, cities and regions.

Several innovations have emerged associated with entrepreneurial movements of bakeries and pastry shops that opened a business unit outside their area of origin. The increased mobility of individuals between European countries, as well as their fixation has contributed to the generation of groups of people outside a given region and, thus, to the creation of a demand for a given product without a corresponding offer. Consequently, several companies producing and commercializing bakery and pastry products have started internationalization processes, either by way of exportation or by way of fixation on the market, increasing the notoriety of the regions of origin, of the products, and introducing in other regions the consumption of traditional products from a region, where these products did not initially exist.

16.4.1 Influence Area of a Bakery/Pastry Shop

The area of activity of a patisserie is related to the size and extent of the population agglomeration in which the business is located. However, the product offering comes from various businesses, and the customer has to choose based on several factors, such as price, quality and convenience.

The digital transition appears as a new paradigm, causing changes in sales behavior and consumption patterns. In the last decade, the internet has penetrated people's daily lives and has become an essential means of communication, both between individuals and between individuals and companies, which has helped in the communication and sale of services and products online (Akram, 2018).

Faced with this new paradigm, companies have invested and bet on the creation and implementation of means of communication and online sales points. Thus, through the use of websites, social networks, delivery applications, marketplaces and online stores, bakeries and patisseries are now more exposed to a higher number of potential customers in addition to individuals who live or visit their surroundings.

In this way, there is an expansion of the area of influence and activity of the businesses, boosting the growth of business volume and placing particular relevance on factors that can distinguish the company from other competitors.

According to the President of AIPAN, the radius of action in Portugal is between 3 and 4 km for the end customer and between 15 km and 20 km in terms of distribution. Regarding the analysis of data from the questionnaire carried out, 21.3% of bakeries and pastry shops assume that the range of action will be between 10–15 km and more than 15 km, and about 40% affirm that the range of action should be up to at 5 km.

16.4.2 Traditional bakery and Pastry Market and Impacts on the Area of Influence

With the new digital paradigm, traditional products find many opportunities to develop and be presented to new consumers. In a context where the importance of authenticity and origin of products is growing, the proliferation of digital commerce presents itself as a key factor for the sustainability and maintenance of tradition. A successful example of this modernization of the traditional was presented in the ninth week of entrepreneurship in Lisbon.

In view of the pandemic situation and the consequent reduction in the number of tourists, a group of companies located in downtown Chiado (Lisbon) joined together to create a marketplace for traditional Portuguese trade and products and, currently, the website has thousands of daily visits, from users from all over the world, contributing to the dynamization of companies and publicize national products, ensuring an increase in sales and consumer knowledge of the products.

In valuing traditional products, bakeries/pastries find ways to revitalize digital communication without losing their identity, investing in an online store that shows the traditional spirit through images that illustrate the culture, colors, fonts suited to past times and product photographs.

Thus, based on the analysis of the responses to the survey carried out, the best way to attract the target audience is through sympathy in the service, with around 95.2%, allied to getting to know the customer better (name, tastes, period of attendance at the store, etc.), which accounts for around 26% and the attribution of discounts and promotions.

16.4.3 New Forms of marketing in the Sector

The sector is constantly evolving and follows the digital transition, with innovation processes such as breads made with various cereals and seeds, new and current formats, for new aspects.

Product differentiation is important for origin identification, quality control, adding even more value to the product.

The product's image, the point of sale and the seller's profile are crucial elements to attract consumers, betting on campaigns to rescue young people for this profession, creating more schools and training centers, generating more preparation for internationalization.

It is important that bakeries and patisseries can convert their traditional offers to digital platforms, which in addition to promoting through a website with availability of contacts, products, essentially with photographs to give visibility and capture the attention of consumers, quick chat so that the customer does not lose interest in the brand and with the business being allocated to social networks it is possible to share experiences.

With the evolution of the pandemic, there was a need to create delivery applications and Marketplaces, which resulted in a sales channel for bakeries and patisseries to market their products, apply new prices and increase their value chain.

16.5 Interconnection Between Digital Marketing and Face-to-Face Marketing

Since marketing is an area of business science, which studies how to capture, retain and interact with customers, it can be developed by any type of company in any sector of activity. One of the first and most important steps to implementing successful marketing campaigns, with the least possible expenditure of resources, is identifying the "persona" where the customers fit. According to Cooper (1999), this persona can be understood as the representation of a character, with characteristics identical to those of the consumers of the company's products.

In recognizing the characteristics commonly found in customers, bakery and pastry companies have a notorious advantage over companies in other sectors since the interaction between customers and the company is direct, routine and informal.

In building a company's online presence, the digital media used must represent a continuation of the physical store, presenting everything the company has to offer, distinctive features of its products and the company's value proposition. Websites, social media pages and streaming platforms become ideal platforms to present new products that contain benefits for the health of consumers, as well as the dissemination of production processes that have drastically evolved, increasing the offer.

With the construction of audiences based on the characteristics of customers in physical stores, omnichannel solutions, the "IoT – internet of things" triggered by 5G, it will be possible to control the consumer experience in the physical store, on the website, in new marketplaces, in the social networks and create differentiating campaigns, relevant content for each consumer, creation of content based on location, profile and unify the activities of the various sales channels and leverage the behavioral data of each consumer in order to reference the steps to follow (Fig. 16.4).

It is essential that presences (online and offline) are aligned, so that a customer who makes a first contact online can get a continuous consumption experience when he goes to the physical store, and that the physical store customer gets an online experience similar to what has already been experienced.

Digital audiences, according to De Kerckove (2013), are the complement to traditional profiles that became interested in the online market, covering personal, social, institutional, political, scientific and technological aspects. As mentioned by De Kerkove, it is necessary to build a solid base for these profiles using social, technological and representative data sciences.

However, it is a constant challenge to create these profiles since they are characterized through algorithms that are not always able to assess the profile and behavior of the consumer correctly. But it is through this characterization that companies develop new products and solutions according to consumer tastes and preferences.

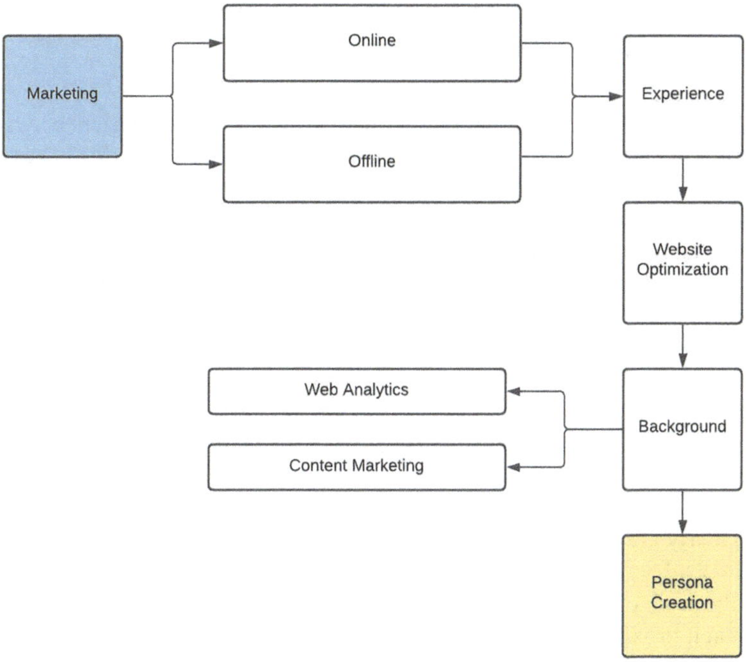

Fig. 16.4 Audience creation. (Source: Own Elaboration)

Traditional marketing often brings more answers than digital marketing due to the physical presence of consumers in bakeries and pastry shops; however, with technological evolution, these establishments must be able to create personal and interpersonal relationships through online communication channels.

Through the analysis carried out on the digital presence of companies removed from Orbis, we can consider that all the companies analyzed have a website, but only 27% of these companies have an online store on their website, and 30% of the companies offer the possibility of placing orders online.

Throughout the analysis, we found that 87%, corresponding to 95 bakeries and patisseries, have incorporated in their offer traditional products adapted to the preferences of local consumers, and about 94 bakeries and patisseries have on their website photographs of the products they sell, one more worth to seduce website visitors.

With technological evolution, the advent of IoT and 5G has become a requirement and an urgency. The technological level change will revolutionize and digitally trigger the means of purchase, making them automatic, immediately triggering an order, which, through a delivery platform, will place an order with the bakery that will deliver it at home without any human influence.

16.5.1 *How to Drive Traffic to the Store Through Digital Marketing*

The omnichannel is already the privileged channel with fewer barriers between the distinction between real vs. digital and vice versa. The best way for bakeries and patisseries to attract website visitors to the physical store is to use a variety of multi-channel digital platforms to build a presence among potential consumers.

As the target audience of most bakeries/pastry shops is quite diverse, whether, in terms of tastes, preferences and demographic characteristics (gender, age, etc.), they must have a digital presence where the various target audiences are present. One of the stimulus to the trend in technology has become the rise of social media and marketplaces, making the role of these disclosures essential for the overcoming and evolution of the bakery/pastry sector.

Based on the developed questionnaire, it is possible to observe that most respondents seek to increase physical store traffic through a digital strategy shared between the website and social networks, creating a close relationship with potential customers, regularly presenting content related to their performance and multimedia material that displays its products.

As the product sold in bakeries is a mix of impulse and demand products, Display and Search proximity ad campaigns are essential to ensure the favorable evolution of the business. From the study carried out on the set of websites of companies at the European level, it is possible to observe that only 7% of companies have, simultaneously, Linkedin, Facebook, Instagram and Youtube. The social network most used by the group of companies in the sample is Facebook (75%), followed by Instagram (49.1%) and the least used is Linkedin (19.4%).

In this way, it is possible to show an effort by companies to be present on social networks, where current and potential consumers are, always considering the characteristics of social networks, observing a preference for more generalist networks that allow for a greater panoply of content (see also Fig. 16.5).

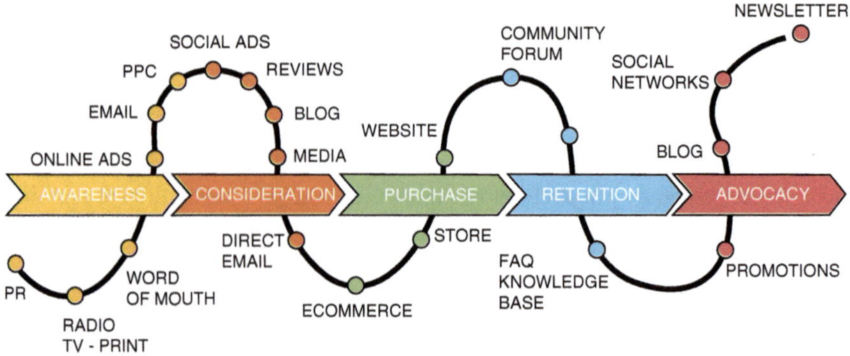

Fig. 16.5 Lead Nutrition. (Source: Social media usage in Europe, 2021)

We can also mention that companies try to serve audiences of different age groups, namely, through the simultaneous use of Facebook and Instagram, wherein the first, there is a population aged over 35 years, while in the second, there is a concentration of the under 35 age group (Statistica, 2021).

The possibility of placing orders through the website with the alternative of picking up the order in the physical store allows taking these users to the center of the business. Through physical presence, bakeries and patisseries nurture relationships with their customers, in different weights, betting on friendliness (93%), and simultaneously with the previous one, they bet on offering promotions and discounts and show interest in getting to know the customer (79%).

16.5.2 How to Attract Digital Traffic Through the Physical Store

There is a significant dependence of society on new technologies, transforming two distinct eras, pre- and post-digitization. Thus, there is a strong need to take consumers from physical stores to digital platforms as a way of encouraging them to follow the brand more closely, making them more active consumers and promoters of bakeries and pastries.

One of the ways that bakeries and patisseries found to promote digitalization was to create a QR Code with the menu, which can take customers directly to the bakery/pastry shop's website and, consequently, attract this traffic to different social networks.

However, in view of the survey carried out, there are still bakeries that do not have current marketing practices in about 26% of respondents. However, 28.57% bet on advertising on social networks, such as Instagram and Facebook, and there is still a small percentage that bet on more traditional methods such as flyers and good service to the public, which suggests that there is still concern about seduce the customer not only for the products, but for the history of the establishment.

16.5.3 Most Appropriate Digital Marketing Tools

Given the above, and after a quantitative and qualitative study of the digital presence of European and North and Central Portugal bakeries and patisseries, the most used network is Facebook and Instagram.

In addition to the construction of relevant content for these two social networks, the focus on video production, either to show processes or in order to draw the attention of the target audience, it is extremely important to create a blog that has a connection to the other digital platforms of the bakery/pastry shop in order to promote and give visibility. Fundamentally, the placement of photos of products, whether traditional or innovative products, helps to create dynamics and interest among website visitors and social networks.

16.6 The Importance of Relational Marketing in the Growth and Sustainability of Small Patisseries

Relational marketing was directly influenced by factors such as the impact of information and technology, the relational and transactional division of markets and customer selectivity. However, the financial factor present in this type of marketing means that many companies are forced to operate in relational markets and in international markets. Relational marketing progressed in three phases: the analysis of the marketing-mix insufficiencies, the evolution of transactional marketing to relational marketing, and the analysis of the main guidelines of the concept.

The main objective of marketing rests on the needs and desires of the customer, as well as their requirements. There are thus differences between these two types of marketing that directly influence the growth of small bakeries and patisseries, such as individual sales and customer retention, discontinuous or continuous contact with the customer, low or high level of commitment.

Thus, if bakeries/pastry shops adopt new forms of relational marketing, they will have several benefits in terms of more customer retention, which leads to an evolution in turnover, based on a close and lasting contact with customers. This type of marketing can be considered a competitive advantage, betting on customer loyalty and satisfaction. It is a relationship that goes beyond the buyer-seller; there is an extra connection with suppliers, competitors, intermediaries, shareholders and investors.

16.7 Conclusion

As evidenced throughout the article, the pastry sector is making efforts to innovate in the way it sells its products and communicates its offer. With the development and proliferation of large commercial surfaces, SMEs in the sector must develop their strengths and mitigate their weaknesses.

We observed that a representative part of companies in the sector are betting on new media (e.g., Facebook and Instagram) to maintain a close relationship with customers.

Content marketing, the development of distinctive products and monitoring of the tastes and needs of the market are essential to guarantee the continuous innovation and future sustainability of the business.

References

Akram, M. S. (2018). Drivers and barriers to online shopping in a newly digitalized society. *TEM Journal, 7*(1), 118–127.

Barberis, C. (1992). Programme de recherche Agrimed. Les micromarches alimentaires: produits typiques de qualite dans les regions mediterraneennes

Cooper, A. (1999). *The inmates are running the asylum*. SAMS. Macmillan.

Kerckove, D. (2013). The skin of culture: investigating the new electronic reality.

OECD. (2018). *Oslo manual 2018: guidelines for collecting, reporting and using data on innovation* (4th ed.). The Measurement of Scientific/Technological and Innovation Activities/OECD Publishing.

Statistica. (2021). *Social media usage in Europe - statistics & facts*. Retrived on: https://www.statista.com/topics/4106/social-media-usage-in-europe/#dossierContents__outerWrapper

Chapter 17
New Business Models in Food Systems for Farmers and Companies, Based on Multi-actor Approach

Aleksandra Figurek, Marinos Markou, and João Miguel Rocha

17.1 Introduction

Due to the general variable conditions that define agri-food production, each action must be carefully planned depending on the successful or undesirable outcomes obtained. Farming is a complex activity, and each of its actions has a variety of unique characteristics that must be observed (processes of biological transformation). Given the significance of biological resources in agri-food enterprises, biological transformation processes can have a significant impact on the attainment and presentation of business outcomes.

Dynamic settings necessitate a high degree of knowledge and commitment to identify effective strategies to adapt to changing conditions in agro - based production operations.

The development of an appropriate business model in the food industry is critical, taking into account the level of subsidies for specific types of production, i.e. products, in order to direct funds both by purpose and volume to those producers who, despite operating on average terms, do not make a profit. Producers make

A. Figurek (✉)
Gnosis Mediterranean Institute for Management Science, School of Business, University of Nicosia, Nicosia, Cyprus

M. Markou
Agricultural Research Institute, Nicosia, Cyprus

J. M. Rocha
Universidade Católica Portuguesa, CBQF - Centro de Biotecnologia e Química Fina – Laboratório Associado, Escola Superior de Biotecnologia, Porto, Portugal

J. M. Ferreira da Rocha et al. (eds.), *Baking Business Sustainability Through Life Cycle Management*, https://doi.org/10.1007/978-3-031-25027-9_17

decisions to adapt to changing conditions, and these choices are influenced not just by economic variables, but also by socioeconomic traits (Gow & Stayner, 1995; Traore et al., 1998; Willock et al., 1999). Farmers are aware of uncertainty and make management decisions with the full understanding that the outcome is unpredictable. The combination of information about their system and previous experiences allows for a subjective judgement of the expected outcome (Meinke et al., 2001).

The significant association between producer qualities and farming decisions is owing to the fact that farms in the Netherlands are primarily conducted as family companies, with the farmer (and his family) serving as entrepreneur, manager, and primary labour force. This also suggests that the farmer chooses his or her own farm business plan, which is based on his or her choices, interests, capabilities, and evaluation of the internal and external environment. Öhlmér (1998), concluded that the decision-making of farmers can be viewed as a process consisting of four steps: problem detection, problem definition, analysis and choice, and implementation. Beach and Connolly (2005) demonstrated the importance of frames to define problems, which is a necessary condition for change. Even though not always rational, previous experiences of both people and firms define the frames through which strategic decisions are viewed and processed. According to the International Institute for Sustainable Development (Bossel, 1999), an indicator quantifies and simplifies phenomena and complex realities into a manageable amount of meaningful information, feeding decisions and directing actions.

Hansson (2008) concluded that information obtained from other farmers and colleagues, which can be interpreted as part of the network, was highly valued by more economically efficient farms.

Using a structural equation modelling methodology, four categories of factors – the decision structure, the farm's business structure, the cognitive structure of the farmer, and the farm's network structure – have been individually analyzed, and are each found to explain the plan to further develop the farm's production, as opposed to continuing without further development or exiting production (Hansson & Ferguson, 2011). Results from their study also have clear implications for farm advisory services, where efforts to support the development of dairy production can become more effective with more conscious targeting of farms that possess characteristics predisposing them to further production developments.

Better educated farmers chose to increase the intensity of the farming system, and cope with the corresponding increase in environmental pressure by improving the production capacity and improving operational management. Ondersteijn et al. (2003) collected financial and nutrient bookkeeping data of 114 farms, over the period 1997–1999 and combined with survey data on farmer characteristics and farm strategies. The potential adopters of alternative farm enterprises may be traced among farm households that pursue a survivalist mode of production. It is argued that the diversity of farm structures observed within this type of farm households cannot be regarded as the decisive factor as far as their mode of survival is concerned (Daskalopoulou & Petrou, 2002).

McCown et al. (1996) illustrated some more advances in increasing model comprehensiveness in simulating farm production systems via reference to with the Agricultural Production Systems Simulator (APSIM).

The facts which pointing the food production system are: the lack of own production of raw materials; insufficient use of natural resources in agriculture and manufacturing capacities in the food industry; inadequate organization of producers and processors, low productivity, extensive production of fruits and vegetables, low average crop yields, the low organization of subsidies, etc.

In order to achieve improvement in the food sector, it is necessary to apply multi-actor approach. Based on that, the relevant bodies, the Ministry of Agriculture, Forestry and Water Management, various institutes and municipal authorities, could gain insight into the state of agricultural production at regional and national level and make rational decisions for its improvement in the future.

It is also necessary to determine whether the agriculture departments by the municipality are capable of adequately and accurately executing the tasks bestowed to them and to measure their performances. Performance measurement – measures the efficiency and effectiveness of municipal services, strengthens municipal accountability, monitors ongoing performance, evaluates past performance, produces information municipalities to make decisions to improve local services, provides information for planning for the future – is one of the tools used to assess how well an organization performs when providing goods (Ministry of Municipal Affairs and Housing, Ontario, 2007).

Performance management and performance measures can help municipalities to develop a continuous system of improvement (GFOA, 2007). Consistent performance measures can help to reveal when a program or service is not being delivered properly or effectively, which can result in insufficient services to the public, and which may cause stagnation in a particular sector. Performance measurements can be integrated to the strategic planning process and budget, which can then help assess accomplishments on a municipal-wide basis. It is crucial, when performance measurement system evolves into a performance management system that establishes a continuum from measurement to management (Plant & Douglas, 2006). This can improve evaluating past resource decisions and facilitate qualitative improvements in future decisions regarding resource allocation and service delivery.

17.2 Multi-actor Approach in Food Sector

Quantitative and qualitative data relating to the food sector, as well as their communication across specific institutions in order to monitor current agricultural and food circumstances, should serve as a foundation for decision-making. Possession of information at both the macro and micro levels is a necessary precondition for appropriately encouraging food producers to improve their efficiency and individual business performance (Diagram 17.1).

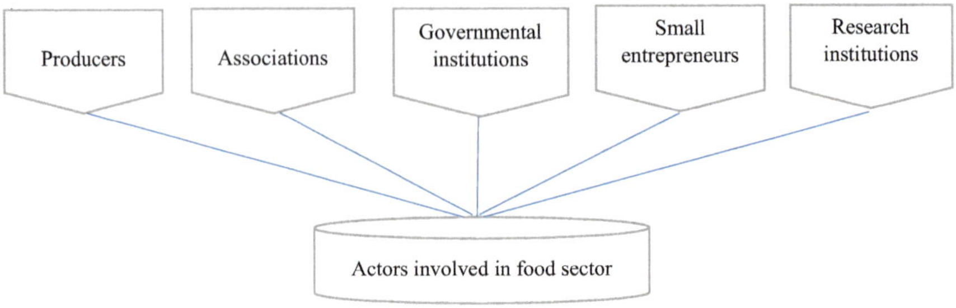

Diagram 17.1 Actors involved

The following activities within the actors are necessary:

- systematically analyze practices and developing a deeper understanding of the collaborative economy phenomenon in agriculture and food sector,
- to identify the technical platforms and infrastructures in agriculture and food sector,
- applying a practice-focused approach, that will support an in-depth examination of the actors, the organizations and the technological tools,
- organizational mechanisms regulating the resource among participants,
- the decision-making processes,
- the nature of the shared resources (e.g. natural, financial, physical, and social or human resources) (Diagram 17.2).

A practice-based approach will consolidate trans-disciplinary perspectives and empirical insights on the human practices of different actors on the individual, collective, organizational, and societal level. This will certainly achieve a great contribution to the development and implementation of the next goals:

- finding aspects of the current practices in food sector,
- collection of existing studies,
- multi-faced view on sharing and caring practices and develop a theoretical framework to interpret and classify the different cases,
- identify sets of technical features of collaborative platforms in agriculture,
- identify the ways how different actors participate,
- identification of the institutions supports the collaborative practices in agri-food sector,
- analyzing of the constraints,
- observing how actors practically overcome limitations,
- identification of the mechanisms to monitor ongoing developments in the sector,
- establishing communication and identification of the networking tools thus strengthening inter-institutional and international linkages among actors.

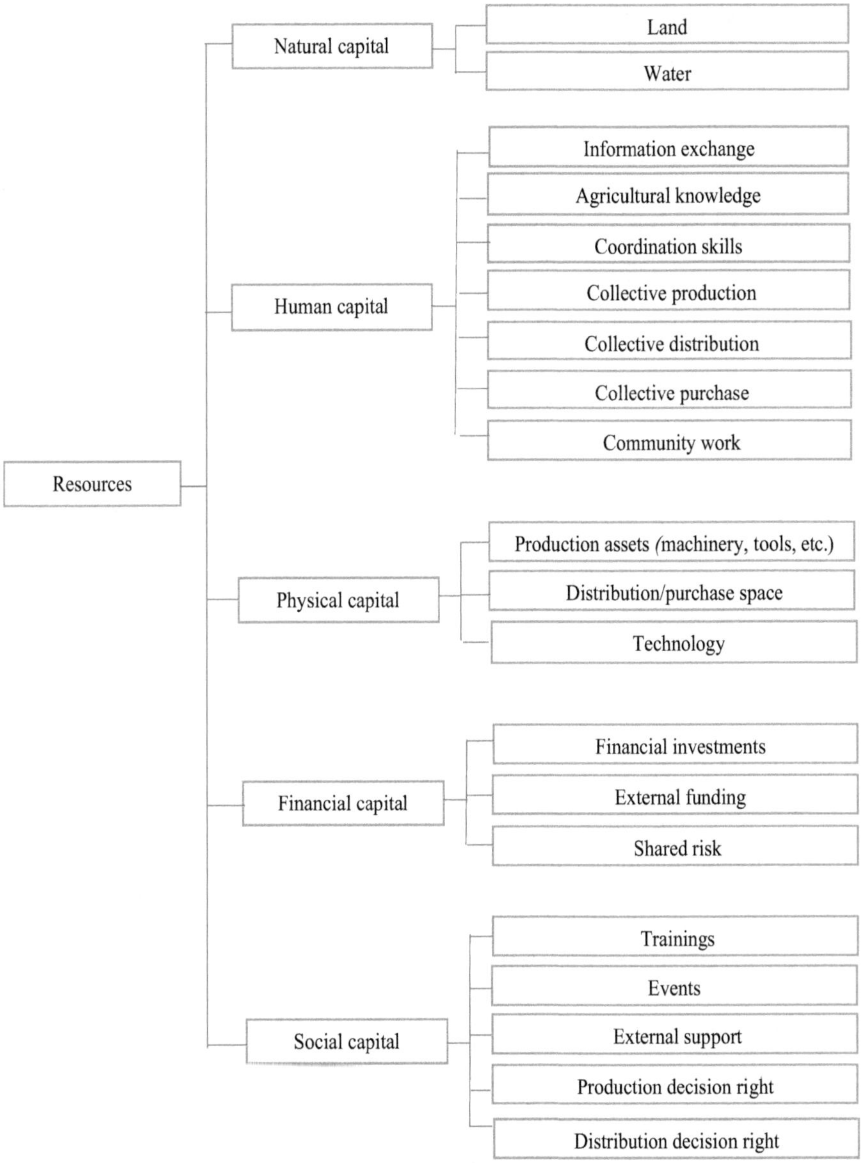

Diagram 17.2 Resources in agri-food sector

17.3 Institutions in the Multi-actors Framework and Their Role

Farms are an important component in the multi-actor approach's framework. Data generated from the continual recording of business events is an important component that is required for the system to function. Farms record business events independently (where there are human resources and preparedness for such operations) or in collaboration with organisational units that are part of the advisory service. The establishment of the individual and aggregate analytical reports and indicators is the essential information basis for other organisations in this sector. A significant number of institutions are interested in owning or gaining access to information reports that contain data on the performance of agricultural holdings; costs incurred in production processes (according to certain categories, years, etc.); the value of realised production by cultures, municipalities, and regions; the method of realisation (internal and external realisation).

Municipal or regional advisory service offices or other local self-government bodies should have statistics on the number of agri-food producers based on their municipal or regional affiliation. Advisory services (or their organisational units and local government offices that perform accounting activities) have a decisive influence in order to make timely and adequate recording of business activities on farms that do not have adequate human resources in terms of the possibility of recording events on farms (farms owned by the elderly or lack of primary education required to compile records). Furthermore, their actions are essential during the processing of obtained data and categorization based on the type of production, farm size, organisational form, regional affiliation, and so on. Data collected from farm holdings by producers or offices, and organisational units created within the advisory service, provide a complete database of agriculture sector implementation operations (Diagram 17.3).

Data relating to the agri-food industry are also useful for manufacturers, suppliers, micro-credit and financial institutions, investors, and other service providers or their customers that want a quality evaluation of opportunities and potential forms of production in the agri-food sector.

Advisory services can utilise this information to advise farmers on the sort of production to use, taking into account the farm's capacity or production resources. This service's aggregate statistics on farm holdings (according to affiliation with a certain area, types of output, and holding size) are critical for the development of future policy actions.

Professional agricultural producer associations (farmers' associations, grain processors, and other associations) are interested on monitoring farm operations in order to obtain timely information on the cost of specific products, funds to be invested in specific productions, standards, market prices, and so on. Consideration of product costs is critical for producers with the same or similarly products (in order to compare them), in order to conduct activities in the future in terms of timely effect on specific costs and improved production process organisation.

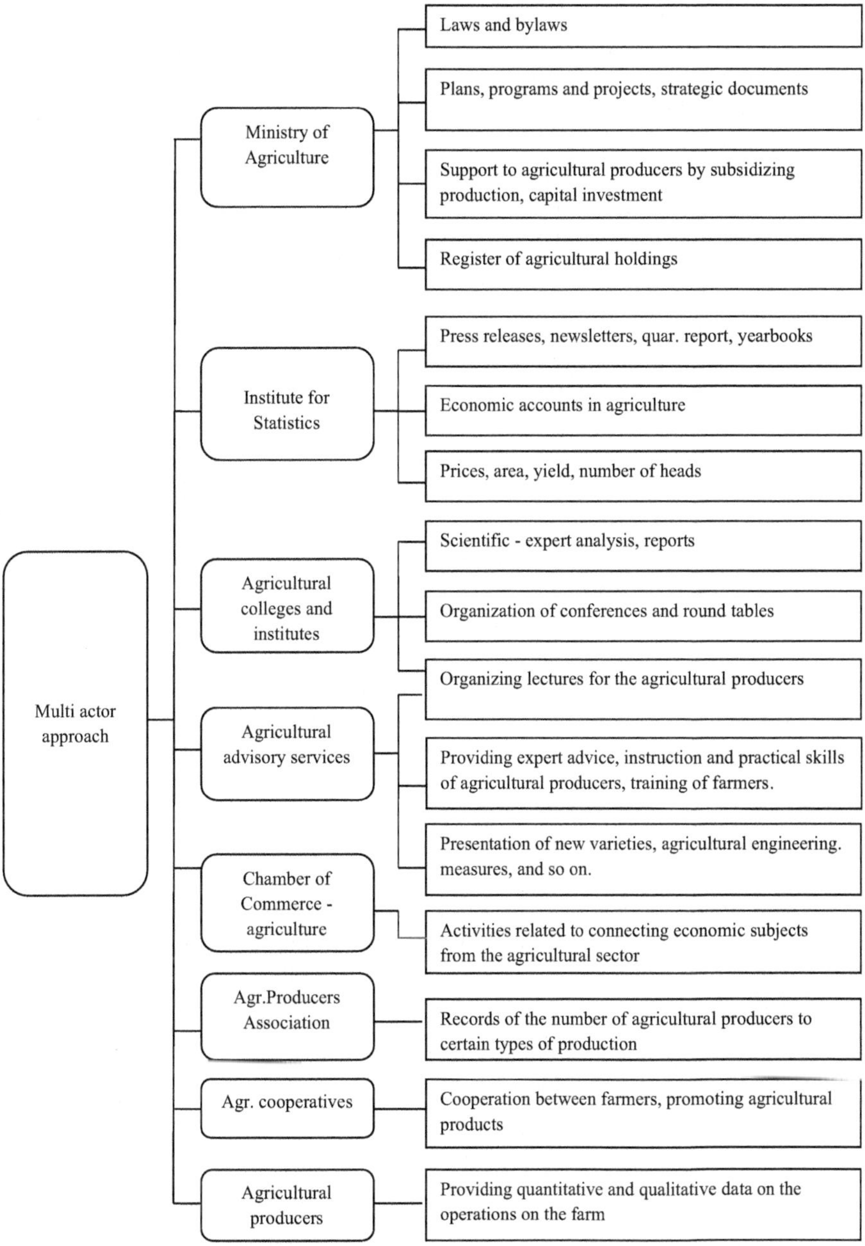

Diagram 17.3 Multi-actor approach

Other types of agricultural producer organisation, such as cooperatives and other groups, provide the same informational benefits. The aforementioned kinds of agricultural producer organisation generate vital information for the micro and macro levels in this industry. Famers can obtain specific information on the means of incentives and other types of assistance that are implemented by the relevant ministry, municipalities, and other institutions at the national or international level through all forms of these and similar groups. Furthermore, records on the number of such associations, the number of agricultural producers gathered around the mentioned associations, and proposals and suggestions originating from such associations (when drafting incentive programmes and other relevant documents and programmes) contribute to the system's better functioning.

Another actor in the stated information flows is the Chamber of Commerce, which should have information on the fundamental capacities that exist in this sector. This would allow for the evaluation of further development modalities and directions, as well as the timely targeting of specific funds originating from domestic and international funds, and, most importantly, the establishment of circumstances for educating potential local and foreign investors. The participation of professional employees from scientific research organisations, institutes, and the like should result in adequate and competent ideas and proposals targeted at strengthening the agri-food sector. These institutions have a crucial role in the structure and operation of this network. The objective of these institutions' utilisation of this information is not just to calculate particular norms and criteria for individual outputs. These institutions can contribute to the development of this sector by using previously established theoretical principles and producing additional information for farmers (organising expert lectures, making suggestions to improve the form and structure of information reports, making suggestions and proposals when drafting strategic documents, and defining certain guidelines aimed at making better decisions).

- The Ministry of Food and Agriculture is the organisation in charge of implementing food policy initiatives, and the presence of this information is critical for future operations and choices. The Ministry, in collaboration with its departments, should:

– Create an acceptable legal framework based on all of the facts and reports given thus far (with appropriate laws and bylaws),
– identifies the appropriate strategic directions that will be focused on the execution of appropriate programmes (by regions, types of production, etc.).
– Adopt steps that have an influence on farm enterprises' operations in terms of enhancing their production activities (which after their implementation will provide better financial effects).

Departments within the ministry or national councils in charge of this sector and agricultural policy should exchange certain information, both at the national level (other ministries, other governmental and non-governmental institutions) and at the international level (Directorate for Agrarian Policy of the EU and other international organizations and institutions) in terms of methods and types of financing, measures taken to improve all types of agricultural production, and innovative approaches used in other countries.

Data provided by consistent recording of business events and documents resulting from the collected and processed data, and information reports (as end products resulting from the collection of such data), then all relevant participants whose main activity relates to or is related to this sector, constitute the backbone of this framework. The collaboration of the above institutions offers additional information, which circulates through the previously specified information flows and enhances the participants' information, both at the micro and macro levels.

17.4 Business Model of Data Collection in Agri-Food sector on the Micro-Level

The quantity and structure of the resources reflect the farm's economic status. It is necessary to explicitly describe a particular economic condition in order to acquire understanding into it. Balances are employed for this purpose because they provide a formal description of the amount and structure of assets and their sources over time. The examination of any business, including the farm, is based on an understanding of the balance sheet and income statement. The balance sheet is a representation of the holding's total assets, as expressed by the current assets available to the holding (assets) and their sources (liabilities). Given that the balance sheet indicates the worth of the farm's assets at a certain point in time (typically the end of the fiscal year), it may be considered a static account of the farm's holdings. However, by comparing balance sheets at the start and end of the accounting period, one may acquire insight into the movement of the holding's assets, capital, and liabilities through time.

The farm's assets include all assets at the farm's disposal (Škarić-Jovanović, 2008): intangible investments, land, forests, long-standing plantations, construction objects, equipment, tools, and inventory, basic flock, other fixed assets, advances and fixed assets in preparation, and long-term financial investments. Investments in real estate, inventory, short-term receivables, and placement and cash equivalents). On the liabilities side, that is, the source of funds, there are the equity of the holding and the liabilities of the holding (liabilities to creditors and creditors).

From the liquidity point of view, the total available assets of the holding are divided into fixed and current assets. Fixed assets consist of land, forests, long-standing plantations, construction objects, equipment, tools and inventory, basic flock, other fixed assets, advances and fixed assets in preparation, long-term financial investments, real estate investments (Scheuerlein, 1997).

Land is one of the basic assets available to agricultural holdings, and is a necessary condition for agricultural activity. In terms of accounting treatment, it is necessary to emphasize three basic characteristics of land that distinguish it from other fixed assets (immovable property, non-lethargy and indestructibility, Finci et al., 1986).

Before assessing the value of the land, it should be emphasized that its value is estimated separately from the biological assets (crops, long-standing plantations) that are on it, or grown (Mihić, 2004). Land can be subject to market turnover, so its value is determined on the basis of the current market price at the balance sheet date. It should be borne in mind that this value depends on the purpose, location, quality of the land itself (which can be influenced by fertilization, land reclamation and other measures), class, but above all supply and demand in the market. In order to determine the value of land as accurately as possible when drawing up the balance sheet, it is necessary to take into account all the above factors.

Households, after initial recognition at cost, are accounted for in accordance with International Accounting Standard 16 (IAS 16) in an alternative manner, that is, at the revalued amount that represents the fair value of the asset at the date of revaluation, less accumulated depreciation and impairment losses. When assessing the value of a building, the year of construction, the type of material, and the depreciation rate for each facility should be taken into account. Machinery and equipment are valued at fair value, less any subsequent accumulated depreciation and impairment losses, on a balance sheet after initial recognition at cost. In calculating the value of equipment and machinery, in addition to the purchase price, the type of equipment or machinery, the year of production and the depreciation rate are taken into account.

Sales costs include commissions to brokers and brokers, agency and commodity fees, customs duties and sales taxes. Sales and shipping costs do not include transportation costs. The group of fixed assets out of function consists of cash intended for investments, receivables for fixed assets, investment material intended for investments in fixed assets, investment works in progress and completed investments out of function. This group of fixed assets is owned by the holding but has not yet been realized and materialized. Cash as fixed assets out of function are assets intended to be used as fixed assets, but only through an adequately usable form. As long as they are in monetary form, their value is out of function, although they belong to fixed assets. An example of this type of asset is e.g. monies earmarked for the purchase of a new tractor. Given that the tractor is not a holding company at the time of drawing up the balance sheet, the funds provided for its purchase have not been realized and are therefore considered to be a fixed asset out of function.

Holding receivables related to fixed assets (sale of fixed assets or compensation for damaged or destroyed fixed assets) are also considered fixed assets.

Inventory of investment material includes all material intended for the purpose of investment in fixed assets, which was found on the holding on the day the balance sheet was drawn up. It is important to emphasize that although they are considered as fixed assets, these assets cannot function as fixed assets in their current state. Examples of these assets are supplies of building materials intended for the construction of farm buildings, various parts for equipment and machinery, uninstalled machinery and planting material intended for the foundation of permanent crops.

Ongoing investments are a form of property, plant and equipment that, due to incompleteness, is not operational at the time the balance sheet is drawn up. This form of fixed assets includes facilities under construction (unfinished barn, warehouse, canal), equipment in the design and installation phase (irrigation or greenhouse heating system) and long-standing plantations in the lifting phase. This category could also be classified as breeding ground in time until it is exploited (Slovic, 2002).

Completed investments out of function represent fixed assets that have been brought to the usable stage and are not yet included in production because the necessary conditions have not been met. In the case of objects, this would e.g. were facilities that had been built but had not yet received the operating permit or had not yet installed the necessary equipment. When it comes to mechanization, the completed investment out of function would be the mechanization that was purchased but has not yet begun to use it. With this group of fixed assets, it is important to emphasize that fixed assets are generally put into operation immediately after the completion of construction, ie procurement. Also, the depreciation period for the assets whose use has begun cannot be interrupted and they are returned to the category of fixed assets out of function.

Working capital is a part of the means of production that, when used in one production process, is physically completely consumed and therefore at once transfers all its value to the business services or products obtained (Andrić, 1998).

This group of assets includes cash held by the holding, supplies of materials, inventories of work in progress, inventories of finished goods, trade and other receivables, and other current assets.

Inventories of materials that may be considered current assets include inventories of food purchased, plant protection products, fertilizers, fuels and lubricants. These inventories are used for current consumption and are measured at fair value at the balance sheet date.

The value of work in progress includes the value of production in progress on the balance sheet date and the value of own products intended for further production. Ongoing plant production is valued according to the investments made in it. Considering that the annual balance sheet is mainly prepared for December 31, the value of crop production in progress in practice represents the value of autumn work (cultivation and preparation of land for sowing, sowing, protection measures implemented).

Trade receivables include the amount of invoices issued for products sold or services rendered that have not yet been charged. This category can be categorized e.g. receivables from the processing industry for surrendered fruit, vegetables or fattened livestock that were not paid on the balance sheet date. Receivables from suppliers represent the value of products or services paid by the holding company that were not realized by the date the balance sheet was drawn up. That would, for example. Was the value of a pre-paid seed or fertilizer that had not been delivered by the balance sheet date. Other working capital is in practice reduced to receivables arising on different bases (unpaid subsidies, miscellaneous contributions, awards, claims for damages and insurance, etc.).

Since the balance sheet represents the current presentation of the value of the assets and liabilities of the holding, so the value of current assets shown in it represents only the current balance and does not allow a more detailed insight into the change in the value of the assets of the holding during the year. The total value of fixed and current assets held by the holding represents the gross value of the holding's assets. When the value of the total liabilities of the holding (loans and the like) is deducted from the gross value of the assets of the holding, the net value of the assets of the holding, i.e. its net capital, is obtained.

The other side of the balance sheet is liabilities, that is, sources of funds available to the farm. The value of the source of funds is generally equal to the value of assets, that is, the assets are equal to liabilities. All assets, that is, all the assets of the farm, have their sources, which determine the origin or manner of acquiring ownership. Funds invested in the creation of assets on the farm or which are the result of the farm's own operations represent own funds, the so-called equity or permanent capital. The funds obtained through loans have a source in the liabilities (debts) of the household. Liabilities can be long-term and short-term, be they liabilities to creditors or creditors.

The Balance Sheet represents the current balance of the value of the assets at the disposal of the holding and the source of the value of those assets. By comparing the balance sheet at the beginning and end of the accounting period, one can only see the difference between one and the other values, but not the movement of cash flows during the accounting period (Diagram 17.4).

In order to monitor the cash flow, it is necessary to analyze the income statement, which shows the performance of the holding's business during the observed period. The income statement shows all income and expenses incurred by the holding during the accounting period. The end result of the income statement is income, or financial expression of the farm business (Krstic et al., 1995). The total household income is represented by the sum of the effective and accrued income that the household generated. Effective revenues represent revenues generated from the sale of products and the provision of services, as well as subsidies for certain types of production. This revenue group is called effective because it is expressed in money received for products sold (wheat, corn, vegetables) and services performed (eg mechanization services). The second group of farm income represents accrued income, not in monetary terms, but as an increase in the material value of production (eg livestock growth), that is, a positive difference in the value of inventories of products or materials and a positive difference in the value of work in progress.

By analogy, incomes and expenditures incurred by the household during the year consist of the above two categories. The category of effective expenditures includes: purchased material (seed and planting material, fertilizer, protective equipment, fuel, lubricants, livestock feed, cattle rug, livestock), paid business services (veterinary services, artificial insemination, natural mooring, counseling services)), paid work (wages, mechanization services).

Expenditure treatment includes depreciation of items, machinery and equipment, a negative difference in the high proportion, a negative difference in initial inventories, finished products, and a material and negative difference in original unfinished production.

The difference between total revenues and total expenditures was represented by the income of the front-runner. Based on the actual version of the holding's income,

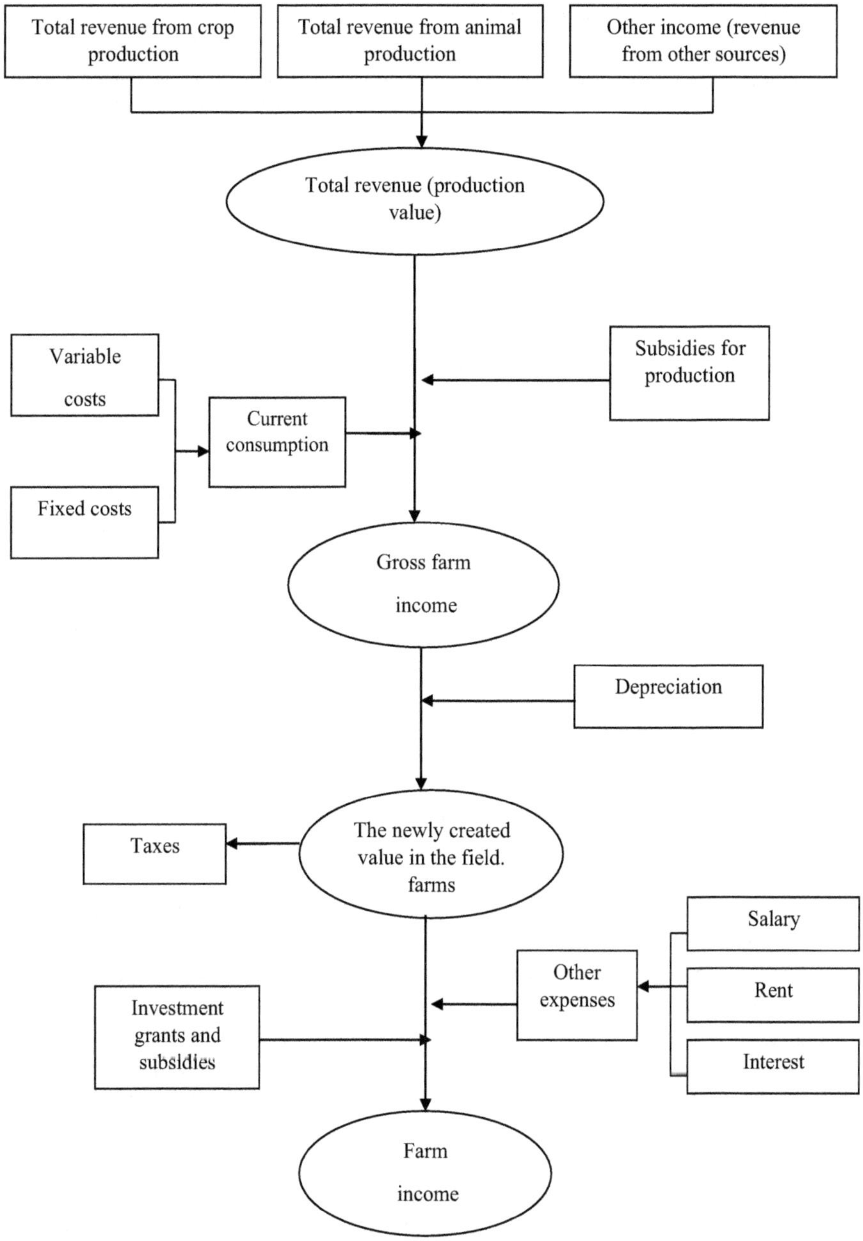

Diagram 17.4 Structure of the income model in the agri-food sector

conclusions are made about agreeing on a substantial management, at the same time whether the management will receive a positive or a negative. Thus, the analysis of the balance sheet made it possible to see the importance of individual income items, unilaterally expanding and assessing the possibilities for improving the dispatch of holdings in the regular period.

17.5 Conclusions

Planning and implementing actions in the food industry need effective organisation and people with necessary expertise in order to carry out activities at all levels. An ideal integrated strategy should encourage producers to use inputs more accurately and to allocate the financial resources created by selling individual and total products in a systematic and complete manner (taking data from both micro and macro levels into consideration).

There are a number of participants in data collection, including research institutes, universities, advisory services, farmers' associations, agricultural cooperatives, NGOs and other institutions, which within the scope of their activities and in the implementation of daily activities have an interest in having data from the agricultural sector.

Based on the data supplied by the suggested model, it is feasible to monitor the operations of agricultural holdings, identify the major variables influencing their profitability, and assess the influence of current agricultural policy action.

The distribution of information between government institutions, agencies, agri-food producers, and other farming industry actors provides to more information and fast solution of specific difficulties.

By achieving greater organization in the realization of production activities, it has enabled agricultural producers to gain insight into their production capabilities and a better evaluation of indicators indicating the success or failure of production activities. Control of work operations combined with a database of past business results is a quality support to the decision-making process.

The economic data through the financial statements reflect the economic picture of the farm and are of great importance for the future management or management of the farm. The objective of the financial statements is to provide qualitative information regarding the implementation of farm activities. They also form the basis for making financial allocation decisions.

The usage of farm activity records offers a clear picture of the farm's financial status and makes it simpler to estimate the farm's potential in terms of future revenue and expenses. In this sense, knowledge represents resources that, when combined with the farm's remaining resources, can contribute to its greater success. The presence of records on farms enables for more accurate forecasting of future activities based on the kind of farm. The financial statements emphasize the results of the management of the farm, that is, establish greater responsibility of the holder of the holding when using both financial and material means at the disposal of the holding.

These reports should emphasize a true picture of the farm business, identify critical points, i.e., production stages and the need to introduce certain modifications through better organization of production activities, which would have the effect of improving the financial result.

References

Andrić, J. (1998). *Costs and calculations in agricultural production*. Contemporary Administration.

Beach, L. R., & Connolly, T. (2005). *The psychology of decision making. People in organizations*. Sage.

Bossel, H. (1999). *Indicators for sustainable development: Theory, method, applications*. A report to the Balaton Group, International Institute for Sustainable Development.

Daskalopoulou, I., & Petrou, A. (2002). Utilising a farm typology to identify potential adopters of alternative farming activities in Greek agriculture. *Journal of Rural Studies, 18*, 95–103.

Finci, Z., Bajcetic, B., & Milosevic, A. (1986). *Organization of Agricultural Holdings*. Svjetlost.

Government Finance Officers Association, Performance Management. Using performance measurement for decision making, approved 2007. http://www.gfoa.org/

Gow, J., & Stayner, R. (1995). The process of farm adjustment: A critical review. *Review of Marketing and Agricultural Economics, 63*(2), 272–283.

Hansson, H. (2008). How can farmer managerial capacity contribute to improved farm performance? A study of dairy farms in Sweden. *Food Economics – Acta Agriculturae Scandinavica C, 5*, 44–61.

Hansson, H., & Ferguson, R. (2011). Factors influencing the strategic decision to further develop dairy production — a study of farmers in Central Sweden. *Livestock Science, 135*, 110–123.

Krstic, B., Andric, J., & Bajcetic, B. (1995). *Models of agricultural holdings focused on livestock production*. Aleksandra.

McCown, R. L., Hammer, G. L., Hargreaves, J. N. G., Holzworth, D. P., & Freebairn, D. M. (1996). APSIM – A novel software system for model development, model testing and simulation in agricultural systems research. *Agricultural Systems, 50*, 255–271.

Meinke, H., Baethgen, W. E., Carberry, P. S., Donatelli, M., Hammer, G. L., Selvaraju, R., & Stockle, C. O. (2001). Increasing profits and reducing risks in crop production using participatory systems simulation approaches. *Agricultural Systems, 70*, 493–513.

Mihić, D. (2004). *Agricultural revision, concepts and methods*, author's edition, Belgrade

Ministry of Municipal Affairs and Housing Ontario, Municipal Performance Measurement Program, 2007

Öhlmér, B. (1998). Models of farmers' decision making. Problem definition. *Swedish Journal of Agricultural Research, 28*, 17–27.

Ondersteijn, C. J. M., Giesen, G. W. J., & Huirne, R. B. M. (2003). Identification of farmer characteristics and farm strategies explaining changes in environmental management and environmental and economic performance of dairy farms. *Agricultural Systems, 78*, 31–55.

Plant, T., & Douglas, J. (2006). The performance management continuum in municipal government organizations. *Performance Improvement, 45*(1), 43–48.

Scheuerlein, A. (1997). *Finanzmanagement für Landwirte*. Verlags Union Agrar.

Škarić-Jovanović, K. (2008). *Financial accounting*. Publishing Center, Faculty of Economics.

Slovic, D. (2002). *Agricultural Accounting*. Fineks.

Traore, N., Laundry, R., & Amara, N. (1998). On-farm adoption of conservation practices: The role of farm and farmer characteristics, perceptions, and health hazards. *Land Economics, 74*(1), 114–127.

Willock, J., Deary, I. J., Edwards-Jones, G., Gibson, A., Dent, J. B., Morgan, O., & Grieve, R. (1999). The role of attitudes and objectives in farmer decision making: Business and environmentally oriented behaviour in Scotland. *Journal of Agricultural Economics, 50*(2), 286–303.

Chapter 18
The Role of Rural Women in Bread Industry – Job Creation and Retention in Rural Areas

Urszula Ziemiańczyk, Anna Krakowiak-Bal, and Aleksandra Figurek

18.1 Introduction

18.1.1 Rural Areas Background

Over the past 20 years, rural areas in Western societies have undergone many changes as a result of socioeconomic restructuring. Increasing demands made by society on rural areas as sites for tourism and recreation as well as quality and regional food production have transformed the countryside from a (predominantly) production to a (predominantly) consumption space (Halfacree, 2006; Slee, 2005). As a result, the countryside has become a multifunctional space for leisure, recreation, working and living (EC, 2007; Marsden, 1999).

The rural areas of the European Union are strikingly varied in terms of social and economic structure, geography and culture. Rural women are not a homogeneous group as well. They have different roles and occupations, on farms and in family businesses, in employment and in community activities. Also, their needs and interests vary, interests also vary, particularly depending on the age group, and depending on the size and composition of the family and the age of the children. The economic and social changes those rural areas are undergoing do not affect all women in the same way: offering opportunities to some and bringing difficult challenges to the others.

U. Ziemiańczyk (✉) · A. Krakowiak-Bal
Faculty of Production and Power Engineering, University of Agriculture in Krakow, Krakow, Poland
e-mail: urszula.ziemianczyk@urk.edu.pl; anna.krakowiak-bal@urk.edu.pl

A. Figurek
Gnosis Mediterranean Institute for Management Science, School of Business, University of Nicosia, Nicosia, Cyprus

J. M. Ferreira da Rocha et al. (eds.), *Baking Business Sustainability Through Life Cycle Management*, https://doi.org/10.1007/978-3-031-25027-9_18

Rural economies, particularly those dependent on agriculture, have been affected by the processes of globalisation, leading to the restructuring and decline of the agricultural sector, the growth of the service sector and increased emphasis on technology. In many areas, this has created unprecedented work and employment opportunities, as well as bringing changes in the role and status of women. These changes have also contributed to further shifts in population, with some rural areas close to towns and cities coming under pressure, while many remote areas continue to suffer a decline in population. In some regions of Europe, economic recession and cutbacks in public services have led to further rural decline, remoteness and poor infrastructure. Young people, especially young women, migrate to towns and cities in increasing numbers.

Responsible for rural areas development throughout Europe are seeking extra sources of income by diversifying their on- and off-farm production and emphasizing local traditions and heritage. For over three decades traditional food products and agritourism have been available as diversification options within a framework of rural development goals. Rural women often prove to be pioneers in taking entrepreneurial initiatives in these sectors, which are often perceived as gendered because these activities (accommodation, food processing and preservation) have traditionally been performed by women in rural areas (Bock, 2004a, 2004b; McGehee et al. 2007a, b).

The significance of rural women in UE countries presents the Fig. 18.1.

The aim of this study is to determine to explore the significance of factors of economic grow on female employment and entrepreneurship. As a background we took Schumpeterian theory of economic growth which consists of three key factors.

Each of these elements will be discussed in the context of the woman's role in the rural environment, with particular emphasis on the bread production.

18.1.2 Schumpeter's Theory of Economic Development – Women's Context

Considering the role of women in socio-economic development, it is worth to use the conclusions from the theory of economic development by Joseph Alois Schumpeter. Despite over a hundred years of history and the changing conditions of the external environment and the mentality of societies, the conclusions of his work are surprisingly up-to-date and in the light of ever faster changes, they are becoming more and more important.

Schumpeter tried to dynamise the static model of the market economy by pointing to the internal causes of economic development. He proved that economic development is mainly caused by internal forces, instead external factors. It is important to meet three conditions that constitute a specific wholeness:

- existence of a creative *entrepreneur* (main market power; innovation power responsible for the surplus maximization);

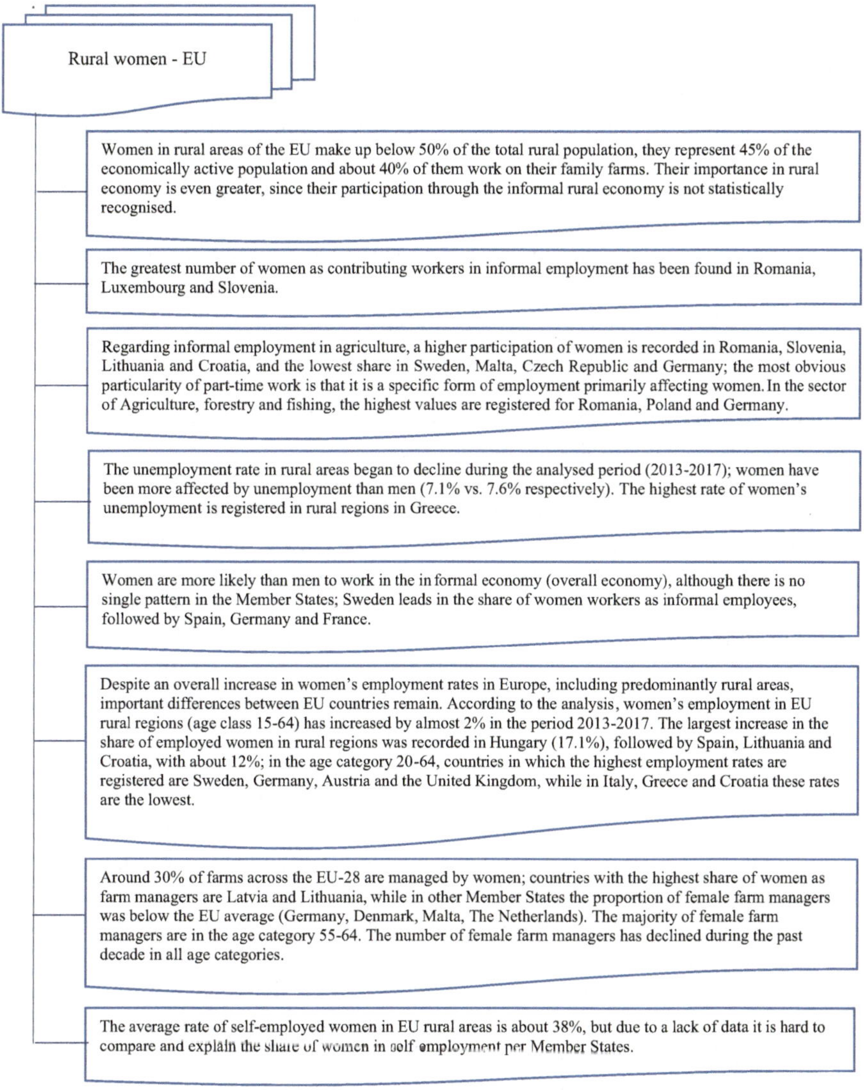

Fig. 18.1 Rural women. (Source: The professional status of rural women in the EU, Policy Department for Citizens' Rights and Constitutional Affairs Directorate General for Internal Policies of the Union, 2019)

- *innovations* (introduced by entrepreneurs, e.g., new products, technologies, new solutions);
- *loan* (the entrepreneur is not obliged to use his own capital, he should use a bank loan).

18.1.2.1 Entrepreneurship

The last few decades in the literature have been a period of lively development and flourishing of numerous economic concepts and models that take into account entrepreneurship. A multitude of scientific works linking these two concepts – entrepreneurship and economic growth (Wach, 2015).

In general, there are no differences in the concept, characteristics, sources' etc. of entrepreneurship regarding gender. Characteristics such as achievement, autonomy, aggression, independence and benevolence between female and male entrepreneurs (Hisrich & Brush, 1984) are equal. Also, no differences were found in risk taking propensity of male and female entrepreneurs. However, women entrepreneurs, as research demonstrates, may do things differently. For example, in comparison to male entrepreneurs, women tend to work more in teams, are less self-centred and personal ego to them is less important than success of the organization or business idea they are pursuing.

It should be stressed that rural women can encounter many constraints when trying to take part in the transformation process. Rural areas tend to be more traditional in regard to the gender issue. This issue is usually a much stronger hindering factor to potential female entrepreneurs than it is in urban areas. Womens' self-esteem and managerial skills are lower in comparison to urban women and access to external financial resources is more difficult than in urban areas. Therefore, special programmes of assistance (technical and financial) to overcome these constraints should be developed and designed to meet the rural women needs in order to allow them an active entrepreneurial restructuring of their communities, to start to develop their own ventures, to expand their already existing businesses, or to function as social entrepreneurs since their number today is still below the potential one. However, it should be underlined that the individuality not quotas, positive discrimination, and equal opportunity' politics is behind the phenomenon of the successful female entrepreneur (Schwager-Jebbink's, 1991).

It is worth to look at the levels of interaction of factors influencing women's employment at geographical, cultural, structural and individual levels (Fig. 18.2).

On a cultural level, local gender ideologies contain dominant norms and values concerning female employment and entrepreneurship, male breadwinning, working mothers and the division of household and family duties (Bak et al., 2000; Knudsen & Waerness, 2001; Little, 1991, 1994). Pfau-Effinger (1994) summarized these issues under the concept of the 'family and integration model', and defines it as the particular combination(s) of cultural norms and values in a society concerning the integration of women and men into different societal spheres, the gendered division of labour, and the societal sphere for caring for children and other dependents. Pfau-Effinger (1994, 2000) uses this model in order to distinguish between nations and their 'gender cultures', but, following Sackmann and Häussermann (1994), a similar model may be used to explain regional differences in female employment behaviour. In both cases it is important for research and analyses to take the historical background of different models into account. Dependent on the specific circumstances of national and regional industrialization processes, specific behavioural

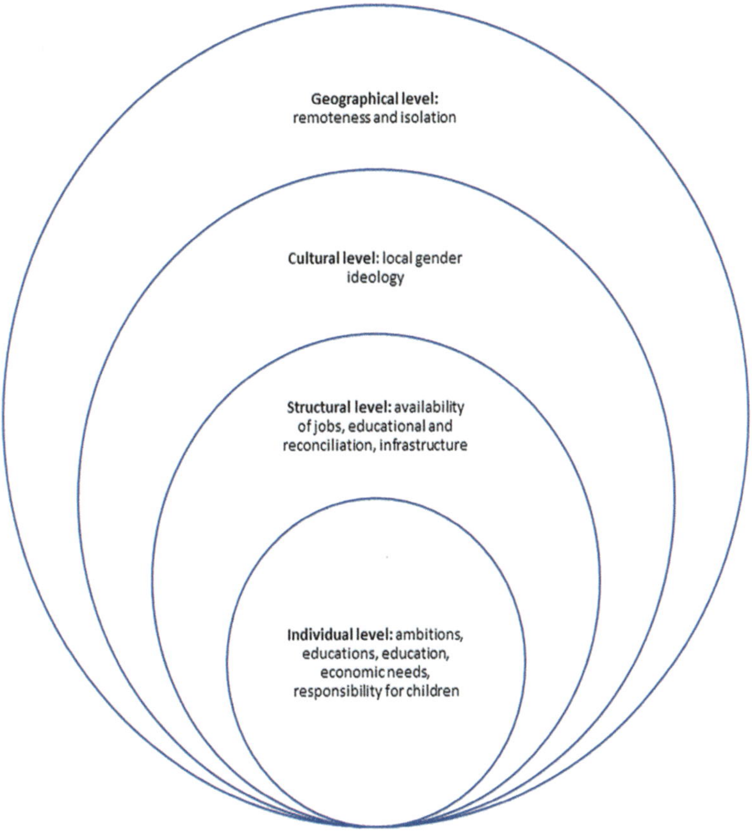

Fig. 18.2 Interaction of factors influencing women's employment at geographical, cultural, structural and individual levels. (Source: Based on Bock, 2004a, b)

patterns are stimulated and supported by norms and values defining what kind of childcare and what kind of female employment is considered appropriate. Such ideologies influence the behaviour of women, but also their identity, their ambition and assessments of the chances of succeeding, along with expectations on likely judgements by others. Even though women differ in their acceptance of dominant local norms and values, they cannot avoid being influenced by them. After all, the same norms and values affect the expectations and behaviour of significant others, among which are local employers, village councils, husbands and parents (and parents-in-law) (Little, 1991, 1994).

At the structural level of the regional economy, the character of the regional labour market and the availability of jobs, especially in the service sector, have an important influence on women's paid work (Dahlström, 1996). The cut-back in public spending and the concentration and reduction of rural services has detrimental effects on the women employment (Commission of the European Communities, 2000; FAO, 1996; Högbacka, 1999; Moss et al., 2000). It is service industries in particular that attract (rural) women and seem to offer good employment opportunities (Overbeek et al., 1998). These are the industries with most important growth levels in recent years and with ongoing growth in employment expected (Terluin & Post, 2000). The actual and future employment opportunities for rural women are thus probably best in areas where the service sector is growing. Moreover, the availability of part-time work, childcare and other services and arrangements are decisive, as these helps support the reconciliation of work and family (e.g. parental leave, emergency leave, flexible working hours (Braithwaite, 1996; Halliday & Little, 2001; Overbeek et al., 1998). Education and training facilities are also important in sustaining the re-entrance of women into the labour market and the updating of their employability (Overbeek et al., 1998). Last but not least, infrastructure and the availability of (public) transport are decisive for enabling women (and men) to accept paid labour outside of their locality (Post & Terluin, 1997).

On an individual level, women's education and professional qualifications, along with their drive and ambition are important factors supporting women, and consequently they influence the extent to which women are able and ready to invest in finding a satisfying job (Moss et al., 2000; Overbeek et al., 1998). Having the right equipment, including means of transport is of course an important resource as well, which is especially necessary to search for and then accept a paid job (Lindsay et al., 2003). Overall, the interaction between aspiration, education and mobility is more effective in influencing women's employment behaviour than are local labour market characteristics. Even when a local labour market has little to offer to those with high education qualifications and considerable ambition, this can stimulate women's willingness to increase their mobility and accept long distances between work and home (Moss et al., 2000; Overbeek et al., 1998). Furthermore, age and household situation are of great importance. The effect of age varies between countries. In some European countries, the gap between young women and men is very small in terms of employment coverage and market share. In 1997, for instance, there was no difference between the labour market participation of women and men in the age group of 15 to 29 years in Greece and men's participation rate was only slightly higher (2%) in Sweden (Commission of the European Communities, 2000). In Finland, France, Ireland and the Netherlands only 6% more men compared to women had paid labour in that age group. In other EU countries, gender differences were remarkable even at a young age. This is true for instance in Italy but also in Denmark, where the participation rate of young men and women differed by 14% and 13% respectively in favour of young men. In almost all European countries women with small children leave the labour market, at least temporarily, or switch to part-time employment (Bock & van Doorne Huiskes, 1995; Franco & Winqvist, 2002; Tijdens, 2002). However, education and ambition do counteract this interrelationship, just as the support of a partner for woman's career and his willingness to take his part in caring for children has an effect. Finally, the economic situation of the individual household is of great importance, as lack of income may force (rural) women to accept whatever work is available as part of the family's survival strategy (Kelly & Shortall, 2002; Perrons & Gonäs, 1998; Plantenga & Rubery, 1999a, 1999b).

The factors described above apply to the situation of women in general. To understand the situation of rural women, it is important to add another level of explanation, which encompasses their geographical context(s). These contexts are affected by population density and distance to urban centres; in short to the 'rurality' of an area. The more remote and geographically isolated an area is, in general the more difficult the situation is likely to be for women.

For one, rurality affects local labour market structures. In many rural areas in Europe, in a last few decades' agriculture was the dominant employer (sector), whereas the service sector, which is more important for female employment, has limited jobs to offer. Long distances to employment concentrations, undeveloped public transport and bad road infrastructure restrict commuting to work (Irmen, 1997; Lindsay et al., 2003). Moreover, husbands or fathers may already be migrating for work purposes, which renders a women's presence at home as indispensable, in order to have someone to look after the family (Halliday & Little, 2001). Women may also be needed to take care of the farm, household food production and extra farm income generating activities. The latter situation is often found in eastern European countries but is also common in the middle of Italy (Bock, 1994) and Norway (Jervell, 1999). As men's engagement in paid work promises a higher income, or men's chances to find work are perceived to be better, it is more likely that male members of households will be the ones to commute or even migrate for work. This pattern is also more acceptable in many countries, as in (rural) Italy, where caring for a home and the family is considered as" natural" woman's job (Trifiletti, 1999).

Moreover, the remoteness of rural areas can adversely affect the resources of individual women and, thus, their ability to find satisfactory employment. Their level of knowledge and experience may be low as there are few possibilities for training and education, or accumulation of work experience. Also the fact that job search channels are often informal (Lindsay et al., 2003), may act against women finding out about (non-advertised) job possibilities. As there is generally little difference in education between young men and women (Looker, 1997), lack of training is probably most constraining for older women. In the old days, rural women had little opportunity to get a professional education. Moreover, their education is most likely outdated today. In this context, it is not surprising that many 'rural' mothers support their children, especially their daughters, in their efforts to get a good education, with the consequence that children are freed from helping in the household and on the farm if it is possible (Bock, 1994; Gourdomichalis, 1991; Laoire, 2001; O'Hara, 1998).

Finally, it is important to recognize that the cultural climate in rural areas tends to be rather traditional and conservative with regard to gender equality, especially in the most remote areas (FAO, 1996; Looker, 1997). Thus, Little (1991, 1994) points out that the typical rural gender ideology is one of the most important constraints on rural women's engagement with paid-work. According to this ideology, rural women should be, above all, mothers and protectors of the community. This rural gender-ideology consequently stimulates and promotes women to stay at home, to care for their family and household and to engage in local community voluntary work

(Little, 1997b). This value disposition also influences local employers, convincing them to offer only temporary, low paid and part-time jobs to women (Little, 1997a). As a result, women in rural areas not only lack the encouragement and support to lead a satisfying professional life but are often actively prevented from doing it.

The geographical character of rural areas (rurality) is not so much a constraint in itself as it strengthens other constraints. It negatively affects the uncertain balance of structural and cultural conditions which are needed morally and practically to enable and empower individual women to make use of their professional capacities and aspirations and overcome the numerous obstacles they face. In many rural areas employment conditions are not as favourable as they were in cities, as the local economy and labour market are less diverse and there is a lack of reconciliation services such as childcare. Additionally, the cultural climate plays a more constraining role, as women and men have fewer opportunities to escape prejudice and gossip, and local social control is more effective in rural than urban areas (Haugen & Villa, 2003).

This interpretation resembles to explaining the position of women through a gender contract or gender system theory (Duncan, 2000; Pfau-Effinger, 1994, 2000). According to this theory, gender-specific norms and values (or 'gender cultures') develop in close relation with structures and institutions that promote gender related employment behaviour. By establishing and re-enforcing, each other, gender culture and structure form one coherent unity, the gender contract. As Duncan (2000) points out, gender contracts are spatially related and vary not only between countries but also within them. Forsberg et al. (2000) reveals how regional difference in female employment patterns in Sweden can be explained by regional variations in the national gender contract. The more equal regions are situated in the North, they are generally larger and offer a lot jobs in the service sector. The more unequal regions are located in the South, which are smaller with an industry-based labour market (Forsberg et al., 2000). Research in Switzerland also reveals how regional gender contracts differ along linguistic boundaries, influenced by ethnic, religious and political factors (Forsberg et al., 2000). Regional variation in female employment patterns may thus be explained as the interaction of regional gender cultures with regional economic and institutional structures (Forsberg et al., 2000). The resulting regional gender contracts define women's roles, opportunities and expectations and, among others, women's employment patterns. However, this determination is not absolute. Above all, women are moving away from dominant gender contracts and acting against social expectations. Secondly, regional gender contracts may and do change over time.

Following Pfau-Effinger (2000) change is the result of a discrepancy between expected and actual behaviour, time lags and variability within groups regarding beliefs and behaviour. Women and men are therefore perceived as important actors of change. By guiding their behaviour according to innovative norms and values, they may introduce new gender practices and change regional gender cultures and structures step-by-step. Young rural women can form such a group, as long as they do not migrate from rural areas in search of better opportunities in cities that have already more equal gender contracts (Forsberg et al., 2000; Ni Laoire, 2001).

Following Duncan (2000), culture is more important in determining gendered employment patterns than economic structures. Depart from this point, it might be expected that change starts from changing cultures and, thus, changing norms and expectations. But only longitudinal regional studies, embedded in comparative research, can reveal whether this is the case or not. Overall, there is most probably to be a continuous interaction between the cultural, structural and individual factors outlined above.

Numerous recent papers discuss the growing share of rural businesses initiated by women in rural areas, including Western Europe. In particular, the pioneering role of women in new businesses in rural areas is emphasized (Anthopoulou, 2010; te Kloeze, 1999), and those women are at the forefront of rural diversification (O'Toole & Macgarvey, 2003) and social revitalization of the rural economy (Little, 2002; Warren-Smith & Jackson, 2004). As highlighted in the European Commission's report (EC, 2000, p. 13), women contribute to the development of rural communities because they often have the added benefit of being aware and knowing about local needs and having special interpersonal and communications skills.

A variety of rural businesses initiated by women can be identified in the literature. First, there are women living on farms who establish a business at the farmhouse (Bock, 2004a, b). These businesses are often driven by the economic necessity to diversify the farm income due to a decline in agriculture and they mainly include activities related to agro tourism (McGehee et al. 2007a, b), food processing, artisanal products and local agrifood production (Anthopoulou, 2010). In general, these women base their business on the farm and supplement the main income of the farm to a considerable degree. Women who are not farmers have also started out-of-home rural businesses (Tigges & Green, 1994) and home businesses (Oberhauser, 1995, 1997). In general, these companies are small-scale and aim to either generate the main source of income for the household or substantially supplement the main income. Oberhauser (1997) and Baylina and Schier (2002), note that working from home is an essential part of household income and Oberhauser (1995) even argues that working from home is a crucial household survival strategy in rural areas, in particular remote rural areas, where high level of unemployment and poverty is observed (Oberhauser, 1995, 1997). In contrast, rural businesses, called a side activity in the Netherlands, are undertaken in the context of greater economic security, and more flexible and desirable work-life balance, especially among women with young children. Perhaps, therefore, the motivation to start a side activity and its role in the context of rural development differs from the companies just described.

Markantoni et al. (2009) define a side activity as a small-scale, home-based business that provides additional income at the household level. The owner of a side activity either combines the activity with gainful employment (full-time / part-time) or unpaid employment, e.g. housework or income from social benefit or retirement pension. In the latter case, the partner provides the main household income. Side activities do not provide full time employment to anyone other than the people who carry out the activities. Most of these activities are initiated and carried out by women. While it is generally believed that small entrepreneurs improve the rural economy and provide a significant source of employment (Atterton & Affleck,

2010), this is not empirically supported for side activities. At the regional level, spillovers affect the rural economy indirectly rather than directly, mainly by diversifying rural (tourism) activities, networking with other entrepreneurs in the region and mobilizing local networks with other spill-over activities in the region. As a result, side activities can play a role in activating and enhancing social vitality in rural areas. Side activities can offer a place for socializing and meeting social needs, not only for initiators, but also for residents and clients outside the region. Since the side activities are small in scale, they can offer a more personal approach to customers. Furthermore, Delfmann et al. (2011) found that side activities can play a key role in building and maintaining social capital in the rural community and thus play a role in rural regeneration. A study by Markantoni et al. (2009) indicated that for those who engage in side activities, their business activities are associated with achieving a modern lifestyle characterized by individualisation and self-realization, while financial motives are of secondary importance (Markantoni et al., 2009). Therefore, it is important to analyze in more detail what motivates female entrepreneurs to take such actions, what their obstacles and problems are, and how they organize their daily lives around them.

Attention should also be paid to the distinction between concepts rural entrepreneurship and entrepreneurship in rural areas. Both notions are related with the creation of firms. "Rural entrepreneurship" means that the organizational needs and behaviors of rural entrepreneurs differ from those in urban areas (Stathopoulou et al., 2004) because they are rooted in the local space. Rural entrepreneurship is embedded in its spatial context trough local resource use. In contrary "entrepreneurship in rural areas" do not accent this separateness (Baumgartner etal., 2013b; Korsgaard et al., 2015; Pato, 2015).

18.1.2.2 Innovation

Theory of economic development by Joseph Schumpeter contributed to the establishment of the Entrepreneurial Theory, in which the main driver of economic development is innovation. According to Schumpeter (1934), the entrepreneur is the innovator who carries out new combinations, which can be:

1. introduction of a new good,
2. introduction of a new method of production,
3. open a new market,
4. use of a new source of supply,
5. creation of some new organisational forms in an industry.

It has been observed that entrepreneurship and innovation, combined, help create new jobs and contribute to improving the economic competitiveness, creating economic growth and new wealth for the rural space, and ultimately improving the quality of life for local residents (Pato, 2015; Vaillant et al., 2012).

It is understood therefore that innovation encompasses internal processes of firms (Pato, 2015; Virkkala, 2007). It may refer to product or process innovations,

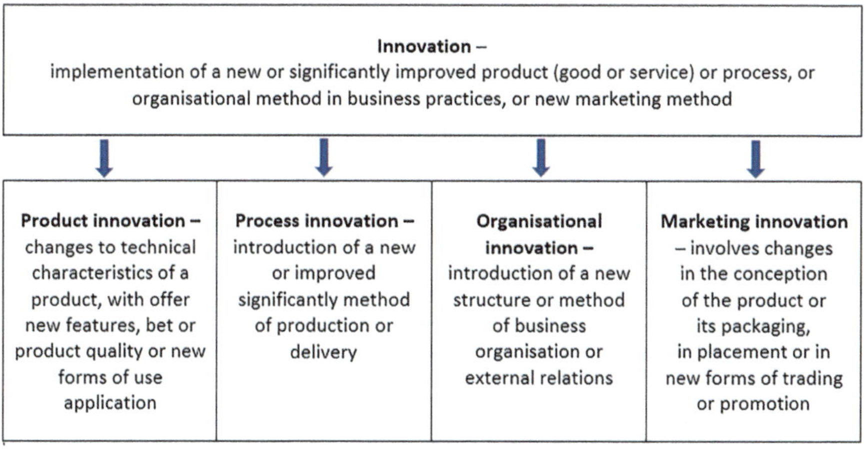

Fig. 18.3 Types of innovation. (Source: Based on OSLO Manual Guidelines 2005)

and marketing or management as well. OECD and Eurostat (2005) use a definition of innovation that is not confined to technological innovation, including the implementation a new or significantly improved product (good or service), or process, or organisational method in business practices (organisational innovation), or new marketing method (marketing innovation) (Fig. 18.3).

Especially in rural areas innovation involves much more than technology. It is about strategy, marketing, management and design.

Innovation – as an ingredient that promotes differentiation, change and/or exclusivity – will create a dominant power in the market, allowing superior and considerable advantages for rural business (Dinis, 2006; Tunney, 2015). In fact, born from the stems of traditional innovation theory (Schumpeter, 1934, Mahroum et al., 2007) observe rural innovation as the introduction of something new (a novel change) to economic or social life in rural areas which adds new economic or social value to rural life. In this sense, it is understandable that the process of rural innovation should involve a plurality of actors, the active participation of local stakeholders and collaborative actions (Brunori et al., 2007; Labianca et al., 2020). Therefore, rural innovation and social innovation are two interconnected concepts. As referred by Neumeier (2012, 2017) social innovation deals with changes of attitudes, behaviour or perceptions of a group of people joined in a network of aligned interests that lead to some kind of tangible improvement for the actors involved or even beyond. In other words, it adds new economic or social value to the rural community (the purpose of rural innovation). It is divided into the 3 following phases (Neumeier, 2012, 2017):

- problematization (an individual or an initial group identifies a problem and a need and seeks a solution);
- expression of interest (other actors are added, seeing an advantage in this cooperation);

– delineation and coordination (a new collaborative form and development strategy).

Important is, that innovations in rural areas are implemented not only by companies on the farm but may involve many actors in the rural space (Brunori et al., 2007). They require an appropriate combination of local knowledge (tacit or implied) with expert knowledge (explicit and formalised) and support of effectively operating networks (Esparcia, 2014). Often the background of the innovations in rural areas is an endogenous approach.

18.1.2.3 Availability of Funds in Rural Areas

Every entrepreneur to start a business must ensure access to financing, e.g., own or borrowed savings, a bank loan, funds from the EU – depending on the type of activity undertaken, it may be different in the number of funds. Lack of funding is a challenge that many women in the business face; however, applying for loans is one way to overcome this common obstacle.

The number of women-led businesses is growing steadily. There are many financial resources that are dedicated to helping female entrepreneurs to set up, maintain and grow their businesses. To stimulate further growth of female entrepreneurship, financial institutions in many countries, including state-owned, private and public banks, have introduced various loan programs that target this segment. Also microfinance institutions have been playing a significant role in promoting women entrepreneurship in developing countries.

Among the credit or loan solutions dedicated to female entrepreneurs there are competitive interest rates, no handling fees, no security or third party guarantees, and the loan repayment period up to 10 years.

However, the average size of a loan for female-owned companies is 31% less than for male-owned companies. Gender differences can discourage you from taking the next step towards financing your business.

Female entrepreneurs are accessing credit that is helping them mostly to grow their revenues, increase their client base and hire additional employees.

18.2 Discussion and Conclusion

Is there still a need to talk specifically about women entrepreneurs. on top of everything that has been already said'? Yes and no. No, because all that has been said about entrepreneurship is directly applicable to women. Women entrepreneurs, as research demonstrates, may do things differently. For example, in comparison to male entrepreneurs, women tend to work more in teams, are less self-centred and personal ego to them is less important than success of the organization or business idea they are pursuing.

There is no difference in characteristics such as achievement, autonomy, aggression, independence and benevolence between female and male entrepreneurs (Hisrich & Brush, 1984). Also, no differences were found in risk taking propensity of male and female entrepreneurs. However, we do need to talk explicitly about women entrepreneurs. It should be stressed that rural women can encounter many constraints when trying to take part in the transformation process. Rural areas tend to be more traditional in regard to the gender issue. In rural areas, the gender issue is usually a much stronger hindering factor to potential female entrepreneurs than it is in urban areas, their self-esteem and managerial skills being lower when compared to urban women and access to external financial resources more difficult than in urban areas. Therefore, special programmes of assistance (technical and financial) to overcome these constraints should be developed and designed to meet the needs of rural women in order to be able to take an active part in entrepreneurial restructuring of their communities, to start to develop their own ventures, to expand their already existing businesses, or to function as social entrepreneurs since their number today is still below the potential one. To this end, based on our experience as well as on the experiences of so many entrepreneurial women we have met across the world in our profession and in business, we very much agree with.

This belief is the one for which we as trainers are responsible to bring to rural women in addition to trying to put in place all factors crucial for rural women to enter into entrepreneurial activities. Without it, entrepreneurial opportunities will not be seen, they will be lost and then the role of women in rural development will be much below their potential.

Fostering entrepreneurship is widely perceived to be a successful rural development strategy for women empowerment leading to positive change.

It is worth mentioning the main needs and motivations, which – from the theoretical point of view – allow women to develop, take business initiatives and cope with sometimes difficult conditions.

Edward Deci and Richard M. Ryan created the Self-Determination Theory, in which they indicate three basic needs that determine human actions: competence (understanding how to achieve various types of external and internal benefits and being effective in the scope of required activities), relationships (development of safe and satisfactory relationships with others) and autonomy (or self-determination and self-regulation of one's own goals, activities, achievements).

In the theory of self-determination, a distinction is made between internally determined behaviors (fully volitional and approved by the individual, so those whose causality is perceived as internal to the self) and those externally controlled (triggered by interpersonal or intrapsychic forces, which are perceived as external in the self). These first-mentioned behaviors are therefore an individual choice, and the latter are the result of submission or obedience, and in some cases resistance. Both types of behavior concern motivation, but their regulatory processes differ (Deci et al., 1991).

Despite trend towards living in towns rather than in rural areas, the latter still offer opportunities and real potential. Surveys show that, in general, rural women have a strong desire to stay in their community and contribute to its development, provided certain requirements are met:

- employment opportunities in the local area, including part-time jobs;
- the possibility of gaining work experience and vocational qualifications;
- local facilities for education and training;
- business services supportive to women's projects and enterprises;
- public transport services compatible with working hours;
- local childcare facilities and social services for the elderly and the sick;
- supportive public and professional organisations.

Women are also seeking a better balance in the division of labour in the domestic household, need encouragement for their personal and professional development and more support in their bid to achieve financial independence, and to participate fully in decision-making.

Empowerment through entrepreneurship leads to self-fulfillment and makes women aware of their status, existence, rights and their position in the society. Many authors indicate that socioeconomic factors such as family support, cultural norms of society, lack of skills, self-confidence, material possession, investment, income and savings etc. discourage women to get into entrepreneurial activities (Itani et al., 2011; Mordi et al., 2010). However, Mishra and Kiran (2014) elucidate that in the modern era, women are becoming socially and economically empowered through business ownership. In social context; women have the power to take decisions, remain autonomous, be self-confident and manage their household independently, through which they can interact within the society (Lemire et al., 2001; Mayoux, 2005); while economic context is their economic stability – they are competent to participate in the economy and make better decisions for their economic position (Golla et al., 2011).

Although women's contributions to local and community development are significant, rural women around the world are a minority in decision-making and planning, especially at the regional and national levels. This is partly due to the multiple roles of women and the workload, but also to the persistence of traditional views on the roles of women and men in society.

The success of entrepreneurship actions depends on the cooperation and creation of networks among the different actors (individuals, agents of supply and organisations) that live in the rural communities.

References

Anthopoulou, T. (2010). Rural women in local agrofood production: Between entrepreneurial initiatives and family strategies. A case study in Greece. *Journal of Rural Studies, 26*(4), 394–403.

Atterton, J., & Affleck, A. (2010). Rural businesses in the North East of England: Final survey results (2009). Centre for Rural Economy Research Report

Bak, M., Kulawczuk, P., & Szcesniak, A. (2000). *Providing assistance to women in Rural Poland: The perspectives of providers and beneficiaries*. Institute for Human Sciences SOCO Project Paper 75.

Baumgartner, D., Pütz, M., & Seidl, I. (2013a). What kind of entrepreneurship drives regional development in European? A literature review on empirical entrepreneurship research. *European Planning Studies, 21*(8), 1095–1127.

Baumgartner, D., Schulz, T., & Seidl, I. (2013b). Quantifying entrepreneurship and its impact on local economic performance: A spatial assessment in rural Switzerland. *Entrepreneurship & Regional Development, 25*(3–4), 222–250.

Baylina, M., & Schier, M. (2002). Homework in Germany and Spain: Industrial restructuring and the meaning of homework for women. *GeoJournal, 56*(4), 295–304.

Bock, B. B. (1994). Female farming in Umbrian agriculture. In L. van der Plas & M. Fonte (Eds.), *Rural gender studies in Europe* (pp. 91–107). van Gorcum.

Bock, B. (2004a). It still matters where you live: Rural women's employment throughout Europe. In H. Buller & K. Hoggart (Eds.), *Women in the European countryside* (pp. 14–41). Ashgate.

Bock, B. B. (2004b). Fitting in and Multi-tasking: Dutch farm women's strategies in rural entrepreneurship. *Sociologia Ruralis, 44*(3), 245–260.

Bock, B. B., & van Doorne Huiskes, A. (1995). The careers of men and women: A life-course perspective. In A. van Doorne-Huiskes, J. van Hoof, & E. Roelofs (Eds.), *Women and the European labour markets* (pp. 72–89). Paul Chapman.

Braithwaite, M. (1996). Women, equal opportunities and rural development: Equal partners in development, *LEADER Magazine* 5. http://www.rural-europe.aeidl.be/rural-en/biblio/women/art03.htm

Brunori, G., Rand, S., Proost, J., Barjolle, D., Granberg, L., & Dockes, A.-C. (2007). *Towards a conceptual framework for agricultural and rural policies*. IN-SIGHT – Strengthening Innovation Processes for Growth and Development.

Commission of the European Communities. (2000). *Gender equality in the European Union: Examples of good practices (1996–2000)*. Office for Official Publications of the European Communities.

Dahlström, M. (1996). Young women in a male periphery – experiences from the Scandinavian North. *Journal of Rural Studies, 12*, 259–271.

Deci, E. L., & Ryan, R. M. (2008). Facilitating optima motivation and psychological well – being cross life's domains. *Canadian Psychology, 49*(1), 14–23.

Deci, E. L., Vallerand, R. J., Pelletier, L. G., & Ryan, R. M. (1991). Motivation and education: The self – determination perspective. *Educational Psychologist, 26*(3,4), 325–346.

Delfmann, H., Markantoni, M., & van Hoven, B. (2011). *The role of side activities in building Rural resilience: The case study of Kiel-Windeweer (The Netherlands)*. Paper presented at the IGU mini-conference: Globalizing Rural places, 20–21 May 2011, in Vechta, Germany Google Scholar

Dinis, A. (2006). Marketing and innovation: Useful tools for competitiveness in rural and peripheral areas. *European Planning Studies, 14*(1), 9–22.

Directorate General for Internal Policies of the Union. (2019). *The professional status of rural women in the EU, Policy Department for Citizens' Rights and Constitutional Affairs*, 2019

Duncan, S. S. (2000). Introduction: Theorising comparative gender inequality. In S. S. Duncan & B. Pfau-Effinger (Eds.), *Gender* (pp. 1–24). Economy and Culture in the European Union.

EC. (2000). EC Women Active in Rural Development. (2000) Assuring the future of Rural Europe Office for Official Publications of the European Communities, Luxembourg, p. 13.

EC. (2007). EC Rural Development Policy 2007–2013 (2007) Agriculture and rural development European Commission, Brussels.

Esparcia, J. (2014). Innovation and networks in rural areas. An analysis from European innovative projects. *Journal of Rural Studies, 34*, 1–14.

FAO. (1996). *Overview of the socio-economic position of Rural women in selected Central and Eastern European countries: Bulgaria, Croatia, the Czech Republic, Estonia, Hungary, Latvia, Lithuania, Poland, Slovakia and Slovenia*. Food and Agriculture Organization

Forsberg, G., Gonäs, L., & Perrons, D. (2000). Paid work: Participation, inclusion and liberation. In S. S. Duncan & B. Pfau-Effinger (Eds.), *Gender, economy and culture in the European Union* (pp. 27–48). Routledge.

Franco, A., & Winqvist, K. (2002). *Women and men reconciling work and family life, statistics in focus – Population and social conditions 9*. Eurostat, Office for Official Publications of the European Communities

Golla, A. M., Malhotra, A., Nanda, P., & Mehra, R. (2011). *Definition, framework and indicators*. International Center for Research on Women (ICRW).

Gourdomichalis, A. (1991). Women and the reproduction of family farms: Change and continuity in the region of Thessaly, Greece. *Journal of Rural Studies, 7*, 57–62.

Halfacree, K. (2006). *From dropping out to leading on? British counter-cultural back-to-the-land in a changing rurality*. Progress in Human Geography, 2006/6, 30/3, pp. 309–336

Halliday, J., & Little, J. (2001). Amongst women: Exploring the reality of rural childcare. *Sociologia Ruralis, 41*, 423–437.

Haugen, M. S., & Villa, M. (2003). *The countryside as a rural idyll or a boring place? Young people's images of the rural*. Paper presented at the XXth congress of the European Society of Rural Sociology, 18–22 August 2003, Sligo.

Hisrich, R., & Brush, C. (1984). *The woman entrepreneur: Management skills and business problems*. University of Illinois at Urbana-Champaign's Academy for Entrepreneurial Leadership Historical Research Reference in Entrepreneurship.

Högbacka, R. (1999). *Women's work and life-modes in rural Finland: Change and continuity*. Paper presented at the conference on 'gender and rural transformations in Europe', 14–17 October 1999, Wageningen University.

Irmen, E. (1997). Employment and population dynamics in OECD countries: An intraregional approach. In R. D. Bollman & J. M. Bryden (Eds.), *Rural employment: An international perspective* (pp. 22–35). CAB International.

Itani, H., Sidani, Y. M., & Baalbaki, I. (2011). United Arab Emirates female entrepreneurs: Motivations and frustrations. *Equality Diversity and Inclusion: An International Journal, 30*(5), 409–424.

Jervell, A. M. (1999). Changing patterns of family farming and pluriactivity. *Sociologia Ruralis, 39*, 100–115.

Kelly, R., & Shortall, S. (2002). 'Farmers' wives': Women who are off-farm breadwinners and the implications for on-farm gender relations. *Journal of Sociology, 38*, 327–343.

Knudsen, K., & Waerness, K. (2001). National context, individual characteristics and attitudes on mother's employment: A comparative analysis of Great Britain, Sweden and Norway. *Acta Sociologica, 44*, 67–79.

Korsgaard, S., Müller, S., & Tanvig, H. W. (2015). Rural entrepreneurship or entrepreneurship in the rural – Between place and space. *International Journal of Entrepreneurial Behavior & Research, 21*(1), 5–26.

Labianca, C., De Gisi, S., Todaro, F., & Notarnicola, M. (2020). Evaluation of remediation technologies for contaminated marine sediments through multi criteria decision analysis. *Environmental Engineering & Management Journal (EEMJ), 19*(10), 1891–1903.

Landström, H. (2010). *Pioneers in Entreprenership and small business research*. Springer.

Laoire, C. N. (2001). A matter of life and death? Men, masculinities and staying 'behind'in rural Ireland. *Sociologia Ruralis, 41*(2), 220–236.

Lemire, B., Pearson, R., & Campbell, G. (2001). *Women and credit: Researching the past. Refiguring the future*. Berg.

Lindsay, C., McCracken, M., & McQuaid, R. W. (2003). Unemployment duration and employability in remote rural labour markets. *Journal of Rural Studies, 19*, 187–200.

Little, J. (1991). Theoretical issues of women's non-agricultural employment in rural areas, with illustrations from the UK. *Journal of Rural Studies, 7*, 99–105.

Little, J. (1994). Gender relations and the rural labour process. In S. J. Whatmore, T. K. Marsden, & P. D. Lowe (Eds.), *Gender and rurality* (pp. 11–29). David Fulton.

Little, J. (1997a). Employment marginality and women's self-identity. In P. J. Cloke & J. Little (Eds.), *Contested countryside cultures* (pp. 138–157). Routledge.

Little, J. (1997b). Constructions of rural women's voluntary work. *Gender, Place and Culture, 4*, 197–209.

Little, J. (2002). Rural geography: Rural gender identity and the performance of masculinity and femininity in the countryside. *Progress in Human Geography, 26*(5), 665–670.

Looker, E. D. (1997). Rural-urban differences in youth transition to adulthood. Rural-urban differences in youth transition to adulthood, 85–98

Madureira, L., Gamito, T. M., Ferreira, D., & Portela, J. (2013). *Inovação em Portugal Rural Detetar, Medir e Valorizar*. Princípia Editora.

Mahroum, S., Atterton, J., Ward, N., Williams, A. M., Naylor, R., Hindle, R., & Rowe, F. (2007). *Rural innovation*. National Endowment for Science, Technology and the Arts (NESTA).

Markantoni, M., Koster, S., & Strijker, D. (2009). *Motivations to start a side activity in rural areas in the Netherlands*. In Geography of Innovation and Entrepreneurship: The 12th Uddevalla Symposium.

Marsden, T. (1999). Rural futures: The consumption countryside and its regulation. *Sociologia Ruralis, 39*(4), 501–526.

Mayoux, L. (2005). *Women's empowerment through sustainable microfinance*. Rethinking Best.

McGehee, Kim, K., & Jennings, G. R. (2007a). Gender and motivation for agri-tourism entrepreneurship. *Tourism Management, 28*(1), 280–289.

McGehee, D. V., Raby, M., Carney, C., Lee, J. D., & Reyes, M. L. (2007b). Extending parental mentoring using an event-triggered video intervention in rural teen drivers. *Journal of Safety Research, 38*(2), 215–227.

Mishra, G., & Kiran, U. V. (2014). Rural women entrepreneurs: Concerns & importance. *International Journal of Science and Research, 3*(9), 93–98.

Mordi, C., Simpson, R., Singh, S., & Okafor, C. (2010). The role of cultural values in understanding the challenges faced by female entrepreneurs in Nigeria. *Gender in Management: An International Journal, 25*(1), 5–21.

Moss, J. E., Jack, C. G., Wallace, M., & McErlean, S.A. (2000). *Securing the future of small farm families: The off-farm solution*. Paper presented at the conference on 'European Rural Policy at the Crossroads', 29 June – 1 July 2000, Arkleton Centre for Rural Development Research, University of Aberdeen

Neumeier, S. (2012). Why do social innovations in rural development matter and should they be considered more seriously in rural development research?–proposal for a stronger focus on social innovations in rural development research. *Sociologia Ruralis, 52*(1), 48–69.

Neumeier, S. (2017). Social innovation in rural development: Identifying the key factors of success. *The Geographical Journal, 183*(1), 34–46.

Nï Laoire, C. (2001). A matter of life and death? Men, masculinities and staying 'behind' in rural Ireland. *Sociologia Ruralis, 41*, 220–236.

O'Hara, P. (1998). *Partners in production: Women, farm and family in Ireland*. Berghahn.

O'Toole, K., & Macgarvey, A. (2003). Rural women and local economic development in southwest Victoria. *Journal of Rural Studies, 19*(2), 173–186.

Oberhauser, A. M. (1995). Gender and household economic strategies in rural Appalachia. *Gender, Place & Culture, 2*(1), 51–70.

Oberhauser, A. (1997). The home as "field": Households and homework in rural Appalachia. *Thresholds in Feminist Geography*, 165–182.

OECD, & Eurostat. (2005). *OSLO Manual Guidelines for collecting and interpreting innovation data*. OECD.

Ogbor, J. O. (2000). Mythicizing and reification in entrepreneurial discourse: Ideology-critique of entrepreneurial studies. *Journal of Management Studies, 37*, 605–635.

Overbeek, G., Efstratoglou, S., Haugen, M. S., & Saraceno, E. (1998). *Labour situation and strategies of farm women in diversified rural areas of Europe*. Office for Official Publications of the European Communities.

Pato, L. (2015). *Rural entrepreneurship and innovation: Some successful women's initiatives*, 55th congress of the European regional science association: "World renaissance: Changing

roles for people and places", 25-28 August 2015, Lisbon, Portugal, European Regional Science Association (ERSA), Louvain-la-Neuve.
Perrons, D., & Gonäs, L. (1998). Perspective on gender inequality in European employment. *European Urban and Regional Studies, 5*, 5–12.
Pfau-Effinger, B. (1994). The gender contract and part-time work by women: Finland and Germany compared. *Environment and Planning, A26*, 1355–1376.
Pfau-Effinger, B. (2000). Conclusion: Gender cultures, gender arrangements and social change in the European context. In S. S. Duncan & B. Pfau-Effinger (Eds.), *Gender, economy and culture in the European Union* (pp. 262–276). Routledge.
Plantenga, J., & Rubery, J. (1999a). Introduction and summary of main results. In J. Plantenga & J. Rubery (Eds.), *Women and work: Report on existing research in the European Union* (pp. 1–12). Office for Official Publications of the European Communities.
Plantenga, J., & Rubery, J. (1999b). *Women and work: Report on existing research in the European Union*. Office for Official Publications of the European Commission.
Post, J., & Terluin, I. (1997). The changing role of agriculture in rural employment. In R. D. Bollman & J. M. Bryden (Eds.), *Rural employment: An international perspective* (pp. 305–326). CAB International.
Sackmann, R., & Häussermann, H. (1994). Do regions matter? Regional differences in female labour-market participation in Germany. *Environment and Planning, A26*, 1377–1396.
Schumpeter, J. (1912/1934). *The theory of economic development*. Harvard University Press.
Schumpeter, J. (1960a). *Teorie rozwoju gospodarczego*. PWN.
Schumpeter, J. A. (1960b). *Teoria rozwoju gospodarczego*. PWN.
Schwager-Jebbink, J. (1991). Views from the top. In *Management education and development for women conference* (Vol. 5). Henley Management College.
Slee, R. (2005). From countrysides of production to countrysides of consumption? *The Journal of Agricultural Science, 143*(4), 255–265. https://doi.org/10.1017/S002185960500496X
Stathopoulou, S., Psaltopoulos, D., & Skuras, D. (2004). Rural entrepreneurship in Europe. A research framework and agenda. *Journal of Entrepreneurial Behaviour & Research, 10*(6), 404–425.
Stevenson, H. H., & Jarillo, J. C. (1990). A paradigm of Entreprenership: Entrepreneruail management. *Strategic Management Journal, 11*, 17–27.
te Kloeze, J. (1999). Family and leisure: Between harmony and conflict. *World Leisure & Recreation, 41*(4), 4–10.
Terluin, I. J., & Post, J. H. (Eds.). (2000). *Employment dynamics in Rural Europe*. CAB International.
Tigges, L. M., & Green, G. P. (1994). Small business success among men-and women-owned firms in Rural areas 1. *Rural Sociology, 59*(2), 289–310.
Tijdens, K. (2002). Gender roles and labour use strategies: Women's part-time work in the European Union. *Feminist Economics, 8*(1), 71–99.
Trifiletti, R. (1999). Women's labour market participation and the reconciliation of work and family life in Italy. In L. den Dulk, A. van Doorne-Huiskes, & J. Schippers (Eds.), *Work-family arrangements in Europe* (pp. 75–102). Thela thesis.
Tunney, E. (2015). Misogyny and marginalization in criminal justice systems: Women's experiences in two post-conflict societies. *Journal of Hate Studies, 12*, 153.
Vaillant, Y., Lafuente, E., & Serarols, C. (2012). Location decisions of new 'Knowledge intensive service activity firms': The rural-urban divide. *The Service Industries Journal, 32*(16), 2543–2563.
Virkkala, S. (2007). Innovation and networking in peripheral areas – Case study of emergence and change in rural manufacturing. *European Planning Studies, 15*(4), 511–529.
Wach, K. (2015). *Przedsiębiorczość jako czynnik rozwoju społeczno-gospodarczego: przegląd literatury*.
Warren-Smith, I., & Jackson, C. (2004). Women creating wealth through rural enterprise. *International Journal of Entrepreneurial Behaviour & Research, 10*(2004), 369–383.

Index

J. M. Ferreira da Rocha et al. (eds.), *Baking Business Sustainability Through Life Cycle Management*, https://doi.org/10.1007/978-3-031-25027-9

Printed by Printforce, the Netherlands